ANIMAL NUTRITION
FROM THEORY TO PRACTICE

This text is dedicated to my wife, Lyn Hynd, who patiently put up with me disappearing for long hours in the study while this passion of mine was realised.

ANIMAL NUTRITION
FROM THEORY TO PRACTICE

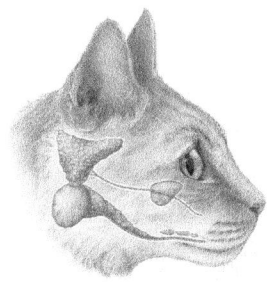

Philip I Hynd

CSIRO
PUBLISHING

CABI

A catalogue record for this book is available from the National Library of Australia.

ISBN: 9781486309498 (hbk)
ISBN: 9781486309504 (epdf)
ISBN: 9781486309511 (epub)

Published in print in Australia and New Zealand, and in all other formats throughout the world, by CSIRO Publishing.

CSIRO Publishing
Locked Bag 10
Clayton South VIC 3169
Australia

Telephone: +61 3 9545 8400
Email: publishing.sales@csiro.au
Website: www.publish.csiro.au

Published in print only, throughout the world (except Australia and New Zealand), by CABI.

ISBN 9781789242911

CABI	CABI
Nosworthy Way	745 Atlantic Avenue
Wallingford	8th Floor
Oxfordshire OX10 8DE	Boston, MA 02111
UK	USA
Tel: +44 (0)1491 832111	Tel: +1 (617)682-9015
Fax: +44 (0)1491 833508	E-mail: cabi-nao@cabi.org
E-mail: info@cabi.org	
Website: www.cabi.org	

Cover images: Photos from Pixabay, illustrations by Christina Rzewucki

Set in 11/13.5 Minion and Helvetica Neue
Edited by Peter Storer
Cover design by Andrew Weatherill
Typeset by Thomson Digital
Index by Max McMaster
Printed and bound by CPI Group (UK) Ltd, Croydon, CR0 4YY

MIX
Paper | Supporting responsible forestry
FSC
www.fsc.org
FSC® C013604

Contents

Preface

Good nutrition underpins the profitability and sustainability of animal production enterprises. It is probably the main single factor impacting on animal welfare across all species, is a key determinant of the impact of animals on the environment, and drives a global feed industry worth more than US$500 billion per annum. The global demand for animal protein is rapidly expanding at the same time as consumers in developed countries are demanding cheaper meat, milk and eggs and pressing for the removal of antibiotics and growth promotants in animal production. Consumers are also demanding higher standards of animal welfare in production systems. Many of these demands can be met through sound nutritional management, but this will only be realised if a scientific, evidence-based approach is taken. Unfortunately, many aspects of nutrition are heavily influenced by marketing claims, often with little, or no, supporting evidence. There are also many self-appointed nutrition 'experts', particularly in pet nutrition, who compromise the lifespan and quality of life of pets by poor nutritional practices.

This text provides an evidence-based approach as the basis for development of practical nutrition solutions for the management of the major companion and production animals.

Most texts on animal nutrition focus in depth on one animal species or class of animal (e.g. ruminants) or cover several species, but at a more superficial level. This text is aimed at senior undergraduate students and graduates working with animals as veterinarians, animal nutritionists or animal production specialists in the field, and provides an in-depth and up-to-date coverage of the nutrition of all the major companion and production animals (horses, cats, dogs, camelids, deer, sheep, goats, pigs, poultry, beef and dairy cattle). Given the importance of grazing animals in global food supply and environmental management, a detailed chapter on grazing animal nutrition is included. The text was developed using an evidence-based approach to nutrition and is heavily referenced to allow further reading and verification of concepts. Readers can garner key concepts quickly by reading the summaries at the start of each chapter or reading the breakout panels, while those wanting more depth can delve into detailed sections and further reading. The text is rich in drawings and flow charts designed to make it easy for reader to grasp often complex concepts in anatomy, physiology and nutrition.

Each chapter contains:

- a background to the animal species, its evolution, current global status and importance, and major nutritional challenges facing the species
- detailed descriptions of the digestive anatomy and physiology of the animal
- the energy, protein, vitamin and mineral requirements of the animal for different life stages and how to formulate practical diets for animals
- current nutritional management systems for the animal
- common nutritional and metabolic diseases of the animal
- supporting peer-reviewed references.

Acknowledgements

The brilliant artwork of Ms Christina Rzewucki is gratefully acknowledged. I was so fortunate to find a student of veterinary biosciences with a unique combination of skills and interest in anatomy, art and computer-design.

About the author

Phil Hynd is Professor Emeritus in Animal and Veterinary Sciences at The University of Adelaide, South Australia, Australia. He holds a degree with first-class honours in Rural Sciences at the University of New England, Armidale, New South Wales, Australia, and a PhD in animal nutrition at The University of Adelaide in South Australia. He was the inaugural JS Davies Postdoctoral Research Fellow at The University of Adelaide, working on the nutrition of grazing beef cattle. He was appointed lecturer in animal nutrition and to the Chair in Animal Science and Head of Department of Animal Sciences at The University of Adelaide in 1997. His research has focussed on the cellular biology and nutritional biochemistry of wool and hair production, and the impact of nutrition on wool fibre characteristics. This work led to the development of alternatives to the mulesing operation to protect sheep from blowfly strike. The impacts of early life environment (particularly fetal programming) on the lifetime production and health of animals became a major focus of his research in later years and has been applied to ruminants and birds. The author has taught animal nutrition and physiology to undergraduate students in veterinary science, animal science and agriculture for 35 years, and has supervised 30 PhD students.

1

Introduction to animal nutrition

Key points

- Nutrition is **the** single major determinant of animal health, welfare, production and the economic viability of animal enterprises.
- Feed costs account for 50–70% of the costs of production in most animal production systems and the global feed manufacturing industry is worth more than US$500 billion per annum!
- Poor nutrition is a major welfare issue, contributing to specific deficiency diseases, metabolic diseases, digestive disorders, obesity, diabetes, orthopaedic problems, cardiovascular problems, and predisposition to infectious and parasitic diseases.
- The global demand for animal feeds and feed additives is rapidly increasing, as the global demand for animal protein expands in line with global per capita incomes.
- Digestive systems fall into two major groups: pre-gastric fermenters (sheep, cattle, goats, deer, macropods, camelids) and post-gastric fermenters (dogs, cats, horses, poultry).
- Energy in animal feed and energy required by animals are measured as heat. The unit of energy is the Calorie or food Calorie (the energy required to raise the temperature of a litre of water by 1°C). The SI unit is the Joule (= Calories × 4.182).
- The energy available to an animal for maintenance, growth, gestation, lactation and activity is called the Net Energy.
- The system used to describe energy availability for most animal species is based on Metabolisable Energy (ME), the energy available after digestion and losses of energy in urine and methane.

- The ME requirement for maintenance is based on the rate of heat production in animals under standard conditions. Basal metabolic rate of animals is best estimated by the general equation $BMR = cW^{0.75}$ (MJ/day) where W is live weight and $W^{0.75}$ is the metabolic bodyweight. The value of 'c' varies with animal species, but is ~0.3 for most animals.
- Essential nutrients are those that cannot be synthesised by the animal in sufficient quantities for physiological functioning.
- The essential nutrients are usually 10 amino acids, 13 vitamins, seven macrominerals and nine microminerals (trace) and two fatty acids, although there are animal species differences in the essential nutrients.
- The law of limiting nutrients means that there will be no animal response to other nutrients until the 'first-limiting' nutrient is provided. Once this nutrient is supplied a 'second-limiting' nutrient limits further response and so on.
- Nutrients interact with one another to reduce absorption and utilisation, often inducing deficiencies.
- The animal response to intake of essential nutrients follows a generalised dose response curve, comprising phases of acute deficiency, subclinical deficiency, homeostasis, subclinical toxicity and acute toxicity. The length of each phase is dependent on: the nutrient; its solubility in fat or water; its mechanism of absorption and excretion; storage sites; and mode of action in metabolism. Some nutrients influence performance at high doses by pharmacological (drug-like) effects.
- Digestion in all animals provides mixtures of simple nutrients that are biochemically interdependent, making animals resilient to wide variations in diet composition.

Why is an understanding of animal nutrition important?
Nutrition drives the profitability of production animal enterprises

More than 60% of the variable costs of production in intensive animal production enterprises are feed-related. Maximising the output of animal products (milk, pork, wool, eggs, sheep meat, beef and chicken meat) relative to the input of feed is therefore critical to the economic profitability of animal enterprises. This is obvious in intensive animal production systems such as pig and poultry enterprises and in dairy systems where the impact of nutrition on profitability is readily apparent. Not surprisingly, when small changes in feed conversion efficiency produce large economic benefits, these intensive enterprises readily applied leading-edge nutritional sciences. It is little wonder that the poultry industry has reduced the amount of feed to achieve a kilogram of chicken from more than 3 kg in the 1950s to less than 1.5 kg in the modern broiler unit. Modern meat birds reach 2.5 kg in 39 days at feed conversion ratios (FCR) of less than 1.6:1

(ACMF 2011). Since the 1960s, broiler growth rates have doubled and the FCR values have halved.

> Achieving feed conversion rates (kg feed consumed per kg live weight gain) of <1.0 seems at first glance to be impossible, but remember that the feed is usually very dry (~90% DM) while animal live weight is ~60% water. A FCR of 1.0 kg feed/kg bodyweight therefore is actually a FCR of dry matter of 1.5 kg dry matter/kg body dry matter.

The efficiency of production of eggs has likewise increased, with only 2 kg feed required to produce 1 kg eggs. While it is clear that the majority of this remarkable progress has been achieved through genetic selection for improved genotypes, the role of nutrition in realising the genetic potential cannot be overstated. Birds fed even slightly unbalanced rations will fail to reach their genetic potential and the profitability of the enterprise will be reduced. This is true for all animal enterprises. Breeding more productive animals puts even more emphasis on correct feeding. The analogy of a Formula 1 racing car on low-octane fuel is a good one. The highest milk yield recorded for a dairy cow is now 35 000 L in a 300-day lactation, with many cows achieving more than 15 000 L per lactation. However, poor nutrition and poor management of her transition from one diet to another means not only that she may not reach this level of production, but she is also susceptible to a suite of metabolic diseases such as ketosis, hypocalcaemia, acidosis, hypomagnesaemia, left-displaced abomasum and laminitis.

Given that the maintenance energy requirement of an animal is a fixed amount at any live weight, maximising production reduces the relative cost of this fixed cost. High-quality feeds are used with greater efficiency than poor-quality feeds because the costs of digestion are lower and the losses of heat in metabolic pathways are lower. So maximising the growth rate and efficiency of gain reduces the relative maintenance costs and importantly the faster growing animals will spend less time on feed to reach a desired market weight. Given feed costs are as much as 60–70% of the variable cost in the enterprise, this could be the difference between profit and loss.

Poor nutrition (underfeeding, nutrient imbalances or overfeeding) is a key determinant of animal health

Poor animal nutrition can take various forms including:

- inadequate dietary energy or protein to meet the requirements for maintenance and good health

- unbalanced diets, in which the levels and ratios of essential nutrients are incompatible with good health and production
- overfeeding, in which the total energy intake over a period of time is incompatible with good health.

All three forms of malnutrition are apparent in animal enterprises, and particularly in companion animals, where overfeeding and inappropriate diets are major causes of health and welfare problems.

Animal nutrition has large effects on the environment

Nearly all of the negative impacts of animals on the environment (overgrazing, enteric greenhouse gas emissions, polluting effects of manures and urine) are related to nutritional management in some way. Attempts are being made to reduce methane production in sheep and cattle using targeted biochemical approaches such as nitrate feeding, reducing the protozoal numbers in the rumen using tannin-containing feeds, and feeding polyunsaturated fatty acids to scavenge hydrogen ions. Better matching of the soluble nitrogen supply to the requirements of the ruminal microbes, by better choice of protein and non-protein sources, will reduce urinary ammonia excretion, hence effluent and atmospheric pollution from feedlots. Optimising the ratios of available calcium to phosphorus for laying birds will reduce phosphorus excretion and pollution. Better grazing and water management systems will reduce overgrazing (Chapter 5). Using animal species better matched to the nutritional environment (e.g. macropods in Australia) may also reduce animal impacts on the environment. All of these solutions to animal problems can be realised through better nutritional practices.

The global animal feed industry is large and growing

As the per capita income of countries increases so too does their consumption of animal proteins. By 2050 it is estimated that the global demand for animal products will have grown by more than 60%. The production of meat (beef, swine and poultry) and dairy products is predicted to double from 2017 to 2050. To meet current demand for animal protein, the global animal feed industry produces more than 1 billion tonnes of compound feed annually and a further 300 million tonnes are mixed on-farm. Global commercial feed manufacturing generates an estimated annual turnover of over US$500 billion. More than 250 000 skilled workers, technicians, managers and professionals are employed globally in the animal feed and sales industries.

The estimated global number of domesticated animals in 2017 are shown in Table 1.1. Chickens dominate world animal numbers but pork is the most consumed meat globally (by weight).

Table 1.1. Global domesticated animal numbers[a]

Species	Millions (2008)	Millions (projected 2030)
Chickens	19 000	24 804
Cattle + buffalos	1500	1858
Sheep + goats	1600	2309
Pigs	800	1062
Cats	600	600[b]
Dogs	525	525
Horses	58	58[b]
Camels	25	25[b]
Alpacas	4	4[b]

[a]Various sources including FAO (2003); <http://faostat.fao.org/default.aspx>
[b]Estimates for cats, dogs, horses, camels and alpacas are very imprecise and forecasts are non-existent

There are now more than 25 billion domesticated animals worldwide. That is, there are 4× more production and companion animals than people on the planet!

To meet the rapidly expanding demand for animal production, the global market for animal feed additives, such as antibiotics, feed enzymes, antioxidants, vitamins and amino acids, is predicted to be worth US$27 billion per annum by 2020. The products are aimed at meeting the companion animal market or producing low-cost meat of high-nutritional value.

The global animal feed manufacturing business is valued at US$500 billion per annum.
The global feed additives market is valued at US$27 billion per annum.
The global pet food market is valued at US$81 billion per annum.

The global pet food market is predicted to be worth US$81 billion per annum by 2020 and is growing at 4% per annum. The dry pet food market alone accounts for more than 23 million metric tonnes of feed.

Digestive systems in animals
Evolution of digestive systems

Animals have evolved a wide array of digestive systems and strategies to obtain nutrients from the feeds available to them. Systems range from the relatively simple digestive tract of the pig, the dog and the cat, to the extreme specialisation of ruminants and camelids for the efficient utilisation of fibrous, low-digestibility feeds, and the enormous large intestinal specialisation of the horse. Animals are broadly characterised as: **(1) post-gastric fermenters** or **(2) pre-gastric fermenters.** The post-gastric fermenters are sometimes called mono-gastrics, or simple-stomached animals, and include pigs, dogs, cats, rabbits, horses, and birds. The pre-gastric fermenters include the ruminants, camelids and macropods (Table 1.2).

The pre-gastric fermenters developed various diverticulae in the simple stomach, which are still evident in animals with intermediate-type systems such as the peccaries (old-world pigs). Even the modern domesticated pig has a cranial diverticulum in their simple gastric stomach, as though they began down the pathway towards pre-gastric fermentation. These diverticulae in the stomach, presumably arose from spontaneous mutation and provided a means of slowing down the passage of digesta, thereby allowing time for the establishment of a large symbiotic microbiota. The advantage of establishing a symbiotic relationship with a microbial population was that poor-quality feeds that were resistant to endogenous enzymes secreted by the stomach and intestines of mono-gastric animals could now be digested by extracellular microbial enzymes such as cellulases. The development of the grasses in the Miocene Epoch (5–23 million years ago) provided an enormous substrate for any animals that could evolve to digest cellulose and hemicellulose, particularly when other nutrients such as protein were deficient in the grasses throughout the year. Cellulose is the most abundant biopolymer on earth (Klemm *et al.* 2005).

Table 1.2. List of selected pre-gastric fermenters and post-gastric fermenters

Pre-gastric fermenters	Post-gastric fermenters
Ruminants Cervidae (deer, moose) Giraffidae (giraffe, okapi) Bovidae (impala, gazelle, springbok, cattle, sheep, goat, wildebeest, antelope, buffalo, bison)	**Carnivores** *Felis catus* (cats) *Panthera* (large cats)
Tylopods (camelids) Camelidae (alpaca, camel, llama, vicuna)	**Omnivores** *Canis lupis familiaris* (dogs)
Diprotodonts Macropodidae (kangaroos, wallabies, quokka) Hippopotomidae (hippopotamus)	**Herbivores** *Equus caballus* (horse) *Equus asinus* (donkey) *Oryctolagus cuniculus* (European rabbit) *Pascolarctus cinereus* (koala)

- Cellulose is the most abundant polymer on the planet.
- Plants produce 180 billion tonnes of cellulose per annum.
- Cellulose is a polymer of glucose molecules joined in a β1–4 linkage.
- Only microbes can degrade these linkages.
- Pre-gastric fermenters use symbiotic microbes to degrade cellulose.

The macropods (kangaroos, wallabies, quokkas) in Australia developed a different means of establishing this relationship. They developed a forestomach comprising a small sacciform compartment and a long tubiform compartment leading to an acid hind-stomach compartment (Hume 1982). The tubiform stomach of the macropods slows the rate of digesta passage with a system of 'baffles' or haustrae that hold back the digesta during contractions of the longitudinal smooth muscle (taenia) lining the tube (Hume 1982).

The camelids and hippopotamus also established a symbiotic relationship with microbes in the first two of three stomach compartments. The horse developed an enormous large intestinal digestive system to allow microbial digestion after gastric and small intestinal enzymic digestion.

The advantages and disadvantages of the pre-gastric and post-gastric fermentation systems are indicated in Table 1.3.

Caecotrophy and coprophagy

Caecotrophy refers to the practice of some small herbivores (rabbits, hares, rats, mice, voles and ring-tailed possum) of consuming special 'soft' faecal pellets produced in the caecum during periods of rest. In wild rabbits this occurs first thing in the morning when rabbits can be seen taking the caecal pellets directly from the anus. These caecotrophes differ from normal faecal pellets in that they are smaller (5 mm in rabbits), higher in protein, lower in fibre and higher in moisture, and they are usually excreted in tight clusters. The caecotrophes are formed in the caecum from solutes and small food particles that are selectively retained in the caecum by antiperistalsis. Bacteria trapped in mucin are returned from the colon to the caecum where they ferment undigested feed to short-chain volatile fatty acids (VFAs). These VFAs are absorbed and contribute significantly to the energy supply of the herbivores that practise caecotrophy. The microbially synthesised vitamins, microbial protein and the mucin coating the faeces are voided and eaten, thereby returned to the start of the digestive tract. This ingenious behaviour allows small herbivores to obtain high-quality nutrients (microbial vitamins B and K and amino acids) from high-fibre diets without

Table 1.3. Advantages and disadvantages of pre- and post-gastric fermentation systems of digestion

Factor	Advantages of pre-gastric fermentation	Disadvantages of pre-gastric fermentation	Advantages of post-gastric fermentation	Disadvantages of post-gastric fermentation
Carbohydrate digestion	Cellulose and hemicelluloses degraded	High-quality carbohydrates, such as starch, degraded to VFAs and lactic acid	High-quality carbohydrates available	Poor extraction of energy from low-quality feeds
Protein digestion	Moderate-quality protein available from poor-quality protein and non-protein nitrogen	High-quality feed proteins degraded to moderate-quality microbial protein	High-quality proteins available for digestion	Cannot produce protein from non-protein nitrogen sources
Digestion rate	Rumination allows digestion while not exposed to predators	Slow rate of passage reduces intake potential (trade-off between intake rate and digestion extent)	Intake not limited by rate of digestion	
Intake of toxins	Toxins and plant secondary metabolites degraded	Some toxins produced in forestomach (e.g. nitrite from nitrate)		Feed toxins and plant secondary metabolites not degraded
Vitamin intake	Vitamins B and K synthesised by microbes before the absorptive site (intestines)	Vitamin supply variable depending on microbial activity and anti-vitamin microbial products (e.g. thiaminases)		Microbial vitamins B and K not available until after absorptive sites (unless coprophagy practised)

the need to carry large amounts of digesta in fermentation vats such as the rumen or camelid stomachs.

Coprophagy refers to the consumption of faeces, which may be accidental (e.g. birds scratching around in dirt or litter) or deliberate, non-nutritional (behavioural) consumption (e.g. pica or depraved appetite in dogs). For poultry, coprophagy can contribute significantly to vitamin supply from bacterial activity in the caecum. The absence of coprophagy in caged birds means B-group and K vitamins must be supplied in feeds.

Structure and function of digestive systems

Detailed descriptions of the anatomy and function of mono-gastric, ruminant and camelid digestive systems are provided in Chapters 2, 3 and 9, respectively.

Analysis of nutrients in feeds

The 'proximate' analysis of feeds

The objective behind feed analysis is to define the feed in terms of its ability to meet the nutrient requirements of the animal. The most common and basic system for broadly defining the feed is called the 'proximate' analysis of the feed, or 'Weende' system (named after the Weende Experiment Station in Germany). The term 'proximate' refers to the fact that most of the measures are 'proxies' for the actual chemical entities being defined. For example, crude protein is estimated, not as protein *per se*, but as nitrogen, which is then converted into a protein value by assuming the protein contains a constant amount of nitrogen (~16%). Hence it is termed 'crude' protein, not 'true' protein. Similarly crude fibre is a 'proxy' for the actual fibre content. Although the proximate analysis system is more than 100 years old, parts of this system are still commonly used in nutrition. The proximate analysis system is summarised in Table 1.4.

Protein analysis in feeds

Note that crude protein in the proximate analysis is calculated using an average conversion factor of 6.25, based on the assumption that, on average, protein is 16% nitrogen. In reality, feeds vary in the nitrogen content of their constituent proteins, so the assumption of 16% is incorrect. To account for this, a set of factors known as Jones' factors have been used to calculate the crude protein content of

Table 1.4. The 'proximate' analysis of feeds (Weende system)

Fraction (mg/kg)	Method	Components
Dry matter	100°C to constant weight	Moisture + some volatiles
Ash	500–550°C until all carbon removed	Inorganic minerals (macro and trace elements)
Crude protein	Kjeldahl digestion to convert all N (except nitrates) to NH_4. Protein is estimated from N content by assuming a constant N content in feed proteins (approx. 16%) so N × 6.25 = CP%	Proteins, amides, amines, amino acids, some vitamins, nucleic acids
Ether extract	Solvent extraction in diethyl ether or petroleum ether until all fats removed	Fats, oils, waxes, pigments, fat-soluble vitamins, sterols
Crude fibre	Acid boiling/alkali boiling	Cellulose, hemicelluloses, lignin
Nitrogen-free extractives	1000 mg (additions of all above = 'x' mg)	Cellulose, hemicelluloses. lignin, sugars, fructans, starch, pectins, tannins, pigments, vitamins, organic acids, resins

Table 1.5. Calculating the crude protein content of feeds from nitrogen percentage or from amino acid content

Feed	Jones' factors	Recommended factors based on amino acid analysis
Milk and dairy products	6.38	5.85
Meat, fish, eggs	6.25	5.60
Cereals (wheat, barley, oats, rye, triticale)	5.83	5.40
Corn	6.25	5.60
Soybean	5.71	5.50
Legume pulses	6.25	5.40
Default value for all unknown feeds	6.25	5.60

Source: Mariotti *et al.* (2008)

specific feeds. More recently these values have also been questioned on the grounds that true protein should be based on the amino acid content of the feed because this is the important nutritional component. Table 1.5 shows the conversion factors calculated from either nitrogen or amino acid content (Mariotti *et al.* 2008).

However, crude protein, regardless of its derivation, is insufficient for describing the availability of amino acids for metabolism. In mono-gastrics, it is customary to determine the total amino acids, 'available' amino acids and 'intestinally digestible amino acids' available in each feedstuff. For ruminants, crude protein is divided into Ruminally Degraded Protein (RDP) and Undegraded Dietary (or intake) Protein (UDP). The metabolisable protein available from microbial protein and UDP is then estimated (see Chapter 4).

Fibre analysis in feeds

The most important change to the proximate analysis of feeds has been to the system for quantifying the fibre component of feeds. Crude fibre in the proximate analysis system was meant to describe the indigestible component of the feed, but non-ruminants can digest some fibre components and ruminants can digest cellulose and hemicelluloses. The van Soest fibre analysis system (van Soest 1967) was introduced to reflect more accurately the digestibility or indigestibility of the various fibre fractions in ruminants and non-ruminants. Figure 1.1 shows the van Soest analysis system.

The van Soest system effectively separates the digestible cell contents (neutral detergent solubles) from the cell wall constituents (neutral detergent fibre). The NDF fraction contains hemicelluloses, cellulose and lignin. Subsequent acid detergent treatment then removes hemicelluloses. Both NDF and acid detergent fibre (ADF) are useful measures of the fibre content of feeds and are therefore related to digestibility, energy density and voluntary intake in mono-gastrics and ruminants.

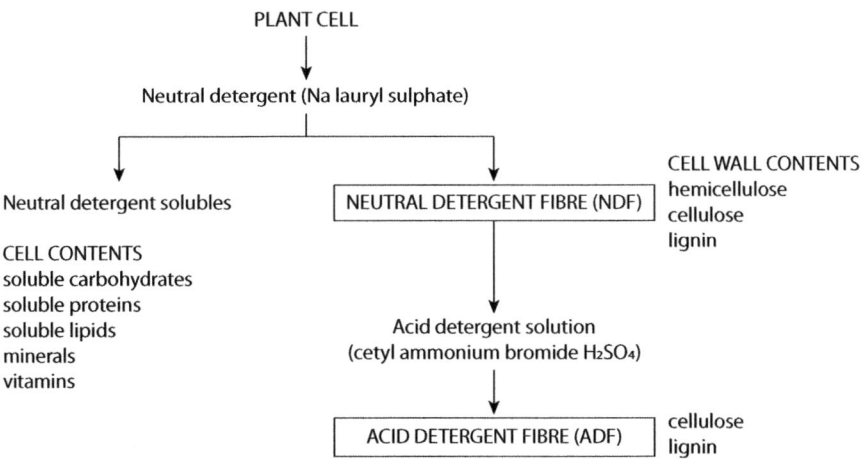

Fig. 1.1. The van Soest (1967) method of fibre analysis.

Near infrared reflectance spectroscopy (NIRS)

Most feed analysis services now provide a combination of wet chemistry (as per the proximate analysis and van Soest systems) together with results from near infrared reflectance spectroscopy (NIRS) analysis. NIRS is a rapid method for predicting the composition of feeds based on relationships between the reflectance spectra of the sample under near infrared light. NIRS spectra are aligned to results from 'wet' chemistry to derive predictive equations that can be readily used to derive feed composition. NIRS data should be viewed cautiously until a large number of samples have been used to generate the calibration equations. Extrapolation beyond the calibrating dataset should be avoided.

Results of feed analysis are used as the basis for formulating feeds for animals as described in Chapter 4. To formulate feeds accurately, however, requires more detailed knowledge than provided by the simple analyses described in the previous section.

Matching the energy requirements of animals with energy in feeds

Energy is the primary limiting factor to animal performance and is the first factor to be considered in formulating rations for animals. Feeding systems for each animal species have been designed to allow nutritionists to match the energy requirement of the animal with the energy supply from the ration (Fig. 1.2).

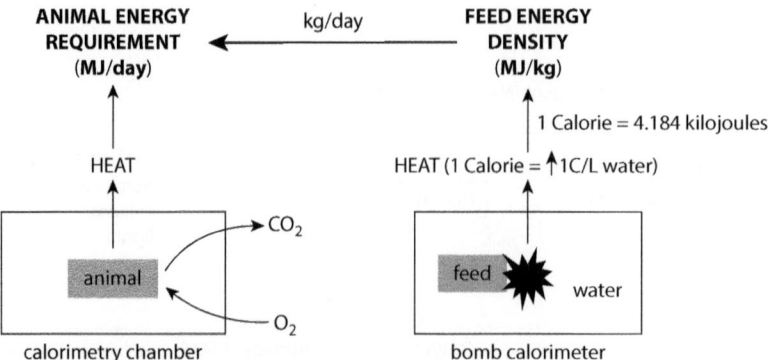

Fig. 1.2. Matching the energy in a feed (measured in a bomb calorimeter) with the energy required by the animal (measured in a calorimetry chamber). The 'bomb' calorimeter (right-hand box) measures the temperature rise in 1 L of water when the feed is burned ('bombed'). This is the number of kilocalories contained in the feed. The animal's energy production (left-hand box) can be measured in a calorimeter as heat produced directly, or indirectly from the rate of oxygen consumption, which is closely related to heat production.

To match the feed energy and the animal energy requirement, the units of energy must be the same. The basic unit of energy is the calorie or kilocalorie (see Box 1.1).

Box 1.1. Energy units used in animal nutrition are confusing!

1. A calorie (called a 'small c' calorie) is the 'the energy required to raise the temperature of 1 mL of water by 1°C (from 14.5°C to 15.5°C) at 1 standard atmosphere in a bomb calorimeter'.
2. However, when nutritionists use the term Calorie ('big C' calorie) in human and pet nutrition, they are referring to a kilocalorie (as defined above).
3. To convert big C Calories to joules (the SI unit of energy): 1 Calorie = 4.184 kilojoules.

The energy unit used for sheep, cattle, horses, pigs, poultry, macropods and camelids is the megajoule (= 10^6 joules).

Estimating the energy content of feeds

The total amount of energy in a feedstuff measured in a bomb calorimeter (Fig. 1.2) is not always available to the animal because the components of the feed may be indigestible, and even those components that are digested and absorbed are not used with 100% efficiency. Some of the absorbed energy is lost in urine due to

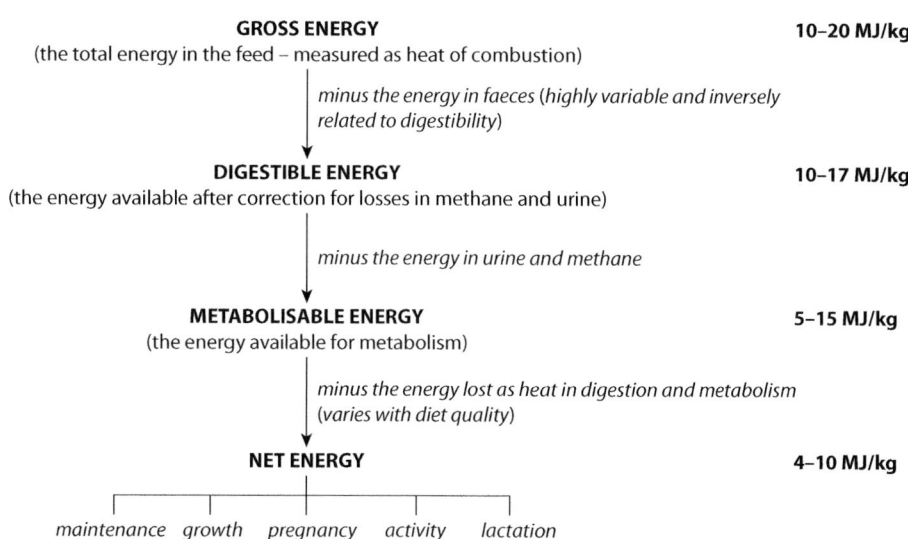

Fig. 1.3. Partitioning of feed energy in animals. Values on the right are typical energy density values at each stage. Feeding systems for animals are based on DE, ME or NE, with corrections applied for losses at each step.

inefficiencies in metabolism, and for ruminants some energy is lost as methane generated in fermentation. The partitioning of feed energy into each of these components is illustrated in Fig. 1.3.

The largest loss of energy in a feed occurs during digestion, where as much as 70% of the dietary gross energy may be lost in faeces. Ruminant feeds commonly fall in a range from 50 to 80% digestibility and for mono-gastric animals 70 to 90%. Losses of energy in urine and as methane range from 12 to 19% in ruminants and 3 to 12% for mono-gastrics. The efficiency with which ME is converted to Net Energy varies with both the diet and the purpose for which the energy is being used (maintenance, growth, gestation or lactation). It is clear from Fig. 1.3 that the best system for estimating an animal's energy requirement and a feed's ability to meet that requirement would be a Net Energy system because this is the actual energy available to the animal. However, measuring Net Energy is difficult because one needs to measure the heat increment, the loss of energy associated with digestion and metabolism. This would need to be done on every feed and for every productive purpose, because the loss of energy between ME and NE is highly variable and depends on both the composition of the feed and the purpose for which the animal is using that energy (e.g. maintenance, growth, pregnancy or lactation). Different systems of expressing energy are used in different animal species, and sometimes between countries within a species (see Chapter 4).

Estimating the digestibility of feeds

Given that digestibility is the main determinant of the energy available to an animal, we need a method for estimating the digestibility of feeds. This can be done in several ways:

1. *In vivo* **feeding trials:** the feed is given to an animal for a minimum period of time (usually >5 days) and then feed intake and faecal output are measured over a 5–7 day period. The feed and faeces are 'bombed' in the calorimeter and energy digestibility calculated. This is expensive and time consuming and has been replaced by other, quicker methods (see points 2 and 4).
2. *In vivo* **feeding trials using markers:** If one knows the concentration of a marker in a feed, the feed intake (total marker intake) and the concentration of the marker in the faeces, one can calculate the faecal output as follows:

$$\text{Faeces DM output (g/day)} = \text{marker intake (g/day)/marker concentration in faeces (g/gDM)}$$

   The marker can be internal (a component of the feed), or an external marker added to the feed or provided to the animal as an infusion or slow-release product.
3. *In vitro* **trials:** the feed is placed into tubes containing enzymes or microbes + enzymes for set periods of time to simulate digestive systems (Tilley and Terry 1963). This system was used for ruminant feeds for many years and provided the basis of calculating the energy content of many thousands of feeds.
4. **Near infrared reflectance spectroscopy (NIRS):** rapid scanning of the feed allows prediction of the digestibility of the feed provided robust equations have been developed for the animal species and the feed type offered.

Having estimated the digestibility of a feed, one can then relate this to the DE requirements of the animal. The problem with this is that the Digestible Energy content of feeds that are extensively fermented in the rumen and/or the hindgut, will be overestimated because there are extensive losses of energy as methane. For the less-complicated digestive systems of cats, dogs and pigs on highly digestible feeds, these errors will be relatively small, but for horses, sheep and cattle on high-fibre diets these errors can be large. The issues around energy systems for the individual species are covered in their respective chapters in detail.

Estimating the energy requirements of animals

The amount of energy an animal needs is equal to its metabolic rate and can be directly measured as the amount of heat the animal produces during metabolism. Heat production is equivalent to metabolic rate because the energy stored in high-energy bonds such as ATP and NADH, becomes heat during metabolic reactions. This can be done directly by placing an animal in a calorimetry chamber, or indirectly by measuring the animal's O_2 consumption and CO_2 production. These can be measured in a respiration chamber but this is expensive and time consuming, and the animal is not behaving normally. Alternatively, O_2 consumption and CO_2 production can be measured indirectly as described in the following sections.

Indirect calorimetry to estimate energy expenditure
Estimating energy expenditure from oxygen consumption or carbon dioxide production
Heat production can be calculated from O_2 consumption or CO_2 output using a face mask and in-line gas meters (Brody 1945). For carbohydrates, each mole of glucose requires 6 moles (134.4 L) of O_2 and produces 6 moles (134.4 L) of CO_2, and this oxidation releases 678 kcal heat (i.e. 5.05 kcal/L O_2 consumed = 21.1 kJ/L). So we can estimate heat production by measuring the total volume of O_2 consumed during carbohydrate metabolism. However, feeds usually comprise a mixture of carbohydrates, fats and proteins. For mixed fats, the heat production is 4.69 kcal/L O_2 consumed = 19.6 kJ/L. For proteins, the value is 4.82 kcal/L or 20.1 kJ/L. Theoretically therefore it is necessary to know the relative amounts of carbohydrate, fat and protein present in the feed before one can accurately estimate heat production from O_2 consumption. If only oxygen consumption is known, it is assumed that an average value of 4.83 kcals/L O_2 (= 20.2 kJ/L O_2) consumed is sufficiently accurate for most purposes (Brody 1945). If the volumes of O_2 consumed and CO_2 produced are known, more accurate estimates can be made by estimating the respiratory quotient (RQ), which is the ratio of CO_2 produced/O_2 consumed. By estimating the RQ, the relative amounts of carbohydrate, protein and fat that were burnt can be estimated (Table 1.6) and hence the heat produced during metabolism as follows.

Respiratory quotient values for carbohydrates, fats and proteins
Burning glucose, fats or a protein (albumen) produces the following reactions:

$C_6H_{12}O_6 + 6O_2 \rightarrow 6CO_2 + 6H_2O$ $\qquad$ RQ = CO_2/O_2 = 6/6 =1.0

$C_{16}H_{32}O_2 + 23O_2 \rightarrow 16CO_2 + 16H_2O$ $\qquad$ RQ = CO_2/O_2 = 16/23 = 0.7

$C_{72}H_{112}N_{18}O_{22}S + 77O_2 \rightarrow 63CO_2 +$ $\qquad$ RQ = CO_2/O_2 = 63/77 = 0.80
$28H_2O + SO_3 + 9CO_2NH_3$

General values for broad mixes of carbohydrates and fats are shown in Table 1.6.

Table 1.6. Energy production can be estimated if the volume of CO_2 produced and O_2 used are measured. The amount of O_2 used/unit of energy produced depends on the composition of the fuel burned. This is estimated from the respiratory quotient (RQ), which is the ratio of CO_2 produced/O_2 used.

RQ (CO_2/O_2)	Fat/CHO (%)	Heat production (Kcal/L O_2)	Heat production (kJ/L O_2)
0.70	100/0	4.69	19.6
0.80	67/33	4.80	20.0
0.90	33/67	4.92	20.6
1.00	0/100	5.05	21.1

Note: No account is made for protein because its heat production falls between fats and carbohydrates and for most mixed feeds the RQ is 0.85, which is similar to the protein RQ of 0.82
Source: Adapted from Brody (1945)

To accurately estimate the volumes of gas produced or used, corrections should be made for standard temperature (0°C) and pressure (100 kPa or 760 mm Hg) conditions (STP).

Estimating oxygen consumption from heart rate
The rate of O_2 consumption is directly related to the heart rate of an animal during the test period, because for any individual animal the amount of oxygen/heart beat (known as the 'oxygen pulse') is relatively constant. Oxygen pulse is 'constant' because stroke volume, haemoglobin content of the blood, saturation of the haemoglobin and uptake of oxygen by tissues tend to be fairly constant in an individual. By calibrating each individual for its own oxygen pulse value using a face mask, one can estimate the metabolic rate (heat production) using a simple heart rate monitor (Brosh *et al.* 1998). Alternatively, an average oxygen pulse value can be used for the animal in question. The value of this method is that it can be used in unrestrained animals, grazing animals and wildlife. An error arises in such estimations because it is assumed that STP values apply at all times and this will not be the case. The errors are likely to be small relative to other errors but corrections of gas volumes can be made if atmospheric pressure and temperature are also measured during gas collection.

Estimating the maintenance energy requirement (basal metabolic rate or fasting heat production) of animals
To estimate the energy requirement of an animal we must start by determining how much energy is needed to maintain the basic functions of life (circulation, excretion, muscle tone, body temperature, secretion, respiration, membrane potentials and so on). The minimum amount of energy an animal needs is termed the basal metabolic rate (BMR), and is defined as the heat production when the animal is maintained under the following standard conditions:

- thermoneutral (20–25°C), so no energy is used to maintain core body temperature

- post-absorptive (depends on the animal species – ruminants require up to 48 h to become post-absorptive and, if pre-testing feed intake was high, this period could be as long as 7 days (Marston 1948)). Digestion, and particularly fermentation, also produces heat that is unrelated to the animal's metabolic rate
- lying posture to reduce heat production associated with muscular activity associated with standing. This is difficult to achieve in animals and so the data can be corrected for this, and this is called **fasting heat production** (FHP) rather than BMR
- calm psychological state. This requires some training of animals to accept the chamber.

This BMR (or FHP) value is then used to provide a platform value on which other activities, such as growth, gestation, lactation, work and exercise, are built. The energy required for these activities and physiological states is then expressed as a multiple of the maintenance or BMR value (e.g. a lactating animal may require 5× the maintenance energy requirement). The BMR or FHP can be measured using the O_2 and CO_2 outputs (discussed earlier) but this is time consuming and difficult to implement as a species-wide feeding standard. Instead, general relationships between heat production and live weight have been derived, which allow the energy requirement of animals differing very widely in live weight to be estimated. This is particularly useful when individuals of one species (e.g. dogs) differ widely in live weight (e.g. from 1.5 kg to 90 kg across breeds).

However, the relationship between heat production and bodyweight is not linear. Smaller animals produce more heat per unit bodyweight than large animals. To allow the energy requirements (metabolic rate or heat production (H)) of animals differing in bodyweight (W) to be compared, the following general equation was proposed:

$$H \text{ (heat production)} = cW^p$$

The value of the exponent (p) has been disputed for decades but the most widely accepted exponent now is 0.75 (Kleiber 1947). The value of the constant (c) varies with animal species, breed, age and sex. Typical values of 'c' are 300–400 kJ (70–95 kcals). The different values of 'c' used to estimate the energy requirements of different species are indicated in the species chapters. Metabolic scaling is particularly important when breeds within a species differ widely in mature body size (e.g. dogs), but is less important when live weights of animals are similar.

Now that we have the basis for estimating the maintenance energy requirement of animals, we can determine the impacts of other physiological states and activities so we can determine the total energy requirement of the animal. The methods used to do this are described in detail in Chapter 4.

Essential nutrients

Essential nutrients are defined as those that meet the following three criteria:

1. The nutrient cannot be synthesised by the body, or cannot be synthesised at a rate sufficient to provide for normal physiological functions, and therefore must be present in the diet, water or the environment (inhaled or absorbed).
2. Clinical signs of a deficiency of an essential nutrient arise when insufficient of the nutrient is provided in the diet, water or from the environment. These signs may be of a general nature (e.g. anorexia or poor growth) or specific to a particular nutrient (e.g. curled-toe paralysis in riboflavin-deficient birds).
3. Provision of the nutrient prevents or removes the clinical signs of the deficiency.

Steps in establishing whether a nutrient is essential

1. Prepare a diet/water/environment complete in all but the test nutrient.
2. Prepare a second diet/water/environment complete in all known essential nutrients including the test nutrient.
3. Feed the deficient ration for a sufficiently long period to establish if the nutrient is essential (Note: for some nutrients such as fat-soluble vitamins this might be for months because they are stored in the liver and fat).
4. Feed a complete ration, including the test nutrient, to 'pair-fed' control animals. Pair feeding means feeding exactly the same amount of a diet containing the test ingredient as is consumed by an animal on the deficient diet. This accounts for the fact that most deficiencies result in depressed intake in addition to the specific effects on metabolism. It is the latter we wish to demonstrate.
5. Quantify the clinical signs (clinical biochemistry, vital signs, general health, specific pathologies) in the test animals and pair-fed control animals.
6. Reintroduce the nutrient and quantify response/recovery.

Creating a diet deficient in a single nutrient can be difficult, particularly if the nutrient is normally found in trace amounts in the diet. Ensuring the water does not contain the nutrient is also difficult because many water supplies contain traces of mineral elements (some arising from the water pipes, such as copper, or water treatment processes). The environment can also contribute to nutrient supply, particularly if animals are housed in metal cages or are in areas where the respired air can contain particles and nutrients (e.g. lead from smelters). A deficiency of nearly all of the essential nutrients produces anorexia to a greater or lesser extent, depending on the nutrient. To determine the effects of the nutrient over and above its anorexigenic effect, it is essential to have pair-fed controls that will account for intake effects. Clinical signs in the test animals but not in the pair-fed controls reflect the specific effects of the deficient nutrient. An example of this test system for essentiality can be found in White et al. (1994) for zinc deficiency in sheep.

Using this approach, the essential nutrients established for animals are: the essential amino acids; essential fatty acids; essential inorganic nutrients (macro- and micro- or trace minerals); and essential organic micronutrients (vitamins).

The law of limiting factors

Before examining the essential nutrients in detail it is necessary to understand the concept of 'the law of limiting factors'. This concept revolves around the fact that animal performance will be 'limited' by the 'most-limiting' nutrient in the diet. This concept was first described in plant biology by a plant biochemist, Carl Sprengel, and then popularised by a chemist, Justus von Liebig. Liebig used a barrel with different length staves to represent the relative supply of nutrients to the plant. This became known as Liebig's barrel and is shown in Fig. 1.4. The quantity of water held in the barrel represents animal performance, which is limited by the shortest stave. This stave represents the first-limiting nutrient. Providing more of any other nutrient will have no impact on performance (barrel volume) until the limiting nutrient is provided. In Figure 1.4, lysine is deficient in the diet, so no matter how well balanced the rest of the ration, animal performance will be limited. Lysine is therefore considered the 'first-limiting' nutrient. Once lysine is provided, animal performance will then be

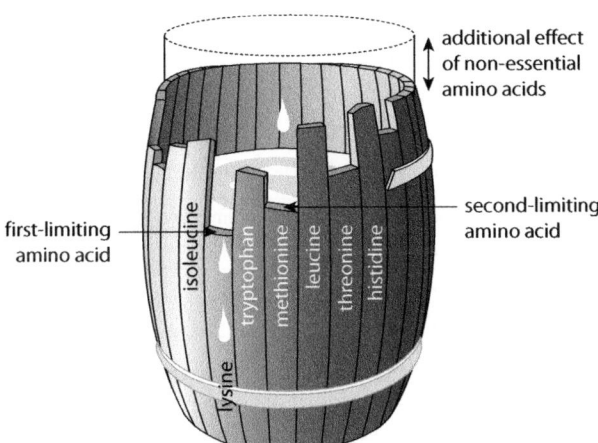

Fig. 1.4. Modified Liebig's barrel illustrating the concept of first, second, third, and so on, limiting essential nutrients in animal nutrition: in this case amino acids. A full barrel represents the potential animal performance if all essential amino acids are present in non-limiting quantities. In this case, lysine is the first-limiting amino acid, followed by methionine, then threonine. New evidence suggests that there is an additional response (dotted hoop) if the non-essential amino acids are provided in addition to the essential amino acids (Wu 2014). Some nutrients present at high levels in feeds or water can induce a deficiency of another essential nutrient (this would be represented as a very long stave reducing the length of another stave). High K levels, for example, induce a deficiency of Mg in ruminants.

limited by the next-limiting nutrient (in this case, methionine), which is referred to as the second-limiting nutrient and so on.

Essential amino acids

Amino acids are required as building blocks for protein synthesis and, for most animal species, ten of the amino acids cannot be synthesised from the carbon skeletons of other amino acids, so they must be provided in the diet. These are termed 'essential', or 'indispensable' amino acids. An easy way to remember them is the mnemonic 'Private Tim Hall' or PVT TIM HALL (Note: cysteine can be synthesised from methionine so strictly speaking is not essential) (Table 1.7).

> Essential amino acids for most animals = Private Tim Hall or PVT TIM HALL

Although the concept of essential amino acids and the concept of first-limiting, second-limiting, third-limiting amino acids, are still valid, there is a growing trend towards including consideration of the entire amino acid 'balance' in the feed protein, because the non-essential amino acids can also limit the performance, health and efficiency of animals (Breuer *et al.* 1964; Rogers *et al.* 1970; Wu 2014). This is represented in Fig. 1.4 by the dotted hoop above the barrel, which in this case represents the impact of non-essential amino acids on performance over and above the impacts of the essential amino acids.

Table 1.7. The essential amino acid requirements in different animal species

Amino acid	Rats	Poultry	Pigs	Ruminants	Horses	Dogs	Cats
Phenylalanine	✓	✓	✓	✓	✓	✓	✓
Valine	✓	✓	✓	✓	✓	✓	✓
Threonine	✓	✓	✓	✓	✓	✓	✓
Tryptophan	✓	✓	✓	✓	✓	✓	✓
Isoleucine	✓	✓	✓	✓	✓	✓	✓
Methionine	✓	✓	✓	✓	✓	✓	✓
Cysteine							
Histidine	✓	✓		✓	✓	✓	✓
Arginine	✓	✓		✓	✓	✓	✓
Leucine	✓	✓	✓	✓	✓	✓	✓
Lysine	✓	✓	✓	✓	✓	✓	✓
Glycine		✓					
Taurine							✓

Measuring protein quality in mono-gastric animals

Protein 'quality' for mono-gastric animals depends on the digestibility of the protein and the balance of the absorbed essential (and non-essential) amino acids. Together these determine the degree to which the amino acids match the amino acid requirement of the animal. The efficiency with which absorbed nitrogen is converted to animal proteins is called the biological value (BV). The biological value of a protein is therefore a measure of the efficiency of utilisation of the amino acids in metabolism. It is estimated by the following equation applied to data from a feeding trial in which the 'true' availability of nitrogen is measured and the 'true' excretion of nitrogen in urine is measured. The 'true' values refer to the correction for nitrogen that appears in the urine and faeces from endogenous sources: these are termed the endogenous urinary nitrogen (EUN) and metabolic faecal nitrogen (MFN):

$$BV(\%) = \left(\frac{\text{absorbed N} - \text{excreted N}}{\text{absorbed N}} \right) \times 100$$

calculated as:

$$BV(\%) = \left(\frac{\text{N intake} - (\text{faecal N} - \text{MFN}) - (\text{urine N} - \text{EUN})}{\text{N intake} - (\text{faecal N} - \text{MFN})} \right) \times 100$$

where N = nitrogen; MFN = metabolic faecal nitrogen and EUN = endogenous urinary nitrogen.

Generally the biological value of animal proteins is greater than that of plant proteins, because they are more closely matched to the 'ideal' protein composition of animals (i.e. contain a better balance of the essential amino acids). The biological value of proteins varies widely, reflecting the variance in the essential amino acid composition of the feedstuff (Table 1.8).

Note the higher values of BV for the animal proteins with the exception of gelatin, which is deficient in several amino acids including lysine, methionine and tryptophan. Maize protein, zein, is deficient in lysine and tryptophan, and linseed is deficient in lysine. One of the great challenges of animal nutrition is to develop

Table 1.8. Typical biological values (BV) for selected proteins for mono-gastric animals

Plant proteins	BV (%)	Animal proteins	BV (%)
Soybean meal	63–76	Fishmeal	75–90
Linseed meal	73–81	Egg protein (albumen)	95–100
Maize	49–61	Milk protein (casein)	95–97
Rice	83	Meat meal	75–85
Wheat gluten	64	Gelatin	0–20

plant proteins that contain high levels of essential amino acids, particularly to replace fish proteins currently fed to aquaculture-farmed animals. Genetic engineering to alter plant protein 'quality' may provide a solution to this problem and make fish farming more sustainable.

Biological value is an index of the value of the protein that is **absorbed**. To get the true value of a feed's protein you need to multiply the BV by the digestibility of the protein. This value is called the **net protein utilisation**, which is a measure of the proportion of the nitrogen intake that is retained by the animal. Protein 'quality' in pigs and poultry is increasingly expressed as 'standardised ileal amino acid digestibilities' relative to an 'ideal protein' (Urbaityte *et al.* 2009; van Milgen and Dourmad 2015).

Measuring protein quality in ruminants and camelids

The complexity of protein digestion in pre-gastric fermenters and the impact this has on amino acid supply are considered in detail in the relevant species-specific chapters.

Essential vitamins (organic micronutrients)

Vitamins are a diverse group of organic molecules ranging from amines, acids, alcohols and aldehydes, some of which are simple organic compounds while others are complexes with essential minerals. All are indispensable to the metabolic processes that sustain life. Some are synthesised within the gut by microorganisms (the B-group vitamins and vitamin K) and one, vitamin D3, can be synthesised in the skin under the action of ultraviolet radiation. Others, such as vitamin A, are synthesised from precursor compounds (in this case β-carotene) in the intestinal enterocytes, and niacin can be formed from the essential amino acid tryptophan.

Vitamins fall into two broad classes of compounds: the water-soluble vitamins and the fat-soluble vitamins. This distinction is important because solubility determines the mode of absorption from the gastrointestinal tract, transport around the body, entry into cells, storage characteristics and modes of excretion. Fat-soluble vitamins are stored in long-term stores in the liver and fatty tissues. To deplete these stores can take months, so deficiencies of fat-soluble vitamins do not present clinical signs for long periods of time after a deficient diet is introduced. Storage also means the fat-soluble vitamins can accumulate to toxic levels over time, and eventually result in pathology and death. Water-soluble vitamins, on the other hand, are rapidly excreted through the kidney tubules, so clinical signs of deficiency become apparent rapidly. For this reason, and also because water-soluble vitamins are stored in labile stores, toxicity of the water-soluble vitamins is unlikely unless very high doses are provided. Initially scientists divided the essential micronutrients they discovered in milk and liver as either fat-soluble (fraction A) or water-soluble (fraction B); hence the lettering nomenclature began.

Numbers were then added to the letters to distinguish the ever-growing discrete B-group and then on to vitamin C, D, E and K (named for 'Koagulation' factor). The preferred nomenclature of the vitamins is to refer to them by their chemical names, especially given the various forms of each of the vitamins depending on side-chain functional groups, methylation, adenylation, hydrogenation and so.

Fat-soluble or water-soluble vitamins?

The solubility of the vitamins in water or fat determines their mode of absorption, transport in blood, entry into cells, storage sites, modes of excretion and time required for depletion to be reflected in clinical signs.

The essential vitamins, their chemical forms, functions and clinical signs of deficiencies and toxicities are summarised in Table 1.9.

Essential minerals (inorganic micronutrients)

The mineral content of an animal's body is the ash remaining after combustion at 550°C for 6 h. The ash comprises ~4% of bodyweight and contains more than 60 mineral elements. In a 650 kg dairy cow this equates to ~25 kg minerals. However, not all of these are essential to the animal's health. Many are simply components of the animal's environment, diet or water. Of the 60 or so minerals present, only 16 are classified as essential as defined earlier. Of these, nine are considered trace elements (microminerals) and seven are macrominerals. Another list of potential essential minerals is also emerging, but yet to be conclusively established. The essential mineral elements and potentially essential elements are listed in Table 1.10.

The macrominerals are required at levels in the feed of the order of grams/day and expressed as g/kg feed DM or per cent of the feed dry matter, while the trace elements are required at levels of milligrams or micrograms per day and typically the feeds contain mg/kg or parts per million. The macrominerals are generally structural components (e.g. Ca and P in bone; S in hair; S in connective tissues), components of buffers (e.g. Na, P in salivary bicarbonates or phosphates), or involved in muscle contraction (Ca, Mg, P) or nerve function (Na/K). The trace elements are generally present as cofactors in enzyme systems (e.g. Cu in lysyl oxidase; Zn in carbonic anhydrase) or are a component of special molecules (e.g. I in thyroxine; Fe in haem). The absolute levels of the trace elements required by animals are remarkably small. As little as one teaspoon of iodine is required for the health of a 60 kg animal for its entire life!

The mineral content of a plant is largely dictated by the soil type (mineral content, soil pH and fertiliser history), plant species (particularly legumes versus

Table 1.9. Essential vitamins for animals: nomenclature, chemical forms, functions and clinical signs of deficiency and toxicity

Vitamin lettering system	Chemical and biochemical forms	Functions	Clinical signs of deficiency	Clinical signs of toxicity
A	**Retinoic acid** **Retinal** **Retinol**	• Epithelial maintenance • Keratinisation • Vision (formation of the bleaching pigment rhodopsin) • Antioxidant • Protection against mutagenesis	• Anorexia • Skin lesions • Roughened hair/feathers • Xerophthalmia (eye pressure) • Abnormal bone growth • Poor vision in low light • Incoordination, paresis • Reproductive abnormalities • Congenital deformities	• Anorexia • Weight loss • Skin thickening • Scaly dermatitis • Haemorrhage • Poor bone strength • Excessive mucus production
D2 (plants) and D3 (animals)	**Calciferol** **1,25 dihydrocholecalciferol** **1,25 dihydroxyvitamin D3** **ergocalciferol (D2 from plants)**	• Calcium absorption (intestinal) • Calcium resorption (renal)	• Rickets (young) • Osteoporosis (adults) • Lameness, sore joints	• Abnormal Ca deposition in soft tissues • Bones break down to provide the soft tissue with Ca • Renal failure (Ca in tubules)
E	**α-tocopherol** **tocotrienols**	• Antioxidant (with selenium) • Gene expression • Signal transduction	• Cardiac and skeletal muscle abnormalities • Reproductive failure • Immune dysfunction • Muscular weakness (paresis) • Mad-chick disease (poultry) • leaky capillaries = exudative diathesis (poultry) • Mulberry heart disease in pigs • Liver necrosis	• Few reports • Haemorrhage • Neuropathies
K	**Menaquinone (animals) K2** **Phylloquinone (plants) K1**	• Blood clotting (synthesis of prothrombin)	• Haemorrhage anaemia • Increased clotting time (prothrombin time) • Contusions (bruising) • Growth disorders	• Anaemia • Skin and respiratory disorders

	Function	Deficiency	Toxicity
B1 **Thiamine** **Thiamine pyrophosphate (TPP)**	• Component of transketolase (pentose phosphate pathway) • Component of decarboxylases (carbohydrate and protein utilisation) • Energy production • Neural function	• Anorexia • Neurological disorders (paraesthesia – tingling) • Polioencephalomalacia (PEM) • Muscle weakness (paresis)	• Unlikely
B2 **Riboflavin** **Flavin mononucleotide (FMN)** **Flavin adenine dinucleotide (FAD)**	• Energy production (component of FAD and FMN) • Blood cell synthesis	• Neural dysfunction (curled-toe paralysis in birds) • Anaemia • Poor growth	• Extremely unlikely due to rapid urinary excretion
B3 **Niacin** **Nicotinamide adenine dinucleotide (NAD)** **NADH**	• Energy production (component of NAD, NADH, NADP and NADPH)	• Diarrhoea • Depression • Dementia • Death • Necrotic enteritis in pigs • Blacktongue • Possibly increased abortions	• Liver damage • Gastric ulceration
B5 **Pantothenic acid**	• Component of CoA (e.g. acetyl CoA) • Fatty acid utilisation • Carbohydrate utilisation • Protein utilisation • Energy production	• Reduced growth rate • Dermatitis in chickens • Haemorrhage • Rough coats • Anorexia • 'Goosestepping' in pigs	• Unlikely
B6 **Pyridoxine**	• Component of 50 enzymes in carbohydrate and protein metabolism • Skin maintenance • Neural function • Red blood cell formation	• Poor growth • Seizures • Anaemia • Depressed immune system	• Very unlikely
B8 (vit H) **Biotin**	• Protein and fatty acid metabolism	• Scaly skin, dermatitis • Impaired keratinisation	• Unlikely

(Continued)

Table 1.9. (Continued)

	Folic acid		
B9	Folic acid	• Methyl group transfers • Protein metabolism • Gene expression	• Anaemia (macrocytic) • Poor growth • Depressed immunity • Death • Neural tube defects	• Unlikely
B12	**Cobalamin** **Cyanocobalamin** **Hydroxycobalamin** **Methylcobalamin** **Adenosylcobalamin**	• Component of methylmalonyl CoA mutase for propionate entry to TCA cycle • Methyl group transfer and folate metabolism • Methionine metabolism • Red blood cell synthesis • DNA synthesis	• Megaloblastic anaemia • Anorexia • Weight loss • Fatty liver • Wool break • Infertility	• Unlikely
Choline		• Component of acetylcholine neurotransmitter • Component of phosphatidylcholine • Methyl group transfer	• Liver damage • Fatty liver	• Vomiting, salivation, sweating • Liver damage

Table 1.10. Essential trace elements and macrominerals on a bodyweight basis (g or mg/kg bodyweight). Several other potentially essential trace elements are also listed.

Macrominerals	g/kg bodyweight	Trace elements	mg/kg bodyweight
Calcium	15	Iron	28–30
Phosphorus	10	Zinc	10–50
Potassium	2	Copper	1–5
Sodium	1.6	Molybdenum	1–4
Sulphur	1.5	Selenium	1–2
Chlorine	1.1	Iodine	0.3–0.6
Magnesium	0.4	Manganese	0.2–0.5
		Cobalt	0.02–0.1
		Chromium	0.08
		Limited evidence for essentiality	
		Boron	
		Silicon	Requirements not yet established
		Aluminium	
		Arsenic	
		Cadmium	
		Tin	
		Vanadium	
		Bromine	
		Lead	
		Lithium	

Source: Underwood (1981)

grasses), stage of plant growth (maturity, leaf/stem ratio) and the climate in which the feed was produced (Fig. 1.5).

Generally legumes contain higher concentrations of minerals than grasses. Mineral content declines with the maturity of the plant. As a general rule, those minerals that are required at higher levels in the plant for plant growth than are required for animal growth are less likely to create deficiencies. An example is Fe, which is required for plant growth at 100 ppm in the plant tissue, but is only required for animal health at 40 ppm, so Fe is unlikely to be deficient for animal growth (unless subjected to haemorrhage or parasitism). On this basis, the most likely mineral deficiencies in animals are P, Cu, I, Se and Co.

Question: how much iodine does a 60 kg animal need in its 10-year life for good health?
Answer: half a teaspoon!

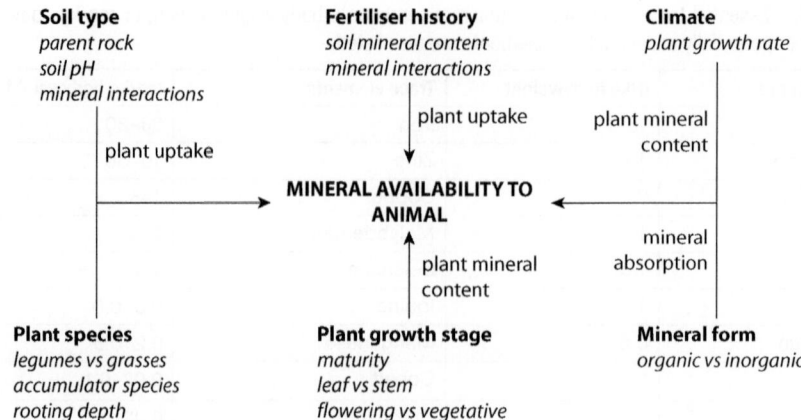

Fig. 1.5. Factors determining the availability of minerals to the animal. Soil type is the major determinant of the mineral status of feeds, hence animal mineral status.

The functions of, and clinical signs of deficiencies of, the essential mineral elements are summarised in Table 1.11.

Essential fatty acids

Rats fed diets deficient in fats develop clinical signs of deficiency, including poor growth, scaly skin, impaired reproduction, poor lactation and ultimately death (see review by Simopoulos 2008). There are two classes of essential fatty acids that must be provided as the parent fatty acid or the derivatives (Fig. 1.6).

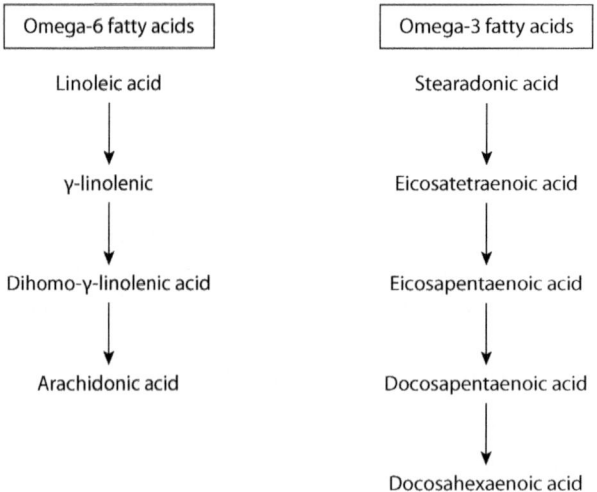

Fig. 1.6. Classes of essential fatty acids for animals.

Table 1.11. Essential mineral elements, their functions and clinical signs of deficiency

Mineral element	Functions	Clinical signs of deficiency
Macrominerals		
Calcium	• Component of bone • Muscle contraction • Nerve function • Coagulation cascade	• **Osteomalacia** (loss of bone mineralisation and density) = rickets in young • **Osteoporosis** (bone density normal but amount of skeletal bone greatly reduced)
Phosphorus	• Component of bone • Component of phospholipids • Component of ATP/ADP and creatine phosphate energy molecules • Component of nucleic acids • Component of buffers	• **Osteomalacia** (rickets in young growing animals) • Depraved appetite (**pica**) • **Haemolysis**
Potassium	• Nerve function with Na • Osmotic equilibrium with Na • Activation of pyruvate kinase	• **Abnormal heart rate** • Growth **retardation/emaciation** • **Muscle weakness**/paralysis • **Diarrhoea**
Sodium	• Neural function • Acid/base balance • Osmotic homeostasis • Absorption of glucose and amino acids	• **Dehydration** • **Poor growth**
Sulphur	• Component of sulphur amino acids cysteine and methionine • Component of insulin, thiamine, biotin and CoA • Major component of wool/hair (as cyst(e)ine)	• Main effects due to S-amino acids deficient • Main effect on **rumen microbes**
Chlorine	• Acid/base balance (chloride shift with bicarbonate) • Component of HCl gastric secretion	• **Alkalosis** • Growth retardation
Magnesium	• Neuromuscular function • Numerous enzyme systems	• **Ataxia** • **Hyper excitability** • Weakness • **Tachycardia** • Osteoporosis
Trace elements		
Iron	• Haem synthesis	• **Anaemia**
Zinc	• Component of carbonic anhydrase for acid/base homeostasis • Component of thymidine kinase for DNA synthesis	• **Anorexia** • Poor growth • Parakeratosis (skin thickening) • Skin lesions • Bone abnormalities
Copper	• Fe transport and haem synthesis • Component of tyrosinase • Component of enzyme linking disulphide bridges	• **Anaemia** • Weak wool with no crimp ('steely' wool) • Enzootic **ataxia** (hind limb paresis or swayback) • **Depigmentation** • Diarrhoea • Weak bones

(Continued)

Table 1.11. (Continued)

Molybdenum	• Component of xanthine oxidase, aldehyde oxidase and sulphite oxidase	• Not observed in nature
Selenium	• Component of glutathione peroxidase • Antioxidant	• Poor growth • **Exudative diathesis** in chicks • Nutritional muscular dystrophy
Iodine	• Component of the thyroid hormones T3 (tri-iodothyronine) and T4 (tetra-iodothyronine)	• Hypothyroidism (**low metabolic rate**, hypothermia); goitre • Hair loss • Reproductive failure • Cretinism in children
Manganese	• Enzyme activator	• **Skeletal deformities** • Poor wound healing (collagen synthesis) • Poor glucose metabolism
Cobalt[a]	• Component of vitamin B12 (cobalamin)	• **Megaloblastic anaemia (see B12)** • **Anorexia** • **Ill-thrift** • **Weak wool** • **Pica (depraved appetite)** • **Rough coat (cattle)**

[a]A fascinating account of the discovery of cobalt as a component of B12 is found in <https://csiropedia.csiro.au/cobalt-deficiency-and-the-cure-for-coast-disease/>

Interactions between nutrients and 'induced deficiencies'

The absolute levels of a nutrient in feedstuffs is not the only determinant of its availability to the animal. The chemical form of the nutrient and its interaction with other nutrients can alter its absorption. Some nutrients 'induce' a deficiency of another nutrient by chemically altering it or by reducing its absorption, usually by competing for the same transporter in the intestines. Other interactions are positive: for example, provision of one nutrient might enhance the absorption of another nutrient or alter the chemical state of the nutrient. Table 1.12 lists some of the major interacting nutrients.

Dose responses of animals to nutrients and mechanisms of nutrient regulation

A generalised relationship exists between the intake of any particular nutrient and the animal response (growth, performance, health) (Fig. 1.7).

Animals have a remarkable ability to maintain the blood concentrations of essential nutrients within tight limits despite widely varying intakes of that nutrient. Homeostatic mechanisms operate to maintain the supply of optimum levels of essential nutrients to biochemical pathways.

Table 1.12. Common nutrient interactions

Nutrient	Interacting nutrients	+ or –	Mechanism
Calcium	Cholecalciferol (D3)	+	↑ Absorption
	Magnesium	–	↓ Absorption
	Zinc	–	↓ Absorption
Zinc	Folic acid	–	↓ Absorption
	Calcium	–	↓ Absorption
	Iron	–	↓ Absorption
	Copper	–	↓ Absorption
	Cobalamin (B12)	+	↑ Absorption
Iron	Calcium	–	↓ Absorption
	Zinc	–	↓ Absorption
	Vitamin A	–	↓ Absorption
	Vitamin C	+	↑ Absorption
Magnesium	Pyridoxine (B6)	+	↑ Absorption/retention
	Calcium	–	↓ Absorption
	Potassium	–	↓ Absorption
	Ammonium	–	↓ Absorption
Copper	Molybdenum	–	↓ Absorption
	Sulphur	–	↓ Absorption
	Zinc	–	↓ Absorption
Manganese	Calcium	–	↓ Absorption
	Iron	–	↓ Absorption
Vitamin A	Zinc	+	↑ Absorption
	Vitamin C	+	Prevents oxidation
	Vitamin E	+	Prevents oxidation
Thiamine (B1)	Pyridoxine (B6)	–	↓ Conversion to active form
Folate	Zinc	–	↓ Absorption
	Vitamin C	+	↑ Retention in tissues
Cobalamin (B12)	Vitamin C	–	Inactivates B12
	Thiamine (B1)	–	Inactivates B12
	Iron	–	Inactivates B12
	Copper	–	Inactivates B12
Vitamin E	Vitamin C	+	Restores oxidised vitamin E
	Selenium	+	Synergistic effects

Next we consider the physiological and metabolic events occurring in each phase.

Phase 1 Acute deficiency

In this zone the concentration of an essential nutrient is insufficient to meet the animal's requirements and the animal shows clinical signs of the deficiency. The signs may be general (poor growth) or specific to the nutrient (neurological, haematological, biochemical, physical). The rate of appearance of the clinical signs

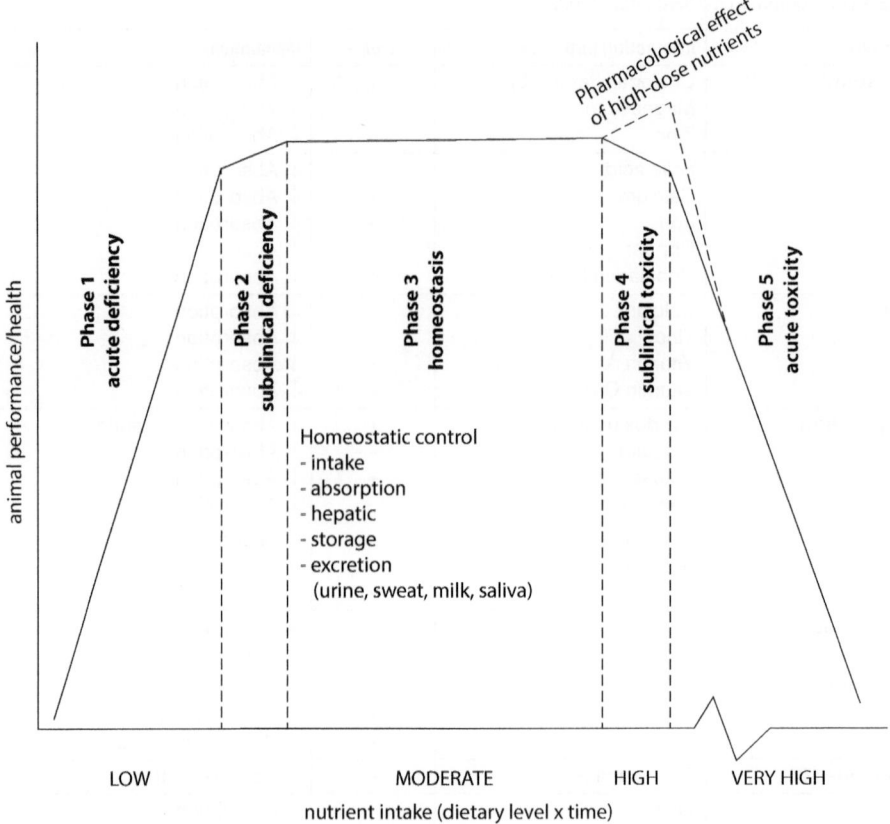

Fig. 1.7. Theoretical general relationship between the intake of a nutrient and an animal's response in terms of performance or health, showing the five phases of response. This general relationship holds for all nutrients, including water and some very toxic compounds. The length of each phase and the nutrient concentration at each boundary differs with the nutrient, the animal species, individual and the length of time the diet has been offered. The pharmacological response to high-dose nutrients (dotted line) is a feature of some nutrients such as zinc, copper, omega-3 fatty acids and vitamin C in some animals.

will depend on the nutrient. For example, a deficiency of a fat-soluble nutrient that is stored in adipose tissue and the liver may not be evident for many months, whereas a deficiency of a B-group vitamin may become apparent over a period of weeks. For non-essential nutrients this phase may not be present.

Phase 2 Subclinical deficiency
This is a common phase for animals in many production systems in which the supply of nutrients is often marginal. It is, by definition, difficult to identify subclinically deficient animals. Carefully designed supplementation trials are required to identify animals in this phase.

Phase 3 Homeostasis

In this phase, the homeostatic mechanisms operate to maintain an optimum level of the nutrient in the tissues in which they are required (e.g. liver, blood, muscle, nerves). By regulating intake, absorption, liver activity, storage (fats, muscles, bones, liver) and excretion (urine, sweat, saliva, faeces, milk), the animal can reduce variations in the nutrient concentration in the feed (Fig. 1.8). If the feed contains low concentrations of the nutrient, the animal may consume more feed, increase the efficiency of absorption of the nutrient from the gastrointestinal tract, mobilise the nutrient from short-term or long-term storage depots, and excrete less of the nutrient from the kidney tubules. If high levels of the nutrient are present in the feed, intake may decline, the efficiency of absorption may decrease, the nutrient may be stored in short-term, medium-term and long-term stores, and the excretion of the nutrient in urine, faeces, milk, sweat and saliva may be increased. The result is maintenance of the nutrient within a range compatible with optimal functioning of the biochemical pathways and tissue requirements.

One of the critical control points for achieving homeostasis of nutrient levels in the body is regulation of absorption from the intestines. Absorption occurs by three main mechanisms:

1. **Passive diffusion:** the movement of the nutrient is directly proportional to the level of intake of the nutrient and its concentration in the digesta.
2. **Facilitated diffusion:** the nutrient binds to a specific transport protein (carrier) protein, which allows it to move down its concentration but does not use ATP to drive against a gradient.
3. **Active transport:** the nutrient attaches to a specific carrier protein, which uses one of two mechanisms to move across the membrane:

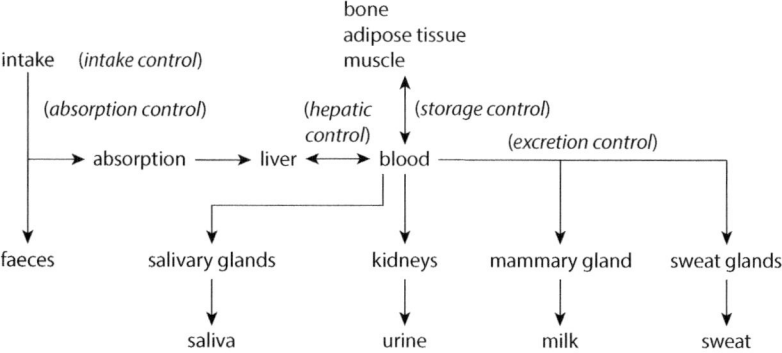

Fig. 1.8. Homeostatic control of nutrient levels in the body. Control mechanisms operate at the level of intake, absorption, excretion and storage to maintain blood levels of nutrients despite varying levels of nutrient intake. Nutrient deficiencies and toxicities occur when these mechanisms are no longer effective.

(a) in co-transport with Na^+ ions, which move down a concentration gradient established by an ATP-dependent pump that pumps $3Na^+$ ions out of the cell and $1Na^+$ in, creating a Na^+ gradient down which the co-transported molecule flows.

(b) in antiport: switches Na^+ for H^+ ions, which then creates a Na^+ gradient to move the nutrient.

Carrier-mediated transport systems are **saturable**; that is, as more of the nutrient is present in the digesta, the carrier becomes progressively saturated with the nutrient and less able to take up more nutrient, in contrast to simple diffusion systems that will continue to absorb nutrient as intake increases. Carrier systems can also be affected by other nutrients if they share the same carrier (see earlier section 'Interactions between nutrients and induced deficiencies').

Phase 4 Subclinical toxicity

If the dietary intake of a nutrient is sufficiently high or provided at an elevated level for a long period of time, the homeostatic control mechanisms (Fig. 1.8) begin to fail and the nutrient begins to accumulate in blood and vital tissues at levels that impair structure and/or function. In the subclinical toxicity phase, this disruption is insufficient to be observed in clinical signs.

Phase 5 Acute toxicity

Continued high level feeding of a nutrient will ultimately cause failure of physiological systems, due to cell death, changes to protein configuration and enzyme activity, changes in blood acid/base balance, neurological failure, dehydration, neuromuscular pathology and ataxia, haemolysis, convulsions, coma and ultimately death.

When some nutrients are present at high levels, they can function as pharmacological agents rather than nutrients. Perhaps the best known example of a nutrient possibly acting as a pharmaceutical agent is high-dose vitamin C. Vitamin C (ascorbic acid) is an essential nutrient for humans and primates with redox functions and as a cofactor for several enzyme systems including those involved in collagen synthesis (Naidu 2003). There is evidence that high-dose vitamin C reduces the growth of various tumours in animal models. Similarly, high doses of copper and zinc improve weaner pig performance, improve intestinal health and have antibiotic effects. However, apart from these examples, and perhaps omega-3 fatty acids (Simopoulus 2008), there is little evidence of positive effects of prolonged use of high-dose nutrients. Given that the nutrient is acting as a 'drug' at these levels, it is highly likely that there will be negative side effects to prolonged high levels of any nutrient. High levels of vitamin C (ascorbic acid), for example, produce renal problems and blood acid/base problems. Prolonged high levels of omega-3 fatty acids have also been associated with blood clotting problems.

Although the generalised dose response curve in Fig. 1.7 holds for all essential nutrients, the precise nature of the relationship differs widely between different nutrients and classes of nutrients. The shape of the curve and the length of each phase will reflect the following characteristics of the nutrient:

- whether or not it is an **essential** nutrient
- its **solubility**; fat-soluble nutrients will accumulate in association with fatty deposits throughout the body. Such nutrients will ultimately reach toxic levels within those tissues. Water-soluble nutrients will be excreted rapidly through the kidneys and therefore are less likely to reach toxic levels in the body.
- its **absorbability**; some nutrients are absorbed with very low efficiency and can be present at high concentrations in the diet without causing problems
- its **absorption mechanism**; some nutrients have specific transport proteins embedded in the intestinal cells. High intakes of these nutrients will overwhelm the capacity of the transporter thereby limiting uptake.
- its **storage**; some nutrients are stored in labile (short-term) stores such as muscles, and others are stored in longer term stores such as bone. Stores with slower turnover will delay the onset of toxicities and deficiencies.

Theoretical relationships between nutrient concentration and animal response to three nutrients differing in the factors above are shown in Fig. 1.9.

Thiamine is water soluble, is rapidly eliminated from the body through the kidney tubules, and is not stored in large quantities. The body can maintain

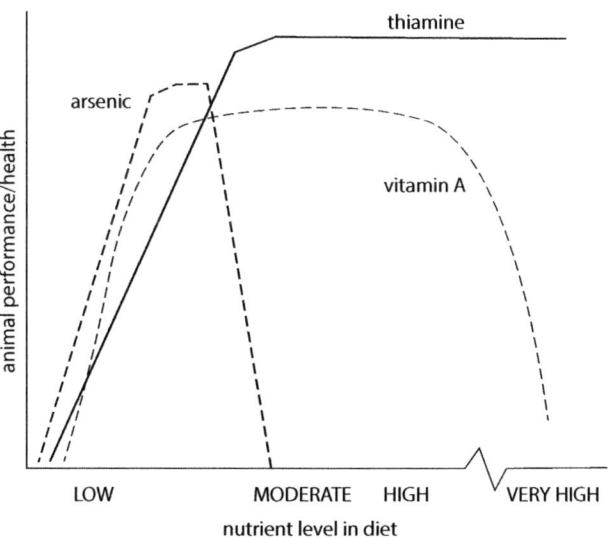

Fig. 1.9. Theoretical dose response curves for three diverse nutrients, arsenic, vitamin A and thiamine. Arsenic has a very tight profile, reflecting high toxicity and limited capacity for homeostasis to control blood levels. Thiamine is excreted rapidly from the kidneys and hence has low toxicity. Vitamin A is toxic but only after prolonged high intakes, and deficient only after prolonged low intakes.

thiamine homeostasis despite very high intakes of the vitamin. Arsenic is water soluble but accumulates in tissues and reaches toxic levels rapidly. A clear requirement for arsenic has not yet been established but positive effects have been demonstrated in several animal species at very low levels (Uthus 1992). The range between these low levels and toxic levels is very small. Vitamin A is fat soluble and stored in the liver in large quantities. Storage and mobilisation as feed levels vary allows a long homeostasis phase but, unlike thiamine, it eventually becomes toxic.

Integrated biochemistry of animal metabolism

Despite the complexity of feeds consumed by animals and the diversity of animal species and digestive strategies, the nutrients available for metabolism are remarkably simple and similar across animal species (Fig. 1.10). In pre-gastric fermenters (ruminants, camelids and macropods), the fermentation products acetic and propionic acid are absorbed across the forestomach epithelium. Butyric acid is metabolised in the epithelial cells to β-OH butyric acid (a ketone body). The epithelial cells of the small intestine absorb amino acids, simple sugars such as glucose, and minerals and vitamins. Fatty acids are absorbed into the lacteals (lymph vessels) of the intestinal villi and are transported to the heart via the lymphatic system. In all species, volatile fatty acids are absorbed from the large intestine (colon and caecum, when present). Importantly, with the exception of the fatty acids and fat-soluble substances such as vitamins A, D, E and K, all absorbed nutrients travel to the liver in the portal vein, so the liver is the first organ to begin the metabolic transformation of the nutrients arising from digestion and absorption. Propionate entering the liver is converted to glucose and stored as glycogen if the animal is in positive energy balance, or it enters the systemic circulation as glucose. Acetic acid and β-OH butyric acid are transferred to the systemic circulation as short-chain fatty acids for energy production or fat synthesis. Amino acids entering the liver can be used to synthesise proteins, are deaminated to produce keto-acids that are used for energy production, or enter the systemic circulation. Vitamins entering the liver are either transferred to the systemic circulation or stored in the liver (particularly the fat-soluble vitamins). Minerals are stored or passed into the circulation. All of the post-liver nutrients are joined in the circulation by nutrients arising from tissue catabolism to become the pool of nutrients available to tissues for metabolism, storage, synthesis or excretion (Fig. 1.10).

Summary of metabolism in animals

Figure 1.11 summarises the major pathways of metabolism of proteins, carbohydrates and fats in animals.

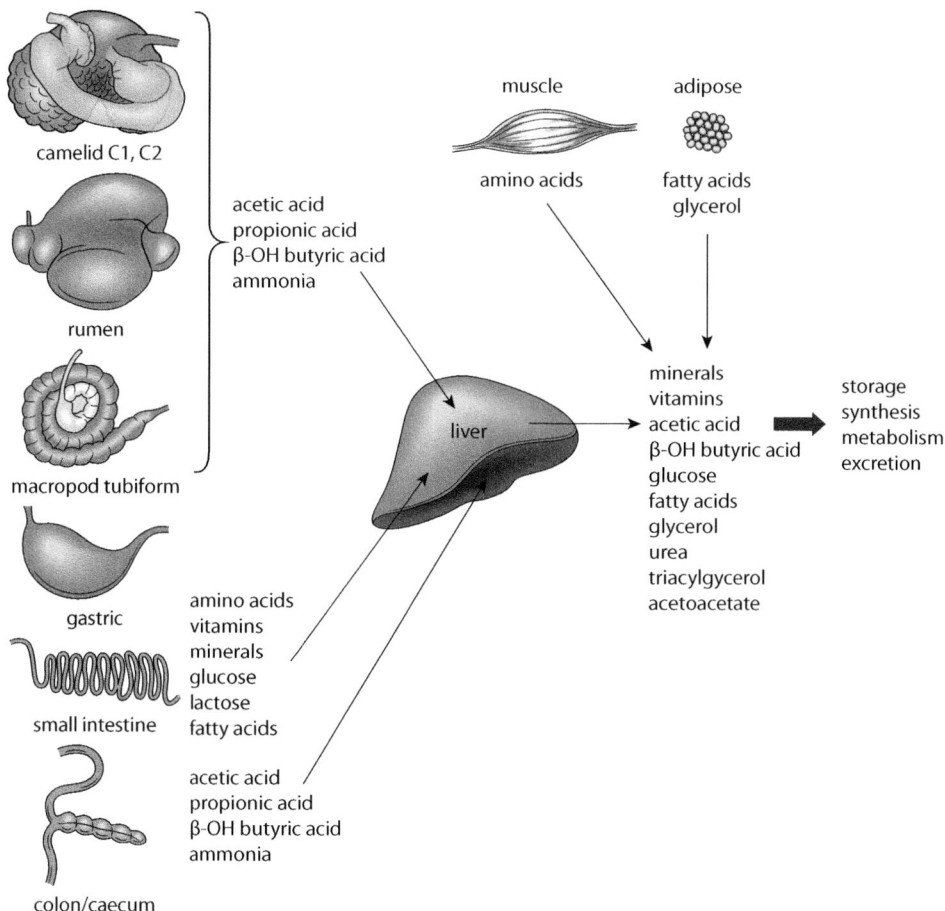

Fig. 1.10. Absorbed nutrients from the gastrointestinal tract. All absorbed nutrients, except fatty acids and fat-soluble compounds, enter the portal vein and travel directly to the liver where they are stored, used for energy production, transformed into other nutrients or pass through into the general circulation via the caudal vena cava. The post-liver nutrients join with nutrients mobilised from tissues or from catabolised tissues to form the available nutrient pool.

Glucose (6C) is broken down in the glycolytic pathway to two molecules of pyruvate (C3) and then either enters the TCA cycle as acetyl CoA or is converted to lactic acid if there is no oxygen available. The first five reactions in glycolysis make up an energy investment phase because ATP is converted to ADP to phosphorylate the sugars. The last five reactions are the energy-generating phase, so overall glycolysis produces pyruvate, two molecules of ATP and two molecules of NADH. The pyruvate under aerobic conditions then forms a 2C unit, acetyl CoA, with the loss of a CO_2 molecule. Acetyl CoA is converted to citric acid (a 6C compound), then isocitrate and a decarboxylation step then produces α-ketoglutarate, which is

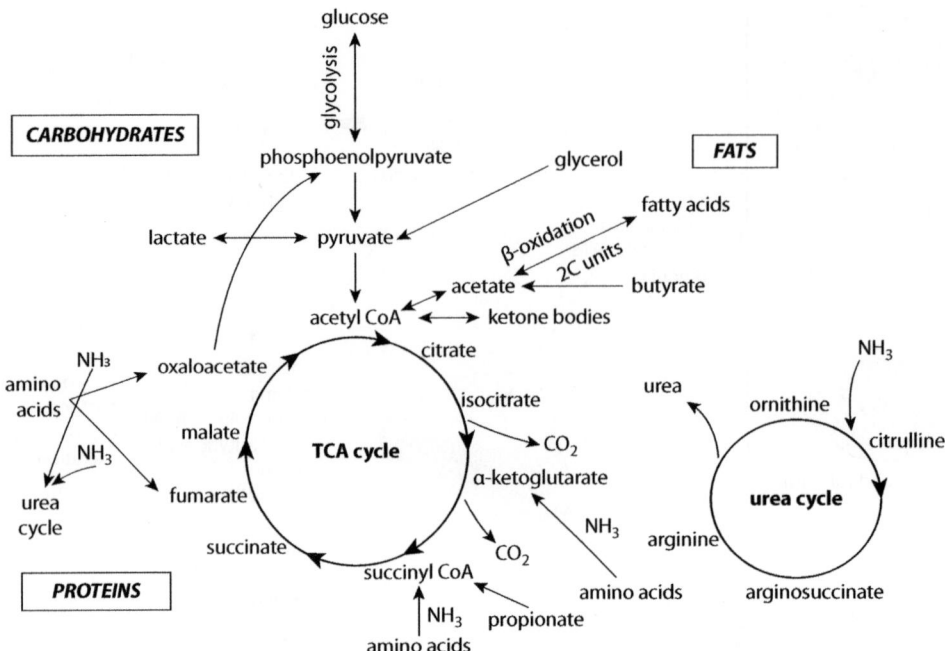

Fig. 1.11. Summary of the major pathways of carbohydrate, protein and fat metabolism in animals. Note the interdependence of the three nutrient classes. The outputs of the TCA cycle are three molecules of NADH, one molecule of $FADH_2$ and one molecule of GTP. The reduced equivalents (NADH and $FADH_2$) are oxidised in oxidative phosphorylation to produce ATP. In total each mole of glucose produces ~36 mols of ATP from glycolysis, TCA cycle and oxidative phosphorylation combined.

converted to succinyl CoA (4C) after another decarboxylation. These two decarboxylation steps are important because the two carbons from β-oxidation (acetyl groups) of fatty acids are now lost rapidly as CO_2. This is why fatty acids, *per se*, which yield two C units from β-oxidation of fatty acids, cannot drive the TCA cycle. Gluconeogenic precursors are also required to burn fats in the TCA cycle ('fats burn in the fire of carbohydrates'). Succinyl CoA is converted to succinate and succinate to fumarate by dehydrogenation. Fumarate is converted to malate and malate to oxaloacetate.

'Fats burn in the fire of carbohydrates'

- 2-carbon fatty acids enter the TCA cycle as acetyl CoA.
- Two CO_2 molecules are lost early in the TCA cycle, so fats cannot drive the TCA cycle alone.
- Fatty acid 'burning' requires glucose precursors to provide TCA intermediates after α-ketoglutarate.

Glucogenic and ketogenic amino acids

Ketogenic amino acids can be converted into acetyl CoA or acetoacetate, the precursors of the ketone bodies. Only two amino acids are strictly ketogenic: lysine and leucine. Amino acids that can be converted to oxaloacetate via pyruvate or from TCA intermediates after α-ketoglutarate can form glucose and are termed glucogenic amino acids. Most of the amino acids are glucogenic (except lysine and leucine) and a few are both glucogenic and ketogenic (phenylalanine, isoleucine, tyrosine, threonine, tryptophan). The mnemonic, PITTT, is a useful way of remembering the ketogenic plus glucogenic amino acids. Figure 1.12 shows the fate of the amino acids in gluconeogenesis and ketogenesis.

Gluconeogenesis is particularly important in pre-gastric fermenters (ruminants, macropods and camelids) because simple sugars such as glucose are rapidly fermented by microbes and are therefore not available for critical roles in neural function and fetal metabolism. Approximately half of the glucose available to ruminants, for example, comes from amino acids; the remainder coming from the conversion of propionate to succinate (Annison *et al.* 1963).

The interdependence of the nutrient classes (carbohydrate, protein and fats) means that, aside from the essential amino acids and fatty acids and an adequate

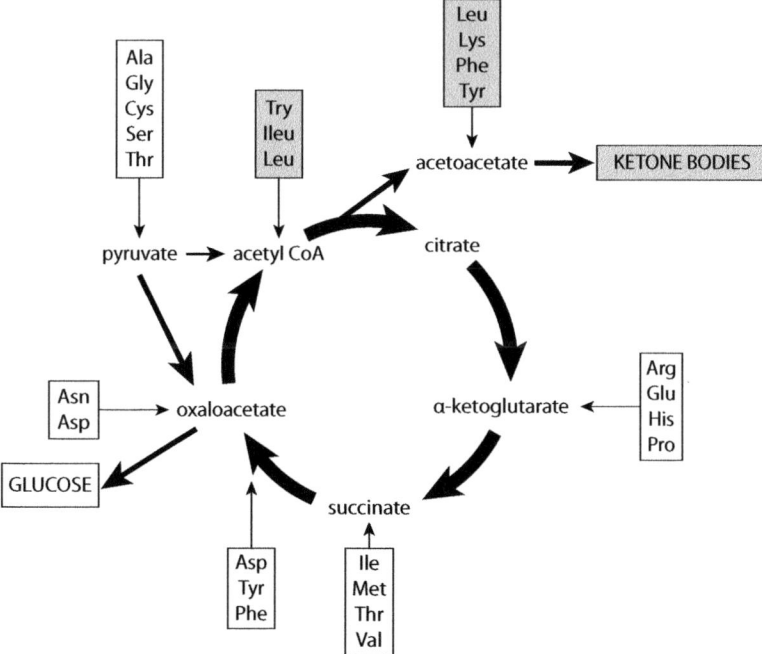

Fig. 1.12. Glucogenic and ketogenic amino acids. Ketogenic amino acids (shaded) produce acetyl CoA or acetoacetyl CoA. Glucogenic amino acids produce intermediates of the TCA cycle after α-ketoglutarate. Some amino acids can produce both ketones and glucose depending on their entry points into the TCA cycle or into pyruvate or acetoacetate.

supply of glucose to drive the TCA cycle, the animal is resilient to changes in the levels of fats, carbohydrates and proteins in the ration. Excess protein results in the amino acids being deaminated, the carbon skeletons used for energy production and the ammonia converted to urea and excreted in urine or recycled to the digestive tract. The fatty acid, propionate, can also be used to make glucose, which is particularly important in milk production (see Chapter 8). Fats can be converted to carbohydrates and enter the TCA cycle as acetyl CoA to produce NADH and $FADH_2$. Energy is generated from these pathways when reduced electron carriers in the form of NADH and $FADH_2$ are transferred to the electron transport chain in the mitochondria. Re-oxidation of these reduced electron carriers creates a proton-motive force that drives H^+ out of the mitochondria. The final complex in the electron transport chain (Complex V) uses this energy gradient to drive ATP synthesis form ADP. Excessive carbohydrate produces acetyl CoA, which can enter fatty acid synthesis pathways to produce fatty acids.

Key features of metabolism in animals

- Energy is produced in animals from the entry of carbohydrates, amino acids and fats into the TCA cycle, forming 3NADH, $1FADH_2$ and 1GTP molecule. The NADH and $FADH_2$ then generate ATP in oxidative phosphorylation in the mitochondria of cells.
- Carbohydrates and some amino acids can produce glucose but fatty acids are very limited in their capacity to produce glucose via ketone bodies.
- Propionate and glucogenic amino acids are important sources of glucose for ruminants and camelids.
- Interconversion of the major nutrients means animals are very resilient to changes in relative amounts of carbohydrates, fats and proteins in diets.
- Despite the wide range and quality of feeds consumed by animals, the nutrients available for metabolism are simple carbohydrates, amino acids, vitamins, minerals and fatty acids.

References

ACMF (2011) *Facts and Figures*. Australian Chicken Meat Federation, Sydney, <http://www.chicken.org.au/industryprofile/downloads/The_Australian_Chicken_Meat_Industry_An_Industry_in_Profile.pdf>.

Annison EF, Leng RA, Lindsay DB, White RR (1963) The metabolism of acetic acid, propionic acid and butyric acid in sheep. *The Biochemical Journal* **88**, 248–252. doi:10.1042/bj0880248

Breuer LH, Pond WG, Warner RG, Loosli JK (1964) The role of dispensable amino acids in the nutrition of the rat. *The Journal of Nutrition* **82**, 499–506. doi:10.1093/jn/82.4.499

Brody S (1945) *Bioenergetics and Growth with Special Reference to Efficiency Complex in Domestic Animals*. Reinhold Publishing, New York, USA.

Brosh A, Aharoni Y, Degen AA, Wright D, Young B (1998) Estimation of energy expenditure from heart rate measurements in cattle maintained under different conditions. *Journal of Animal Science* **76**, 3054–3064. doi:10.2527/1998.76123054x

FAO (2003) *World Agriculture: Towards 2015/2030: an FAO Perspective*. Earthscan Publications, London, UK.

Hume ID (1982) *Digestive Physiology and Nutrition of Marsupials*. Cambridge University Press, New York, USA.

Kleiber M (1947) Body size and metabolic rate. *Physiological Reviews* **27**, 511–541. doi:10.1152/physrev.1947.27.4.511

Klemm D, Heublin B, Fink H-P, Bohn A (2005) Cellulose: fascinating biopolymer and sustainable raw material. *Angewandte Chemie International Edition* **44**, 3358–3393. doi:10.1002/anie.200460587

Mariotti F, Tome D, Mirand PP (2008) Converting nitrogen into protein – beyond 6.25 and Jones' factors. *Critical Reviews in Food Science and Nutrition* **48**, 177–184. doi:10.1080/10408390701279749

Marston HR (1948) Energy transactions in the sheep I. The basal heat production and heat increment. *Australian Journal of Biological Sciences* **1**, 93–129. doi:10.1071/BI9480093

Naidu KA (2003) Vitamin C in human health and disease is still a mystery? An overview. *Nutrition Journal* **2**, 7. doi:10.1186/1475-2891-2-7

Rogers QR, Chen DM, Harper RG (1970) The importance of dispensable amino acids for maximal growth in the rat. *Proceedings of the Society for Experimental Biology and Medicine* **134**, 517–522. doi:10.3181/00379727-134-34826

Simopoulos AP (2008) The importance of the omega-6/omega-3 fatty acid ratio in cardiovascular disease and other chronic diseases. *Experimental Biology and Medicine* **233**, 674–688. doi:10.3181/0711-MR-311

Tilley JMA, Terry RA (1963) A two-stage technique for the *in vitro* digestion of forage crops. *Journal of the British Grassland Society* **18**, 104–111. doi:10.1111/j.1365-2494.1963.tb00335.x

Underwood EJ (1981) *The Mineral Nutrition of Livestock*. 2nd edn. Commonwealth Agricultural Bureau, Slough, UK.

Urbaityte R, Mosenthin R, Eklund M (2009) The concept of standardised ileal amino acid digestibilities: principles and application in feed ingredients for piglets. *Asian-Australasian Journal of Animal Sciences* **22**, 1209–1223. doi:10.5713/ajas.2009.80471

Uthus EO (1992) Evidence for arsenic essentiality. *Environmental Geochemistry and Health* **14**, 55–58. doi:10.1007/BF01783629

van Milgen J, Dourmad J-Y (2015) Concept and application of ideal protein for pigs. *Journal of Animal Science and Biotechnology* **6**, 15. doi:10.1186/s40104-015-0016-1

van Soest PJ (1967) Development of a comprehensive system of feed analysis and its application to forages. *Journal of Animal Science* **26**, 119–128. doi:10.2527/jas1967.261119x

White CL, Martin GB, Hynd PI, Chapman R (1994) The effect of zinc deficiency on wool growth and skin and wool follicle histology of male Merino lambs. *British Journal of Nutrition* **71**, 425–435. doi:10.1079/BJN19940149

Wu G (2014) Dietary requirements of synthesisable amino acids by animals: a paradigm shift in protein nutrition. *Journal of Animal Science and Biotechnology* **5**, 34. doi:10.1186/2049-1891-5-34

2

Digestion in the mono-gastric animal

Key points

- Animals with a single-chambered gastric stomach fall into one of three categories: the mono-gastric carnivores; the mono-gastric omnivores; and the mono-gastric herbivores.
- The anatomy of the jaws, dentition, gut length and size of the caecum and colon reflect the preferred diet and the digestive strategy employed by the animal.
- The mono-gastric carnivores (e.g. cats) have an absolute requirement for pre-formed vitamin A, the amino acid taurine and the fatty acid arachidonic acid. They also require high dietary levels of arginine and niacin.
- The gut wall of mono-gastrics comprises an external serosa, muscularis externa (longitudinal and circular muscles) and a mucosa (muscularis mucosa, lamina propria and mucus membrane). The enteric nervous system is extensive and together with hormones, coordinates the digestive processes.
- The entero-endocrine cells in the gut make up only 1% of the gut cells but comprise the largest endocrine organ in the body.
- The gastric stomach comprises four regions: the cardia at the entrance of the oesophagus; the fundus; the body; and the pyloric antrum at the entrance to the duodenum through the pyloric sphincter.
- The gastric stomach mucosa is folded into numerous gastric 'pits' that are lined by simple columnar epithelium containing specialised cells for producing pepsinogen (chief cells), mucus (neck cells), hydrochloric acid (parietal or oxyntic cells), histamine (entero-endocrine cells) and gastrin (G cells).
- The small intestinal epithelia comprises simple cuboidal or columnar cells lining finger-like projections (villi) with numerous microvilli on their luminal side.

Between the villi are crypts containing the intestinal stem cells and transit-amplifying cells migrating from stem cell division. Goblet cells produce mucins to coat the mucosa, the epithelial cells produce digestive enzymes, Paneth cells produce antimicrobial compounds and entero-endocrine cells produce a suite of hormones. The epithelium of the large intestine is smooth and has no villi.

- The enzymes of the pancreas (peptidases, proteases, lipases, amylases) and bile from the liver are usually secreted into the duodenum via the common bile duct.
- Simple products of digestion (amino acids, simple sugars, vitamins, minerals) are absorbed by simple diffusion, facilitated diffusion or active transport. Many interactions (positive and negative) between nutrients occur at the level of transporter activity.
- Gut-associated microbes (the microbiota) number about the same number as host cells and up to 100 times the levels of gene expression (the microbiome). This gut microbiome is the subject of enormous research effort to determine its effect on animal health and disease.
- Microbial activity in the large intestine of mono-gastric animals produces short-chain volatile fatty acids that can contribute significantly to the animal's energy supply (particularly in large herbivores such as horses).
- Common pathologies of digestion in mono-gastrics include gastric ulceration (horses) and exocrine pancreatic insufficiency (dogs).

Introduction

Digestive strategies of mono-gastric animals

As discussed in Chapter 1, the digestive anatomy of animals can be broadly divided into those with a mono-gastric (simple-stomached) system and those with more than one stomach compartment. The animals with multi-chambered stomachs include the macropods (kangaroos, wallabies, tree kangaroos, pademelons, quokkas, potoroos, bettongs), the camelids (camels, alpacas, llamas, vicunas), the ruminants (cattle, bison, sheep, goats, deer, antelopes, giraffes) and the hippopotamus. These animals are pre-gastric fermenters that consume diets of varying quality (differing in fibre content, protein content, physical form and chemical content). Some are considered 'grazers' (e.g. sheep and cattle), some are 'browsers' (e.g. goats) and some with a combination of the two are called 'intermediate' feeders (e.g. some deer), but all are herbivores, living exclusively on plant-derived materials (leaves, shoots, roots, petioles, seeds, nuts and fruits). The simple-stomached, mono-gastric animals, in contrast, have more diverse diets. Broadly, they can be divided into: mono-gastric carnivores, which require meat for all, or the majority, of their diet; mono-gastric omnivores, which vary widely in the relative proportions of meat and plant components; and mono-gastric herbivores, which consume only plant-derived materials.

The mono-gastric carnivores

Some mono-gastrics, such as the cats (members of the *Felis* and *Panthera* genera), are considered obligate or strict carnivores, although they do have the capacity to digest some carbohydrates (see Chapter 11). Carnivores have a wide muzzle allowing a wide gape for grasping prey. The muscles used for chewing (the masseters and pterygoids) are reduced in cats, and the temporalis muscle is greatly enlarged to provide a strong bite force for capturing and holding struggling prey. The dentition of carnivores comprises long, sharp canines and blade-like molars, spaced and overlapped to provide a slicing action on muscles of their prey. The temporo-mandibular joint of carnivores is broadly in the same plane as the teeth, which provides a simple hinged jaw. This anatomical arrangement has two main consequences: first the bite produces a 'back-to-front' scissors motion slicing the food; and second, it provides a very stable and strong vertical bite with little lateral movement. This structure prevents dislocation by struggling prey. There is no salivary amylase detectable in cats (McGeachin and Akin 1979) and little opportunity for mixing of the feed with saliva because chewing is minimal. The digestive tract of carnivores is short and simple with a small gastric stomach, short small intestine, short large intestine and small caecum. This reflects a diet high in protein and energy comprising muscle, fat, connective tissue, blood and small quantities of highly digestible carbohydrates (muscle glycogen). The stomach of the large cats (*Panthera* genus) can expand to hold some 60–70% of the total tract contents when large meals of captured prey are consumed. Domestic cats, on the other hand, appear better suited to many smaller meals, perhaps reflecting greater social solitude and less pack-hunting. The carnivores have a very small caecum and a smooth, non-haustrated (no 'baffles') colon that has a similar diameter to the small intestine. Little is known of the bacterial population inhabiting the large intestine of cats. Garcia-Mazcorro and Minamoto (2013) reported that the large intestine of cats contained 10^{11} bacteria/g of contents and that 10 phyla of anaerobic bacteria were present. The contribution of anaerobic bacterial end products to nutrient supply in cats on different diets is not known, but is likely to be small.

Carnivores have evolved to rely on the nutrient content of their prey to provide, for example, pre-formed vitamin A (synthesised in the prey animal from carotenoids). The high levels of ammonia generated by proteolysis in carnivores increases the activity of the urea cycle (see Fig. 11.9) and thereby the requirement for arginine. The amino acid taurine, synthesised in non-carnivores from L-cysteine, cannot be formed in carnivores because they lack the enzyme cysteinesulphinate decarboxylase. Taurine is therefore an essential dietary amino acid for cats. Cats also require high levels of dietary niacin, because they have a limited capacity to convert the essential amino acid tryptophan to niacin. The unsaturated fatty acid, arachidonic acid, is essential in cats because, unlike other animals, they cannot synthesis it from linolenic acid.

The mono-gastric omnivores

The omnivores are a very diverse group of animals that consume a wide variety of feeds. The diet of omnivores can vary from close to 0% plants/100% animals (Polar Bear) to 100% plants/0% animals at certain times of the year (Grizzly and Black bears). This evolutionary variation is reflected in wide anatomical variations in dentition, jaw anatomy, gut length and caecal volume. Grizzly and Black bears in Yellowstone National Park, USA, for example, consume a diet that varies throughout the year and, at times, includes moths, berries, fish, elk carcases, mushrooms, ants, weeds and earthworms (Herrero 1978). To accommodate periods of animal consumption, bears have a low plane of the temporo-mandibular joint relative to their teeth, to provide stability and strength to the jaw. They lack a caecum and, unlike caecal and ruminal fermenters, cannot digest low-quality plant materials. Instead, bears have adopted a behavioural strategy of selecting plant diets carefully to ensure only highly digestible parts are consumed.

The mono-gastric herbivores

The mono-gastric herbivores range in bodyweight from a few kilograms (rodents) to more than 5000 kg (elephants). All are post-gastric fermenters (i.e. fermentation activity mainly in the large intestines). They are broadly classed as 'colonic' fermenters or 'caecal' fermenters. This distinction seems to depend on bodyweight. The colonic fermenters are large animals (>50 kg) and caecal fermenters are small (<5 kg) (Hume 1982). The jaws of these plant eaters have well-developed chewing muscles (masseters and pterygoids), and the temporo-mandibular joint is positioned above the plane of the teeth, thereby allowing sideways movements of the jaw to allow grinding of the plant parts between broad, closely spaced molars. Canines, if present, are usually small and used mainly for fighting rather than feeding. The saliva of the mono-gastric herbivores may contain amylase, which has sufficient time to act on carbohydrates released from the plant cells during mastication.

A major disadvantage of being a mono-gastric herbivore is that, unlike the pre-gastric fermenters, the microbial synthesis of protein and vitamins B and K takes place **after** the absorptive small intestinal epithelium. Lagomorphs (rabbits, hares) and some rodents have adopted a strategy of caecotrophy in which special pellets produced in the caecum overnight are consumed in the early morning. The caecal pellets are smaller than normal faecal pellets and are covered in a slimy mucous containing high levels of vitamins B and K.

Table 2.1 summarises the digestive anatomy and special features of nutrition in the major classes of mono-gastric animals.

Table 2.1. Selected list of major mono-gastric animal species, with a summary of the characteristic anatomical and nutritional features of different mono-gastric digestive strategies

Digestive strategy	Anatomical features	Nutritional features
Mono-gastric carnivores Domestic cats (*Felis catus*) Large cats (*Panthera* spp.) 	• Wide muzzle for grasping prey • Large *temporalis* muscle for grasping prey; reduced *masseters* and *pterygoid* muscles • Sharp, elongated canines • Blade-shaped molars that overlap to slice muscle • Temporo-mandibular joint in same plane as teeth allows back-to-front scissor-like motion of jaws to slice muscle • Simple hinge-type jaw attachment with little lateral movement • Short digestive tract • Large expandable stomach volume (60–70% of total GIT) for large meals • Short, narrow caecum if present	• No salivary amylase • Digestion dominated by small intestine (high-nutrient feeds) • Require pre-formed vitamin A • High requirement for niacin, arginine • Taurine is required in diet • Arachidonic acid is required in diet • High protein requirement • Lack glycolytic activity (glucokinase) so limited carbohydrate digestive capacity
Mono-gastric omnivores Bears (*Ursus* spp.) Dogs (*Canids* spp.) Pigs/swine (*Sus* spp.) Primates (*Homo, apes*) Rodents (*Rattus* spp., *Mus* spp.) Raccoon (*Procyon* spp.) Some poultry (*Gallus* spp.) 	• Combination of sharp canines (grasping, tearing) and broad molars (grinding plants) • Intermediate variations between carnivores and herbivores on masseters, pterygoids and temporalis muscles for grasping and chewing • Varying temporo-mandibular plane relative to teeth plane for varying slicing/chewing motions (e.g. bears have a flatter plane to allow stable grasping) • Varying caecum volume	• Highly variable ratio of plants/animal diets

Table 2.1. (Continued)

Mono-gastric herbivores (caecum and colon fermenters)		
Horses (*Equus caballus*) Rabbits (Oryctolagus spp.) Guinea-pigs (*Cavia* spp.) Some poultry (Gallus spp.) Koalas (*Phascolarctos cinereus*) Wombats (*Vombatus* spp.) Elephants (*Loxadonta* and *Elephas* spp.) Pandas (*Ailuropoda melanoleuca*) Rhinoceros (*Rhinoceros* spp., *Diceros* spp.)	• Well-developed masseters and pterygoids for chewing • Thick muscular tongue • Fleshy lips • Temporo-mandibular joint above teeth plane for lateral chewing/grinding • Canines (if present) are small and used mainly for fighting • Teeth closely spaced for grinding • Saliva may contain amylase	• Digestion dominated by large intestines • Depend on symbiotic relationship with microbes in large intestine for energy (VFA) supply • Caecotrophy can provide microbial protein and vitamins B and K in caecal fermenters • More selective than most ruminant herbivores; higher protein, more digestible diets

Caecal fermenters (<5 kg)

Colonic fermenters (>50 kg)

Anatomy and histology of the mono-gastric digestive tract
General anatomy of the simple mono-gastric gastrointestinal tract

Figure 2.1 shows the simple relatively simple gastrointestinal tract of a typical mono-gastric animal, in this case the pig. The stomach has three main regions: the fundus, body and pylorus leading to the duodenum. The stomach of mono-gastrics varies between species in its relative size and capacity to expand. The duodenum loops around the pancreas. The jejunum enters the ileum, which joins the caecum and colon at the ileo-caeco-colic junction. The size of the caecum varies between species as does the size of the colon. Detailed descriptions of these regions and their epithelia are now considered.

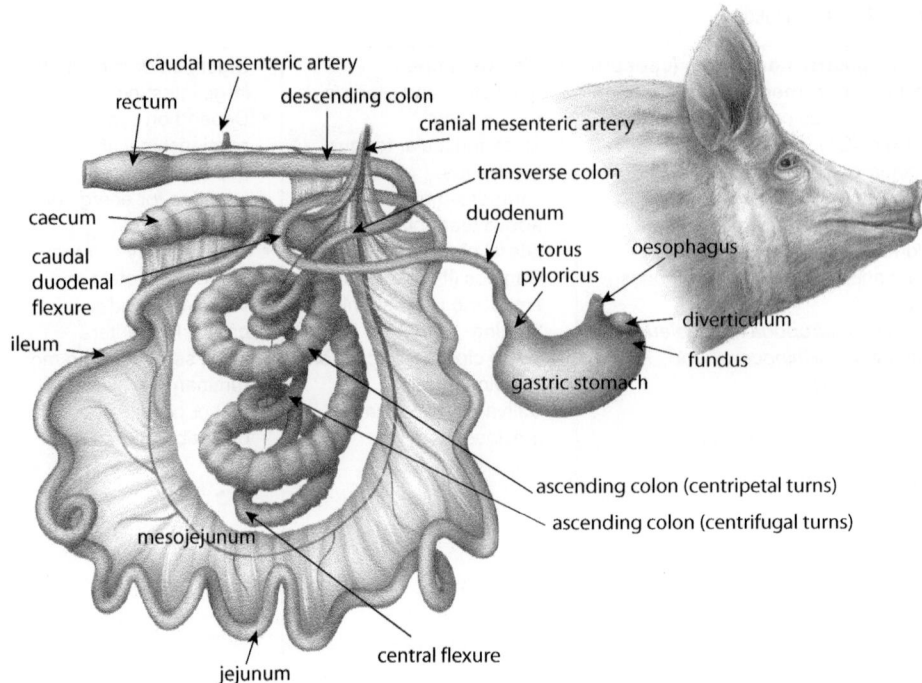

caudal mesenteric artery
rectum
descending colon
cranial mesenteric artery
transverse colon
caecum
duodenum
caudal
duodenal
flexure
torus
pyloricus
oesophagus
ileum
diverticulum
fundus
gastric stomach
ascending colon (centripetal turns)
ascending colon (centrifugal turns)
mesojejunum
central flexure
jejunum

Fig. 2.1. The simple, mono-gastric gastrointestinal tract of a pig, a general model for all mono-gastric animals.

Generalised structure of the gastrointestinal wall

The cross-section of most parts of the gastrointestinal tract reveals an outer layer of connective tissue (the serosa), two layers of smooth muscle running at 90 degrees to one another (the muscularis externa), another layer of loose connective tissue (the submucosa) containing blood vessels, and the mucosa comprising the muscularis mucosae, lamina propria and the mucous membrane (Fig. 2.2).

The mucous membrane contains epithelial cells that vary in morphology throughout the tract. Most epithelial cells are absorptive but some are specialised to produce mucus (goblet cells) or hormones (entero-endocrine cells). The lamina propria contains loose connective tissue, blood vessels, but also lymphoid tissue arranged in nodules called Peyer's patches. The thin layer of smooth muscle (the muscularis mucosae) contracts the mucosae to maximise contact between the digesta and the mucous membrane. The submucosa contains thick connective tissue allowing elasticity of the gut. Blood vessels and lymph vessels are also present in the submucosa. The gut is extensively innervated by an enteric nervous system comprising a submucosal plexus called the Meissner's plexus and a myenteric plexus between the muscle layers in the muscularis externa (also called the myenteric or Auerbach's plexus). The two muscle layers control the motility of the gut. The serosa

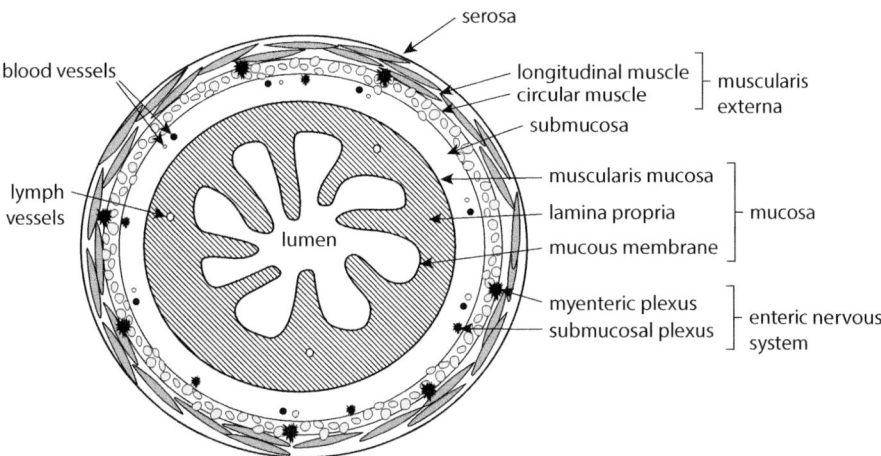

Fig. 2.2. Generalised structure of the gastrointestinal tract showing the concentric layers from outside to the lumen (serosa, muscularis externa, submucosa and mucosa). Different sections of the tract differ in the degree of folding of the epithelium and the nature of the epithelial cells in the mucosa.

comprises an inner layer of fibrous connective tissue and an outer layer of epithelial tissue called the mesothelium, which secretes a watery fluid. This allows the gut and other organs to slide easily passed one another. The mesothelium is continuous with the mesentery: the thin layer of connective tissue containing the blood vessels and nerves. The mesentery is continuous with the peritoneum and attaches the gut to the abdominal wall. The greater omentum is a fold of peritoneum hanging from the stomach and encapsulating the intestines. It is thin, appears perforated, and contains adipose tissue and extensive vasculature. The omentum appears to 'float' over the intestines. It contains macrophages and may play a role in limiting peritoneal infections.

Stomach anatomy and histology

The gastric stomach in most mono-gastric animals is a simple, expandable sac immediately behind the diaphragm. It has four anatomical regions: the cardia, fundus, body and pyloric antrum. The entry of ingesta to the stomach is via the lower oesophageal (cardiac) sphincter and the exit to the duodenum is via the pyloric sphincter, the sphincters being rings of thick smooth muscle. The wall of the fundic region is thin and easily expandable. The wall of the body region is folded into longitudinal pleats called 'rugae', which flatten when the stomach expands during a meal. The pyloric antrum has thick muscular walls which contract forcefully to mix the gastric contents and to empty the contents into the duodenum.

The lining of the stomach contains numerous gastric 'pits' or invaginations, which are lined by various secretory cells (Fig. 2.3). The cells at the top of the gastric pits secrete a bicarbonate-rich mucus. Those in the 'neck' of the pits, the

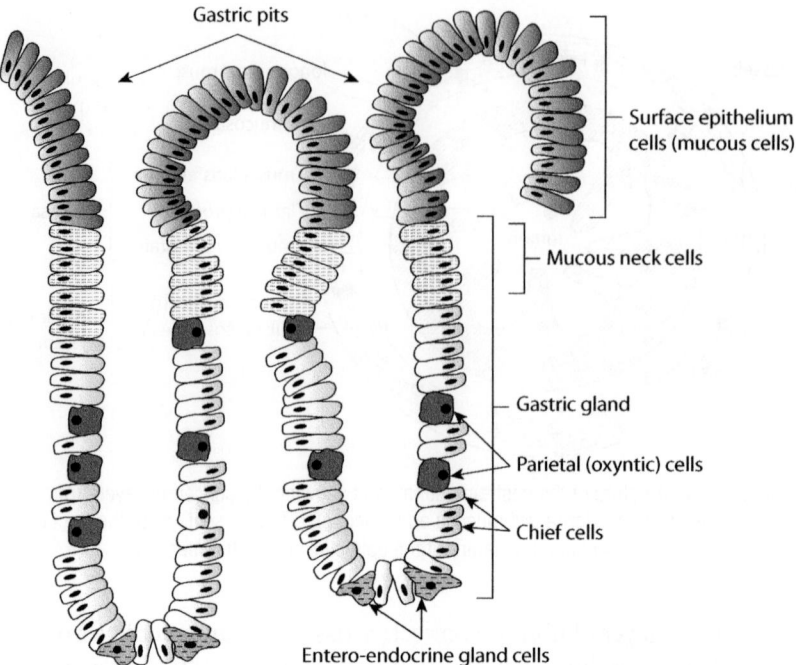

Gastric pits

Surface epithelium
cells (mucous cells)

Mucous neck cells

Gastric gland

Parietal (oxyntic) cells

Chief cells

Entero-endocrine gland cells

Fig. 2.3. Diagrammatic representation of the epithelium of the simple gastric stomach. The stomach wall has a mucus-lined epithelium with invaginations (gastric pits) containing various secretory cells. The cells at the top of the pits produce a bicarbonate-rich mucus; neck cells also produce mucus. The parietal or oxyntic cells produce hydrochloric acid and intrinsic factor. Chief cells produce pepsinogen and the entero-endocrine cells produce histamine. G cells present in the pyloric antrum produce the hormone gastrin.

neck cells, also produce mucus but without bicarbonate. The gastric gland forming the bottom half of the pits contains chief cells that produce pepsinogen and, in young animals, the protease rennin (chymosin), which coagulates milk. Parietal cells produce hydrochloric acid and also intrinsic factor: a glycoprotein necessary for absorption of vitamin B12 in the small intestines. In the base of the pits, the enterochromaffin-like cells are entero-endocrine cells that produce and release histamine in response to gastrin, a hormone produced by the G-cells of the pyloric antrum (Håkanson and Sundler 1991). Histamine, in turn, regulates acid production by the parietal cells (see 'Gastric secretions' later).

Intestinal anatomy and histology

The small intestinal epithelium comprises numerous villi, finger-like projections ~1 mm long, with crypts of Lieberkühn between each (Fig. 2.4). The cells lining the villi and crypts are called enterocytes. Enterocytes are usually simple columnar cells and those involved in absorption have numerous microvilli (as many as 3000–6000/cell). The villi increase the intestinal surface area for absorption and

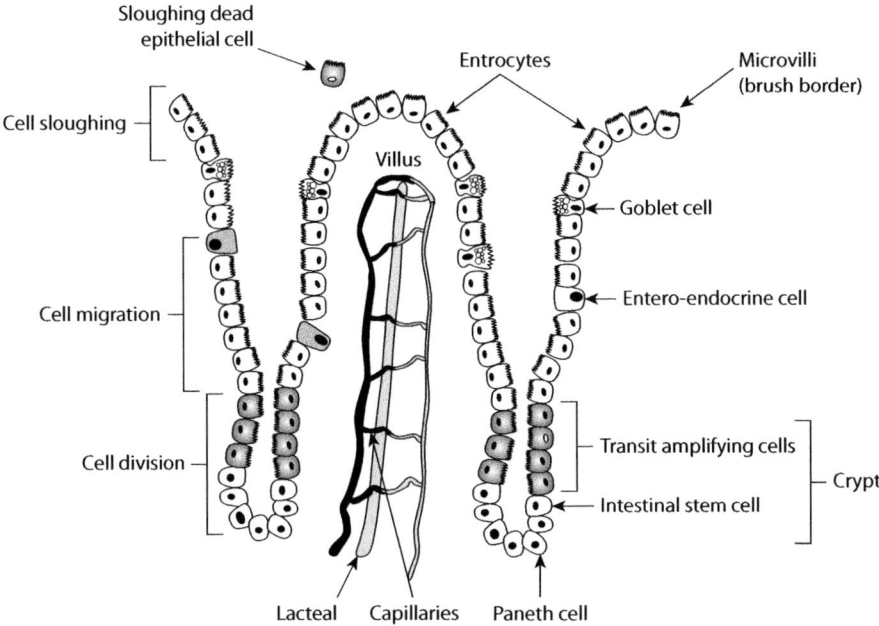

Fig. 2.4. Diagrammatic representation of an intestinal villus and two crypts of Lieberkühn. Stem cells in the basal crypts divide and become transit-amplifying cells, which migrate up the villus and eventually slough off into the intestinal lumen. The apical membrane of the villus epithelial cells contains numerous microvilli, which greatly increase the absorptive surface area and produce brush border enzymes. Goblet cells contain mucin in vesicles and release these to form a glycopolysaccharide layer with important protective functions. Entero-endocrine cells produce a wide range of hormones controlling gut motility, gut secretions and feeding behaviour.

secretion by 10-fold, and the microvilli increase the surface area a further 20-fold. The surface of the cells with microvilli is called the 'brush border' and the enzymes produced by these cells are called the brush border enzymes (see 'Gastric secretions' later). Goblet cells are long, thin cells scattered throughout the epithelium. They produce mucin, which is stored in secretory vesicles. Paneth cells, which produce antimicrobial proteins and peptides, such as defensins, are found in the lower crypts scattered among the intestinal stem cells. The stem cells produce transit-amplifying cells, which in turn rapidly divide and migrate towards the lumen to produce the intestinal cell populations. Entero-endocrine cells reside scattered in the upper half of the villi (Fig. 2.4). Although entero-endocrine cells account for only 1% of the entire intestinal epithelium, they nevertheless are the largest endocrine organ in the body. They produce hormones that regulate gut function, feed intake (see Chapter 4), blood flow and the immune system.

Entero-endocrine cells comprise only 1% of the intestinal cells but are the largest endocrine organ in the body.

Colon and caecum anatomy and histology

The wall of the colon has a mucosa comprising low columnar to cuboidal epithelial cells lining straight tubules and resting on a thin basement membrane. There are no villi in the colon or caecum. The epithelium contains absorptive cells and large numbers of mucus-secreting goblet cells. The base of the crypts contains the proliferative cells, which migrate up the crypt and slough off into the lumen. The lamina propria contains capillaries, lymphatics and inflammatory cells. The muscularis mucosa is thin and regular. The submucosa comprises loose connective tissue and Meissner's plexuses. The inner layer of the muscularis externa contains the plexus of Auerbach.

Secretions of the mono-gastric gastrointestinal tract

The remarkable feats of the digestive tract are that: it can digest food but doesn't digest itself; it harbours as many bacteria as there are cells in the human body (Sender *et al.* 2016), but is not overwhelmed by them; it produces hydrochloric acid and powerful enzymes but doesn't dissolve itself; and it is exposed to toxins in foods but isn't killed by them. On top of this, the majority of the body's immune system resides in the gut epithelia. Many of these feats are achieved by the secretion of mucus, which plays a multifaceted role in housing commensal bacteria, preventing overwhelming immune responses and protecting the enterocytes from hydrolysis. Much of the protective activity of the mucus is a function of the presence of complex carbohydrates (oligosaccharides), which cover the proteins and prevent the proteolytic enzymes reaching the peptide bonds (Pelaseyed *et al.* 2014). The mucus glycocalyx and the unstirred water layer at the intestinal surface form a diffusion barrier through which nutrients must pass before entering the absorptive cells. Other protective mechanisms include secretion of enzymes as inactive pro-enzymes, which are only activated when secreted into the gut lumen. Bicarbonate (salivary, intestinal and pancreatic) plays an important role in buffering gastric acids, and acid secretion is regulated by hormones to prevent ulceration and gastritis.

The digestive tract:
- digests feed using powerful enzymes and hydrochloric acid but does not digest itself
- harbours as many bacteria as there are cells in the body, but is not overwhelmed by them
- is exposed to food toxins but is not killed by them
- is the largest endocrine organ in the body
- contains more nerves than the central nervous system.

Entero-endocrine cells in the stomach and intestines

Entero-endocrine cells present in the epithelium of the stomach and intestines secrete more than 50 different peptides, which generally affect the secretory and motor function of the gut. Some of these hormones, such as CCK, ghrelin, PYY and GLP-1, are generally regarded as important factors that affect the short-term regulation of food intake, acting both at the CNS level and locally.

Gastric secretions

There are six major secretions of the gastric epithelium: (1) **hydrochloric acid** from the parietal cells; (2) **pepsinogen**, an inactive zymogen from the chief cells; (3) **rennin** or chymosin, a protease that coagulates milk caseins; (4) the hormone **gastrin**, which regulates gastric emptying and acid secretion; (5) **bicarbonate-rich mucous**, which covers the luminal surface, protecting and lubricating the stomach wall from damage by acid, pepsin and irritating chemicals in foods; and (6) **intrinsic factor**, a glycoprotein required for absorption of cobalamin (vitamin B12) in the small intestine (see 'Absorption of vitamins' later). In some species, gastric lipase is also present.

Mechanisms of acid secretion by the stomach

The secretion of hydrogen and chloride ions into the gastric lumen is an active process working against concentration gradients (the concentration of hydrogen ions in the lumen is ~3 million times higher than in blood (Yao and Forte 2003). A K^+/H^+ ATPase pump (the proton pump) pumps hydrogen ions into the lumen in a swap with potassium ions, using energy from ATP (Fig. 2.5). The hormone acetylcholine acts on the muscarinic receptor, histamine works on the H2-type receptor and gastrin on the gastrin receptor to increase acid secretion.

Pepsinogen secretion and activation

The main proteases in the acid stomach are the pepsins, which are members of the aspartic protease family. Pepsinogens are synthesised and stored as inactive pro-enzymes (zymogens) in chief cells of the gastric epithelium. On exposure to acid (pH 1.8–3.5 optimum) the activation peptide is cleaved, unmasking the active proteolytic site of the enzyme.

Rennin (chymosin) secretion in the stomach

If milk proteins consumed by young animals remained soluble and liquid, they would pass rapidly out of the stomach, reducing their digestion and availability. The enzyme rennin coagulates these milk proteins by degrading κ-casein. Because κ-casein keeps the other caseins in solution, its degradation by rennin results in the caseins becoming insoluble and 'clotting' into a semi-solid curd like soft cheese, which remains in the stomach. This then allows proteolysis of the caseins under the action of the pepsins.

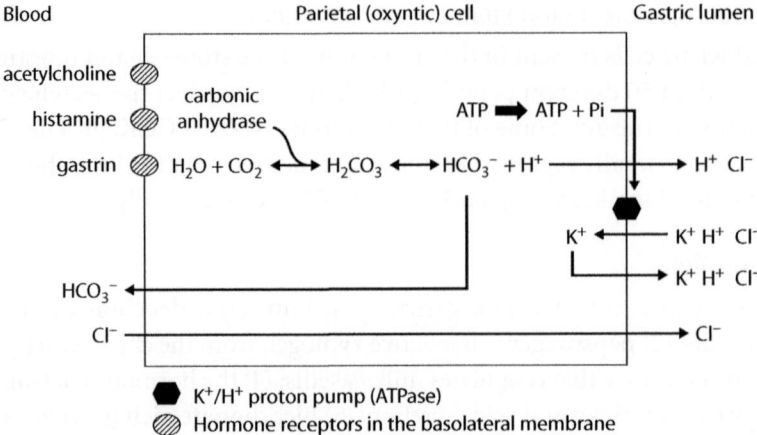

Fig. 2.5. Mechanism of gastric acid secretion into the gastric lumen. The K$^+$/H$^+$ proton pump uses ATP to actively pump H$^+$ ions into the lumen in an electrochemical 'swap' with K$^+$ ions. Chloride ions, which have swapped with bicarbonate ions in a chloride 'shift' on the basolateral membrane, enter the lumen through the apical membrane. The hormones acetylcholine, histamine and gastrin interact with receptors on the blood side to alter the activity of the proton pump, hence acid secretion. Inhibitors of this pump (e.g. omeprazole) are used to prevent gastric ulceration (see Chapter 12, page 333).

Secretion of intrinsic factor and absorption of vitamin B12

The gastric parietal cells secrete intrinsic factor (IF), a glycoprotein that is required for the absorption of vitamin B12 (cobalamin) in the intestines. Congenital deficiencies in the synthesis of intrinsic factor produce 'pernicious' anaemia as a direct consequence of B12 deficiency and subsequent failure of red blood cell synthesis. Binding of the IF to B12 takes place in the small intestine when pancreatic proteases digest other proteins bound to B12 in the stomach. The free B12 can then bind to IF, which binds to ileal receptors facilitating absorption (see 'Absorption of vitamins' later).

Intestinal secretions and absorption

The duodenal epithelium contains numerous Brunner's glands that secrete an alkaline, bicarbonate secretion to neutralise the gastric acid (pH 2.0–3.5). This raises the digesta pH closer to the pH optima of the intestinal and pancreatic enzymes (6.0–7.0). The total volume of fluid moving across the intestinal epithelium in pigs is ~6 L/day and this movement of fluid is a secondary consequence of the activity of transporters of Na$^+$, Cl$^-$ and HCO$_3^-$. Digestion of nutrients in the small intestine is brought about by the activity of secretions from the enterocytes (intestinal cells) and the bile and pancreatic juices secreted into the intestines. In pigs (45 kg), ~1.8 L of bile and 1.8 L of pancreatic juice enter the duodenum daily. The enzymes on the surface of the microvilli are called 'brush border' enzymes. They comprise six carbohydrases (lactase, trehalase, sucrase, isomaltase, maltase II and maltase III) and a large number of peptidases. The luminal phase of digestion is carried out by the salivary, gastric and pancreatic enzymes, which degrade nutrients into smaller substrates that are then further degraded by the enzymes of the brush border (the mucosal phase of digestion).

Paneth cells at the base of the crypts of Lieberkühn are thought to play an important role in defending the stem cells around them. They produce α-defensins and other antimicrobial peptides **such as** lysozymes. These substances are stored in secretory granules and released into the intestinal lumen when bacterial antigens are present.

Pancreatic proteases and peptidases

Although proteolysis begins in the gastric stomach under the action of pepsins, the majority of protein breakdown occurs under the action of pancreatic proteases in the intestines. These are synthesised and stored in vesicles as inactive trypsinogen and chymotrypsinogen. When these are released into the intestinal lumen, trypsinogen is converted to trypsin by the enzyme enterokinase embedded in the intestinal mucosa. The trypsin, in turn, converts chymotrypsinogen to the active protease chymotrypsin. Trypsin and chymotrypsin break proteins down to small peptides, but not to amino acids. Pancreatic carboxypeptidase can achieve this, but peptidases in the mucosal wall carry out the majority of the final cleavage to absorbable amino acids.

Pancreatic lipase

Although there may be some gastric lipase in the stomach, pancreatic lipase is the major source of lipolysis. Triglycerides are broken down to a monoglyceride and free fatty acids by pancreatic lipase. Bile salts from the liver, which are present in the pancreatic juice entering the intestines, are then required for fatty acid absorption.

Pancreatic α-amylase

Pancreatic α-amylase hydrolyses starch to maltose, maltotriose, α-limit dextran and glucose. The brush border enzymes degrade these to simple monosaccharides. Fibre and resistant starches pass to the large intestines for microbial fermentation.

Major digestive secretions

- Gastric stomach: HCl acid, pepsinogen, rennin (chymosin), bicarbonate-rich mucus, intrinsic factor.
- Intestinal epithelium: bicarbonate, brush border enzymes (carbohydrases, peptidases, enterokinase), antimicrobial peptides.
- Pancreas: pancreatic amylase, pancreatic lipase, pancreatic proteases (trypsin, chymotrypsin).
- Liver/gall bladder: bile.

Absorption of digestion products in mono-gastric animals
Absorption of carbohydrates in the intestines

Simple, absorbable molecules from the breakdown of complex carbohydrates, proteins and fats are absorbed by simple diffusion or active transport across the brush border membrane of the enterocyte in the upper third of the small intestinal villi. Other transporters then operate in the apical membrane to release the nutrients into the venous drainage of the gut thorough the portal vein to the liver (Levin 1994). The absorption mechanisms for glucose, galactose and fructose are shown in Fig. 2.6. Glucose absorption occurs via a Na symport transporter (SGLT1), which drives glucose into the cell against a concentration gradient (Kiela and Ghishan 2016).

Absorption of amino acids in the small intestine

The small intestine is the primary site of amino acid absorption. Proteins are digested by proteolytic enzymes including gastric pepsin, which cleaves proteins at the aromatic amino acids producing smaller proteins, peptides and amino acids. These are delivered to the duodenum where the pancreatic pro-enzymes, activated by enterokinase from the brush border, further break them down to amino acids, dipeptides and tripeptides. Amino acids enter the epithelial cells by simple diffusion, facilitated diffusion or active transport. Most of the amino acids enter the cell by active transport. The amino acid transport systems (Table 2.2) are broadly based on the electrochemical charge of the amino acid (anionic, cationic, basic). Further characterisation of the transporter activity is then based on their dependence on Na^+, Cl^-, or H^+, their affinity (K_m) for the amino acid, their velocity of action (V_{max}), and the extent to which they are inhibited by specific amino acid

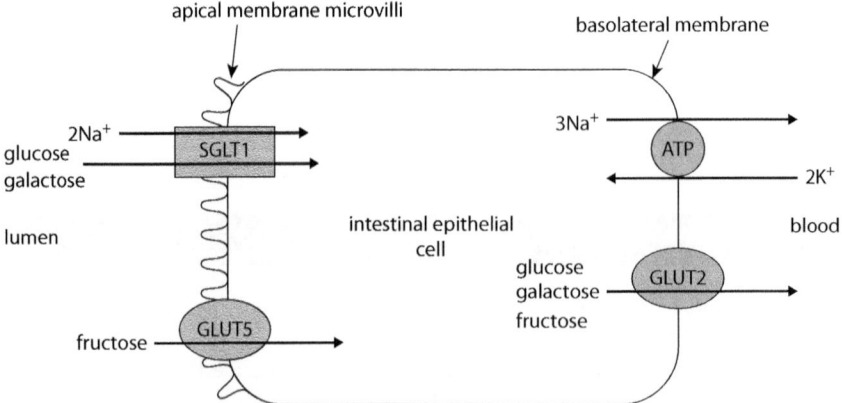

Fig. 2.6.   Absorption of glucose, galactose and fructose across the enterocytes of the small intestine. SGLT1 is a Na-dependent transporter in the apical membrane of the enterocyte. Na follows its concentration gradient downhill because the ATPase pump drives Na out of the cytoplasm (Kiela and Ghishan 2016).

Table 2.2. Intestinal amino acid transport systems

System	Details	Amino acids
B⁰	• Na-dependent neutral amino acids	Arginine, histidine, lysine
B⁰,⁺	• Na and Cl⁻ dependent • neutral and aliphatic amino acids	Arginine, histidine, glycine, alanine, valine, leucine, isoleucine
b⁰,⁺	• Na-independent, neutral, cationic and cysteine	Arginine, histidine, lysine, cystine
IMINO	• Na and Cl⁻ dependent	Proline, hydroxyproline
β	• Na and Cl⁻ dependent	Taurine, β-alanine
X⁻$_{AG}$	• Na and H⁺ dependent • anionic amino acids	Aspartate, glutamate
ASC	• Na-dependent, neutral amino acids	Lanine, serine, cysteine
N	• Na and H⁺ dependent	Glutamine, asparagine, histidine
PAT	• H⁺ coupled electrogenic • short-chain amino acids	Glycine, alanine, proline

Source: Adapted from Kiela and Ghisham (2016)

analogues. Most of the essential amino acids are neutral and are absorbed through systems ASC and B⁰. They then leave the epithelial cell by antiport, in an exchange with glutamine, which then acts in the cell as an intestinal fuel.

Peptides longer than four amino acids are not absorbed by the small intestine. Di-and tri-peptides are efficiently absorbed by a transporter called PEPT1, which is linked to a Na⁺/H⁺ exchanger (NHE3). The peptides taken up by this transporter across the apical membrane are rapidly broken down to amino acids by cytoplasmic proteases.

Absorption of the products of fat digestion

Some fat digestion occurs in the mouth of some mono-gastrics (lingual lipase) and in the stomach (gastric lipase), but the majority occurs in the duodenum under the action of pancreatic lipase and bile acids. Monoglycerides, long-chain fatty acids, phospholipids, cholesterol and fat-soluble vitamins form micelles with the emulsifying bile salts. The contents of the micelles enter the enterocyte where they are re-esterified into triglycerides. These then combine with protein, phospholipid and cholesterol to form chylomicrons. The chylomicrons enter the intestinal lymphatics in the lacteals of the villi (Fig. 2.4) and are transported to the thoracic duct at the heart to enter the systemic circulation. Medium-chain fatty acids can directly enter the enterocytes without being hydrolysed and they can then enter the portal vein directly.

Absorption of vitamins

The water-soluble vitamins are absorbed by a combination of simple diffusion, facilitated diffusion and carrier-mediated active transport (Said 2004). Most water-soluble vitamins are absorbed in the jejunum. Fat-soluble vitamins are absorbed by the same mechanisms as fats in the intestines under the action of bile salts to form chylomicrons. Zinc is essential for the formation of chylomicrons, so zinc deficiency reduces fat-soluble vitamin absorption. Vitamin A is formed from the precursor carotenoids, which are absorbed by simple diffusion and cleaved by dioxygenases to retinaldehyde in the mucosal cells. It is this step that cannot be carried out by carnivores, such as cats, and is the reason they require pre-formed vitamin A.

Biotin is absorbed by the Na-dependent carrier-transporter, SMVT (sodium multivitamin transporter). Absorption of cobalamin (B12) is complicated by the requirement to bind to a factor produced in the stomach, the gastric intrinsic factor (GIF). Cobalamin (B12) binds in the stomach to a transporter, transcobalamin-1 (TC1). Trypsin in the duodenum releases cobalamin from this transporter and it binds to GIF. This B12/GIF complex then binds to a receptor called 'cubam' on the apical membrane. The B12/GIF complex moves to the lysosome where GIF is degraded, releasing the B12 for transport across the basolateral membrane by MRP-1 (multidrug resistant associated protein-1). Folic acid is transported across the membranes by three specific transporters (reduced folate carrier, proton-coupled transporter and the GP-1 anchored folate receptor). The carrier for niacin has not been characterised, but it acts like organic anion transporter 10 (OAT10). Transporters for pyridoxine (B6) have also yet to be identified and characterised. Riboflavin (B1) is transported by riboflavin transporters 1 and 2 (RVT1 and 2) and two thiamine (B1) transporters (THTR1 and 2) have been identified in humans (Said 2004). Similar transporters probably operate in pigs as in humans.

Absorption of minerals

The mechanisms of absorption of minerals are highly variable due to large differences in their electrochemical charge, carrier transport systems, which are variably regulated, and varying interactions between the mineral element and organic compounds in the digesta (Suttle 2011). Regulation of the absorption of minerals is critical to the protection of the animal from potentially toxic levels of the mineral and distinguishing chemically similar elements that are not essential nutritionally. Regulation of absorption must operate to ensure continuing supply of the mineral for metabolic pathways despite low levels of the mineral in the feed (see nutrient homeostasis Fig. 1.8).

Calcium absorption

Calcium absorption occurs by two mechanisms: one is non-saturable and the other is carrier-mediated and therefore saturable (i.e. once the carrier is saturated with calcium no further transport can occur by this mechanism). At high levels of calcium intake, the calcium crosses the epithelium between the cells (paracellular) through tight junctions (which join the cells). Several types of tight junction exist and some increase calcium absorption and others reduce it. Claudins, proteins making up the tight junctions, vary in their selectivity for calcium. Claudins-2, -12 and -15 increase calcium absorption, while Claudin-5 reduces it. Vitamin D3 (1,25(OH)$_2$cholecalciferol) increases calcium absorption from the intestines probably by up-regulating expression of Claudin-2 and Claudin-12 (Fujita *et al.* 2008). The second mechanism is saturable and transcellular (the calcium goes through the enterocyte). It allows absorption of calcium against a concentration gradient and therefore can operate at low levels of calcium in the lumen (low calcium intakes). The mechanism involves calcium channels TRPV-5 and -6, which operate in the apical (lumen side) of the membrane of the enterocyte (Kiela and Ghishan 2016). The calcium enters the enterocyte and binds to calbindin-D9k. The calbindin carries the calcium to the basal membrane and exits via another transporter (NCX1). Both the channels and transporter are up-regulated by vitamin D3 (Suttle 2011).

Phosphate absorption

Phosphate is absorbed in the intestines by a saturable transporter (NaPi-IIb), which is Na- and vitamin D (1,25(OH)$_2$D3)-dependent. At high levels of P intake, passive diffusion of P occurs via the paracellular (between cells) route. The P leaves the basal membrane via facilitated diffusion through unknown transporters.

Magnesium absorption

Magnesium is largely absorbed from the distal small intestine by three mechanisms: passive diffusion, solvent drag and carrier transport (see review by Schweigel and Martens 2000). Passive diffusion through the paracellular route accounts for the majority of absorbed magnesium (Suttle 2011). Magnesium is a good example of an element whose absorption is inhibited by the presence of other minerals and compounds, creating an induced deficiency. The mechanisms of absorption of magnesium in ruminants and the mechanisms of induced deficiency of magnesium are shown in Fig. 7.2.

Absorption of the trace elements

The mechanisms of absorption of the trace elements are diverse and depend heavily on their chemistry, and particularly their electrochemical charge. Free cations are toxic to cells, so cation absorption involves complexation of the element with organic compounds and transport via a carrier protein to protect the intestinal mucosa. There are two processes for trace element absorption: (1) an active, specific transport, which is up-regulated when the element is in low concentration in the feed; and (2) a non-specific transport by the divalent metal transporter DMT-1, which is down-regulated when mineral intakes exceed requirements (Suttle 2011).

It is this transporter that is responsible for the adverse interactions among minerals in absorption. For example, high levels of manganese cause down-regulation of DMT-1 and subsequently poor absorption of iron and zinc. Homeostasis of iron levels in pigs must be tightly regulated at the level of intestinal absorption because there is no excretion mechanism and most iron is retained by recycling mechanisms except during haemorrhage. Iron is absorbed in the ferrous state (Fe^{2+}) and transported across the apical membranes attached to the divalent metal transporter 1 (DMT-1), which also transports magnesium, copper, zinc, cobalt and cadmium. Iron deficiency up-regulates DMT-1. The iron leaves the basolateral membrane of the enterocyte attached to a transporter ferroportin (FPN1). The absorbed iron then attaches to transferrin for blood transport. The majority of zinc is absorbed in the duodenum and proximal colon. There are several zinc transport proteins including Zip4, which is the major apical transporter. Zip5 is the basal transporter of zinc. Zinc is also transported by DMT-1. Copper is absorbed from the stomach and duodenum by a saturable (carrier) process. The carrier (CTR1) has a high affinity for copper. Iodine is mainly absorbed by simple diffusion through the gastric mucosa.

Absorption of fats, amino acids, minerals and vitamins

- Most amino acids are absorbed in the intestines attached to carrier proteins specific for their electrochemical charge (cation, anion, basic).
- Fats join with proteins in the enterocytes to form chylomicrons, which enter the lymphatics in the villi and enter the general circulation at the heart.
- Mechanisms of absorption of minerals are highly variable depending on the charge, valence, solubility and chemical complexing with organic compounds. Most minerals attach to carrier transport proteins, which are common to other minerals, resulting in competitive inhibition of absorption.
- Water-soluble vitamins are absorbed by a combination of simple diffusion, facilitated diffusion and carrier-mediated active transport mechanisms.
- Fat-soluble vitamins are absorbed in the same way as other fatty compounds.

Control of intestinal digestion

The presence of partially digested fats and proteins (peptides and amino acids) in the duodenum stimulates the release of the hormone cholecystokinin, which interacts with receptors in the acinar cells of the pancreas to stimulate the release of pancreatic enzymes into the duodenum. Cholecystokinin also stimulates contraction of the gall bladder, hence the supply of bile to the duodenum. Cholecystokinin is also present in neurones in the intestinal wall and the brain, but the neural role of this hormone is unclear. Secretin, the first hormone discovered by Starling and Bayliss in the early 1900s, is stimulated by the presence of acid in the duodenum. It stimulates the release of a secretion rich in bicarbonate from the bile and pancreas to neutralise the acid.

Digestion in the large intestines

There are no villi in the large intestine, and the mucosa comprises simple columnar epithelium. There are no digestive brush border enzymes. Goblet cells produce large quantities of mucus, which lubricates the digesta passage. About 40% of the total gastrointestinal tract contents reside in the large intestine in many mono-gastrics. The retention time of this material is 20–38 h, allowing time for microbial fermentation of carbohydrates that have escaped small intestinal digestion, and time for reabsorption of water and electrolytes. The reabsorption of water in the colon follows the active uptake of electrolytes, particularly sodium. Sodium and magnesium are absorbed efficiently from the large intestine. Carbohydrates and proteins that have escaped degradation in the small intestine are fermented by microbes in the large intestine. Dietary fibre is digested by the extracellular enzymes (cellulases, hemicellulases and pectinases) of the microbes, producing short-chain volatile fatty acids (VFA) as end products. Methane is also produced but in small quantities, accounting for perhaps 1–2% of total gross energy intake. The main cellulolytic bacteria in the large intestines are *Fibrobacter succinogenes, Ruminococcus flavefaciens, R. albus, Butyrivibrio* spp. and *Clostridium herbivorans*. The short-chain fatty acids include acetic, propionic and butyric acids, which are absorbed across the large intestinal epithelium. The acetate is used directly by peripheral tissues for energy, the propionate is converted to glucose in the liver and the butyrate is converted to ketone bodies by the colonic enterocytes. VFA provide ~20–30% of the total energy supply in mono-gastrics such as pigs, depending on the fibre content of the diet (Bergman 1990). The nitrogen required by the large intestinal bacteria is largely undigested dietary protein, microbial protein from the small intestines, ammonia, and endogenous protein from intestinal sloughing and digestive secretions from the pancreas. The microbial activity in the large intestine greatly influences the total output of faecal nitrogen. This is important when estimating the protein and amino acid requirements, because simply using faecal output as an indicator of the amino acid availability to the pig would provide false and varying values. For this reason, the availability of amino acids to the pig is estimated on the basis of ileal output, not faecal output. There is little absorption of microbial amino acids from the large intestines.

The gut microbiome

The population of bacterial cells in the gastrointestinal tract (the gut microbiota) is enormous. Early estimates, still commonly quoted, were that there are ten times more microbial cells than host somatic cells, but recent estimates indicate there are similar numbers of microbes as host cells (Bull and Plummer 2014). This huge population is referred to as the gut microbiota and the total expression of the microbial genes is referred to as the gut microbiome. The bacteria belong to two major phyla (Bacteroidetes and Firmicutes), but there are more than 1000 microbial species-level organisms. The total expression of microbial genes (the microbiome) is ~100 times the expression of the host genome. This population of microbes has been implicated in gut health (bowel diseases), obesity, Type 2 diabetes and atopy (Bull and Plummer 2014). This population of gut bacteria can protect the host from pathogens and can modulate the immune system. Dysbiosis (an unbalanced population of bacteria) appears to be associated with a suite of metabolic syndromes, but it is difficult to separate cause and effect in these cases. Recent studies showed that the microbes in obese mice had a greater capacity to extract nutrients from feeds than the microbes in non-obese mice and that transplantation of the different microbiota to germ-free mice generated the donor phenotype in the recipient (Turnbaugh *et al.* 2006). Advances in our understanding of the gut microbiome and its interaction with the host will have enormous implications for animal nutrition in the future.

References

Bergman EN (1990) Energy contributions of volatile fatty acids from the gastrointestinal tract in various species. *Physiological Reviews* **70**, 567–590. doi:10.1152/physrev.1990.70.2.567

Bull MJ, Plummer NT (2014) Part 1: The human gut microbiome in health and disease. *Integrative Medicine (Encinitas, Calif.)* **13**(6), 17–22.

Fujita H, Sugimoto K, Inatomi S, Maeda T, Osanai M, Uchiyama Y, *et al.* (2008) Tight junction proteins claudin-2 and -12 are critical for vitamin D-dependent Ca2 absorption between enterocytes. *Molecular Biology of the Cell* **19**, 1912–1921. doi:10.1091/mbc.e07-09-0973

Garcia-Mazcorro JF, Minamoto Y (2013) Gastrointestinal microorganisms in cats and dogs: a brief review. *Archives of Veterinary Medicine* **45**, 111–124. doi:10.4067/S0301-732X2013000200002

Håkanson R, Sundler F (1991) Session 5: Trophic effects of gastrin. *Scandinavian Journal of Gastroenterology* **26** (Supplement 180), 130–136. doi:10.3109/00365529109093183

Herrero S (1978) A comparison of some features of the evolution, ecology and behavior of black and grizzly/brown bears. *Carnivore* **1**, 7–17.

Hume ID (1982) *Digestive Physiology and Nutrition of Marsupials.* Cambridge University Press, New York, USA.

Kiela PR, Ghishan FK (2016) Physiology of intestinal absorption and secretion. *Best Practice & Research: Clinical Gastroenterology* **30**, 145–159. doi:10.1016/j.bpg.2016.02.007

Levin RJ (1994) Digestion and absorption of carbohydrates— from molecules and membranes to humans. *The American Journal of Clinical Nutrition* **59**, 690S–698S. doi:10.1093/ajcn/59.3.690S

McGeachin RL, Akin JR (1979) Amylase levels in the tissues and body fluids of the domestic cat (*Felis catus*). *Comparative Biochemistry and Physiology. B, Comparative Biochemistry* **63**, 437–439. doi:10.1016/0305-0491(79)90274-8

Pelaseyed T, Bergstrom JH, Gustafsson JK, Ermund A, Birchenough GMH, Schutte A, *et al.* (2014) The mucus and mucins of the goblet cells and enterocytes provide the first defense line of the gastrointestinal tract and interact with the immune system. *Immunological Reviews* **260**, 8–20. doi:10.1111/imr.12182

Said HM (2004) Recent advances in carrier-mediated intestinal absorption of water soluble vitamins. *Annual Review of Physiology* **66**, 419–446. doi:10.1146/annurev. physiol.66.032102.144611

Schweigel M, Martens H (2000) Magnesium transport in the gastrointestinal tract. *Frontiers in Bioscience* **5**, d666–d667. doi:10.2741/A542

Sender R, Fuchs S, Milo R (2016) Revised estimates for the number of human and bacteria cells in the body. *PLoS Biology* **14**, e1002533. doi:10.1371/journal.pbio.1002533

Suttle N (2011) Absorption of minerals and vitamins. In *Encyclopedia of Dairy Sciences.* 2nd edn. (Eds J Fuquay, PF Fox and PLH McSweeney) pp. 996–1002. Academic Press, San Diego CA, USA.

Turnbaugh PJ, Ley RE, Mahowald MA, Magrini V, Mardis ER, Gordon JI (2006) An obesity-associated gut microbiome with increased capacity for energy harvest. *Nature* **444**, 1027–1031. doi:10.1038/nature05414

Yao X, Forte JG (2003) Cell biology of acid secretion by the parietal cell. *Annual Review of Physiology* **65**, 103–131. doi:10.1146/annurev.physiol.65.072302.114200

3

Digestion in the ruminant animal

Key points

- Ruminants are even-toed ungulates (hooved animals) and include: deer, giraffes, sheep, cattle, antelopes, bison, ibex, goats, duikers, wildebeest, gazelles, dikdiks, oryx, bighorn sheep and buffalos.
- Ruminant use more than 55% of the total land mass area of the planet, making them one of the most successful groups of mammals on Earth.
- Ruminants can obtain energy from cellulose, the most abundant polymer on Earth (800 billion tonnes) and can obtain protein from non-protein nitrogen sources (such as ammonia) by hosting a microbial population in the first three stomach compartments.
- The ruminant digestive tract has: four stomach compartments (reticulum, rumen, omasum and abomasum); a small intestine (duodenum, jejunum and ileum); and large intestine (large caecum, spiral colon, rectum). The relative capacities and functioning of the compartments vary with feeding habit (grazers, browsers or intermediate feeders).
- The epithelium of the reticulo-rumen and omasum is stratified squamous and contains honeycomb ridges (reticulum), filiform and fungiform papillae (rumen) or leaf-like laminae (omasum).
- The reticulo-rumen contracts in two major movements: Type A movements are mainly backward-moving, mixing movements that average ~1/minute; Type B movements are forward-moving and average ~1 every 2 minutes in both sheep and cattle). Eructation (belching) occurs during some Type B movements and rumination occurs during some Type A movements.

- The onward flow of feed particles from the reticulo-rumen depends on their size, density and shape, which determines their probability of presentation to the reticulo-omasal orifice.
- Ruminal fluid has the following characteristics: pH 5.0–7.4 (acidic); redox potential −350 mV (anaerobic), osmolarity 200–400 mOsm/kg (isotonic);and the gas phase is 65% CO_2, 27% CH_4, 0.6% O_2 and 7.9% N_2.
- Each millilitre of rumen fluid contains approximately 10^{10}–10^{11} bacteria, 10^6 ciliated protozoa and 10^4 anaerobic fungi.
- Microbial fermentation produces short-chain volatile fatty acids, (acetic acid (C2), propionic acid (C3) and butyric acid (C4)), microbial vitamins B-group and K, microbial protein and microbial lipids.
- The microbial nutrients join with Undegraded Dietary Protein, fermentation-resistant carbohydrates, plant lipids, feed vitamins and minerals.
- Many health issues in ruminants relate to poor regulation of ruminal pH (acute lactic acidosis and subacute ruminal acidosis), inadequate eructation of gases (bloat), physical damage to the reticulo-rumen ('hardware' disease), and poor nutrition of the ruminal microbes.

Introduction

The ruminants are one of the most successful groups of terrestrial animals on the planet. They inhabit a wide diversity of environments globally and contribute significantly to many ecosystems, agriculture, recreation and cultures. The success of this group of animals is largely based on their ability to survive and reproduce on low-quality, low-protein and high-fibre plant materials. To produce high-protein, high-value products such as meat, milk, fibres, blood, work, utensils, weapons and skins from such low-quality resources have made them highly valued by humans for millennia. There are ~3056 million domesticated ruminants and 310 million wild ruminants on Earth (van Soest 1994). The ability to digest cellulose in a symbiotic relationship with gut microbes lies at the heart of the success of this group of animals. Plants produce ~180 billion tonnes of cellulose each year, making cellulose the most abundant biopolymer in Earth. No mammal is able to produce cellulase enzymes, so forming a symbiotic relationship with microbes was a major breakthrough in accessing this abundant resource. The total area of land that could only be effectively used by grazing animals is shown in Table 3.1.

Ruminants use ~55% of the total land area globally and only ~5% of that is on land that could be used for cropping. The remainder is rangelands containing vast quantities of low-nitrogen, high-cellulose materials. This is important because

Table 3.1. Total land areas and proportion of land in each land-use category globally

Land use	Area (billion hectares)	Proportion of total (%)	Proportion utilisable by ruminants (%)
Urban/industrial	0.48	3.5	0
Cultivated	1.37	10	5
Unproductive	2.06	15	0
Forests	4.11	30	10
Rangelands	5.48	40	40
TOTAL	**13.7**	**100**	**55**

Source: Adapted from Church (1988)

comparisons are often made between the efficiency of food production by animals and plants. Although animals are significantly less efficient than plants at producing food energy or protein/hectare, they can do so from otherwise unusable landscapes such as rangelands and non-arable areas.

Ruminants play a unique role

Ruminants play vital roles in producing high-quality meat, milk and fibres from low-quality forages on rangelands and from crop residues. Ruminants are less efficient at producing food than plants and mono-gastric animals, but they produce high-quality food from otherwise unusable land areas and substrates.

This chapter covers the digestive anatomy and physiology of digestion in ruminants, providing the background to the chapters on the nutrition of sheep and goats, deer, beef cattle and dairy cattle.

Gross anatomy of the ruminant digestive tract

Figure 3.1 shows a generalised digestive tract of a ruminant. The relative capacities of each stomach compartment and the dynamics of flow of liquid and roughage particles from the reticulo-rumen vary between ruminant species (e.g. deer versus sheep versus goat versus cattle). Variations reflect differences in feeding behaviour and diet preference. Grazers, intermediate feeders and browsers therefore differ in the capacity and turnover rate of material in the reticulo-rumen.

The four 'stomachs' in ruminants derive embryologically from a simple stomach and not the oesophagus as once thought (Warner 1958). It is therefore more accurate to refer to them as four **compartments** of the stomach, a convention followed in this book. The first two compartments are the reticulum ('honeycomb')

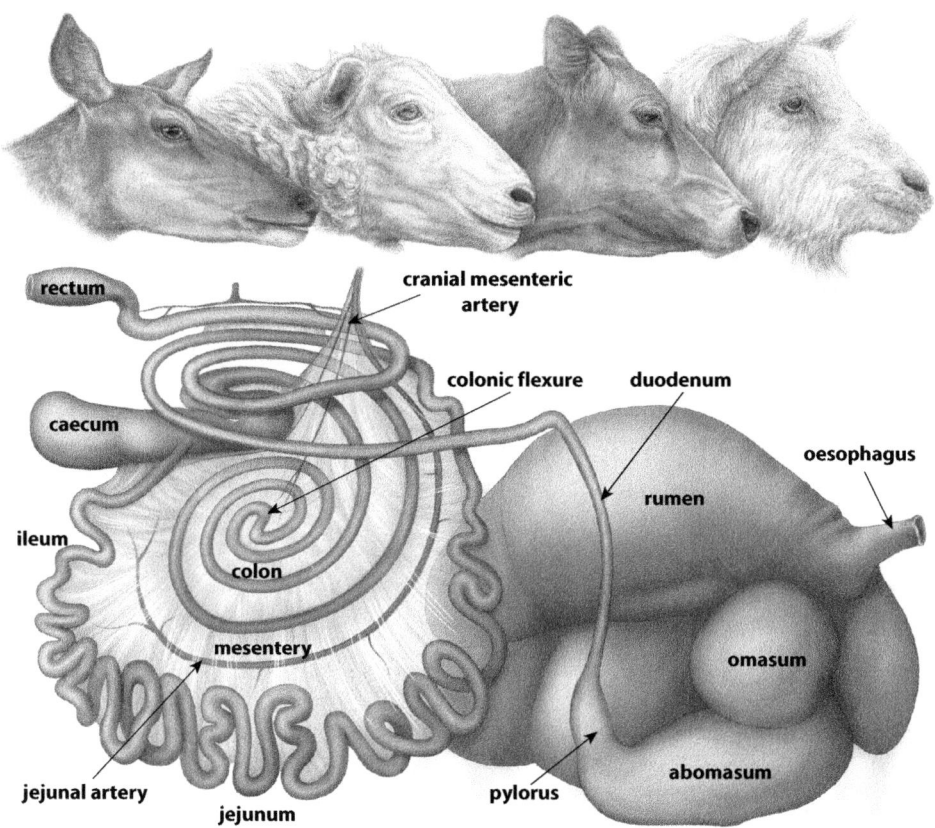

Fig. 3.1. The generalised gastrointestinal tract of ruminants (note the stomach compartments are viewed from the right side; the remainder is schematic to show all regions clearly).

and rumen ('paunch'), often collectively referred to as the reticulo-rumen because they are anatomically and functionally closely linked (see 'Reticulo-rumen motility, eructation, rumination and digesta movements' later). The third compartment is the omasum ('bible') and the fourth, the abomasum ('true' gastric stomach). The abomasum is the ruminant equivalent of the glandular 'gastric' stomach. The four compartments occupy 75% of the abdominal space and 100% of the left half of the abdomen is taken up by the rumen (Fig. 3.2).

The rumen lies in the sub-lumbar fossa, the large cavity beneath the lumbar vertebrae and extending from the diaphragm to the pelvis. The reticulum lies cranial to the rumen on the left of the median plane and compressed against the diaphragm. The omasum and abomasum lie to the right of the median plane. The abomasum is an elongated sac lying largely on the abdominal floor, then extending dorsally to the pylorus and duodenum.

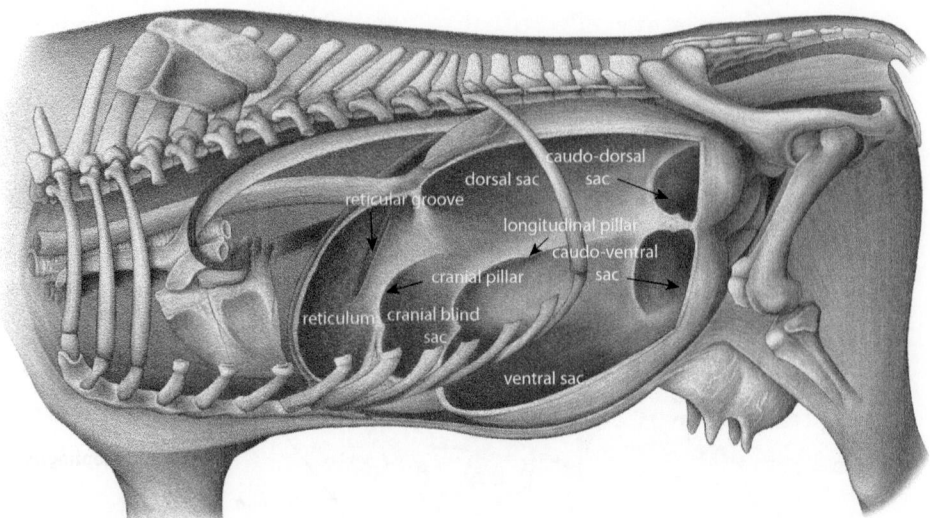

Fig. 3.2. The reticulo-rumen occupies the majority of the left abdominal space in the region known as the sub-lumbar fossa (the large space below the lumbar vertebrae).

Capacity of the ruminant compartments

The relative capacities of the ruminant compartments change with age as the animal develops from a milk-consuming pre-ruminant to a forage-consuming ruminant (Fig. 3.3). In sheep and cattle, the capacity of the reticulo-rumen increases from ~30% at birth to 70% in the adult animal. The capacity of the reticulo-rumen increases from 35% at birth to 60–65% as adults. The capacity of

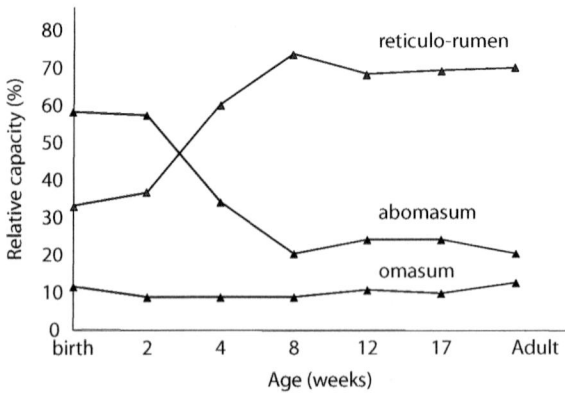

Fig. 3.3. Changes in the relative capacities of the compartments of ruminants with age. The change from milk digestion in the abomasum to forage digestion in the reticulo-rumen is apparent. The compartments of adult ruminants are as follows: reticulo-rumen (60–70%); abomasum (15–20%) and the omasum (10–20%). Source: Adapted from data of Church (1988).

the abomasum declines from 60% at birth to 20% as adults in sheep and from 50% at birth to 14% as adults in cattle. The contribution of the omasum remains fairly constant at ~5–9% in sheep and 14–24% in cattle.

> The relative capacities of the stomach compartments in adult ruminants are 60–70% (reticulo-rumen), 15–20% (abomasum) and 10–20% (omasum).

Anatomy of the reticulo-rumen

The rumen is broadly divided by the longitudinal pillar into a dorsal sac and a ventral sac. These two major sacs are further divided into a cranial blind sac, which lies below the reticulum, and two caudal blind sacs called the dorsal blind sac and the ventral blind sac (Fig. 3.4). The sacs are delineated by internal pillars called the cranial pillar, longitudinal pillar, ventral coronary pillar and dorsal coronary pillar.

A single cranial blind sac, the anterior blind sac, is formed by the cranial pillar. So the rumen effectively has a dorsal sac, ventral sac, cranial blind sac, a

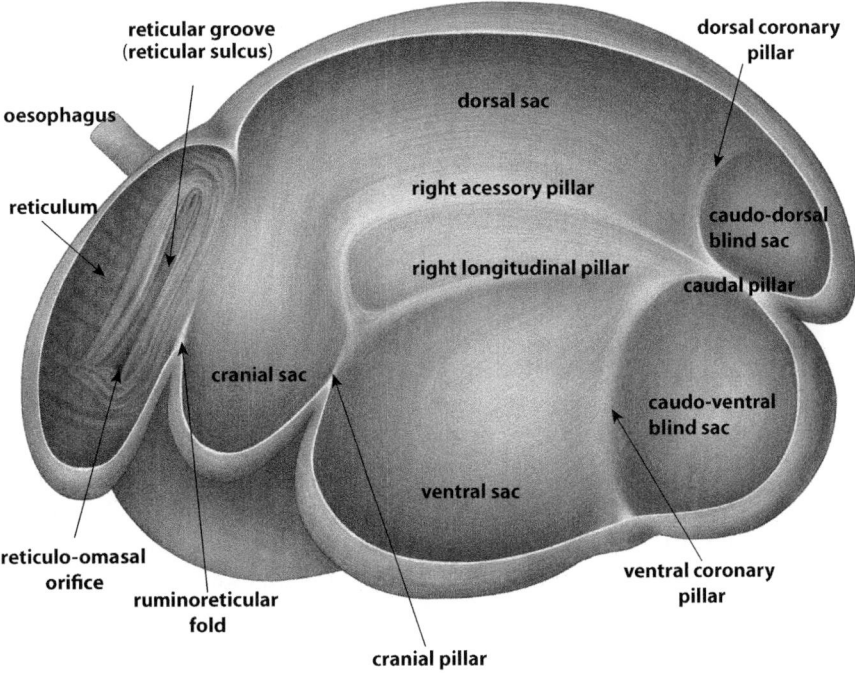

Fig. 3.4. Internal structure of the reticulo-rumen showing the pillars and folds that separate the rumen from the reticulum and that form the major dorsal and ventral sacs, the cranial sac, caudo-ventral blind sac and caudo-dorsal blind sac. The pillars on the inside of the rumen are seen as grooves from the outside of the rumen.

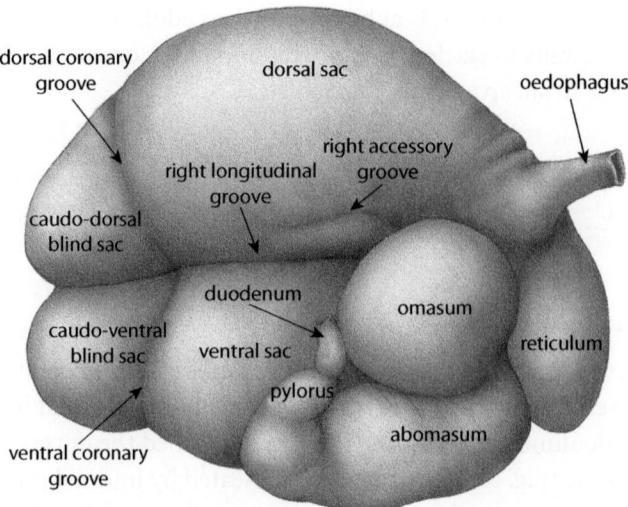

Fig. 3.5. External appearance of the four stomach compartments (reticulum, rumen, omasum and abomasum) of the ruminant animal (right-side view). The major sacs (dorsal sac, ventral sac, caudo-dorsal sac and caudo-ventral sac) are shown (Note: the cranial blind sac is not shown). Note the grooves correspond to the internal pillars (shown in Fig. 3.4).

caudo-dorsal blind sac and a caudo-ventral blind sac. These divisions are apparent from the exterior of the rumen as 'grooves' in the rumen wall, the longitudinal groove and the dorsal and ventral coronary grooves (Fig. 3.5).

The reticulo-rumen occupies the majority of the left side of the animal in the sub-lumbar fossa (the space between the ribs, vertebrae and left hipbone). Externally the rumen is divided by grooves, which correspond to internal pillars that create a large dorsal sac, large ventral sac, cranial blind sac and two caudal blind sacs (ventral and dorsal).

The oesophagus enters the reticulum at the cardia and at the top of a groove known as the oesophageal groove, reticular groove or reticular sulcus (Fig. 3.6). The groove extends to the opening of the omasum and when its smooth muscular walls contract they form a channel through which the ingested liquids (usually milk) pass directly into the omasum. This bypassing of milk to the omasum without rumen fermentation is stimulated by a vagal reflex known as the oesophageal groove reflex, and is stimulated in the young ruminant by sucking on the teat with the head in an extended position. This reflex allows milk to be digested efficiently by the abomasum and intestines and prevents accumulation of milk in the rumen, which causes maldigestion and scouring. Solid feed consumed

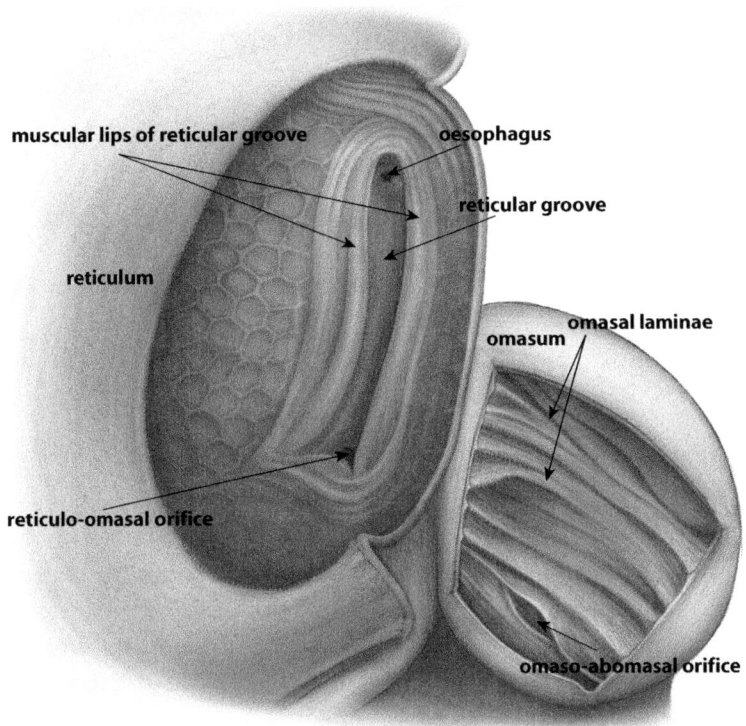

Fig. 3.6. The reticular groove (reticular sulcus) extending from the base of the oesophagus to the omasum. Solid feeds pass directly into the reticulum and then into the rumen in the second reticular contraction of the primary mixing contraction (see Fig. 3.10). During milk sucking, a vagal reflex closes the smooth muscle lips of the reticular groove forming a channel to the reticulo-omasal orifice, forcing the milk to bypass the rumen.

by ruminants normally passes directly into the reticulum and is quickly forced into the rumen by frequent contractions of the reticulum and rumen (see 'Reticulo-rumen motility, eructation, rumination and digesta movements' later).

The ruminal epithelium is a stratified squamous, keratinised epithelium divided into long, finger-like (filiform) or fungi-like (fungiform) projections called papillae. These papillae present a large surface area for absorption of the volatile fatty acids (VFA), which are the primary energy source of ruminants (see Fig. 3.21). Papillae vary greatly in their size, shape and colour depending on the species of ruminant, feeding habits of the animal, season, diet digestibility, the concentration of VFA in the rumen fluid and position within the rumen. Papillae development is stimulated by VFA mixtures containing butyrate and propionate (Lane and Jesse 1997). Starvation or drought feeding results in a reduction in papillae size and a tendency towards the production of filiform (circular) papillae shapes, rather than fungiform (mushroom-shaped papillae). The density of papillae also changes with diet quality (Fig. 3.7).

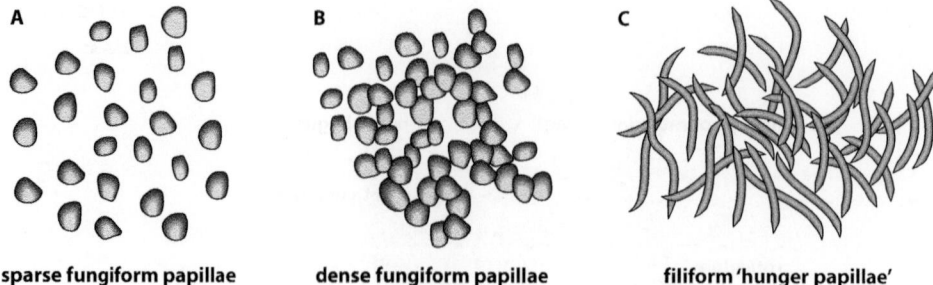

Fig. 3.7. Rumen papillae from animals on different diets. A = low quality and low quantity of hay; B = high-quality mixed ration; C = filiform papillae during a drought.

Development of the ruminal epithelium

Ensuring optimal development of a fully functional ruminal epithelium is vital to the transition from milk feeding to solid feeds. At birth the ruminal epithelium is thin, has a poor blood supply and is devoid of papillae. If the animal is maintained on milk only, the epithelium remains in this undeveloped state, but provision of fermentable substrates, such as concentrates and hay, stimulate the growth of papillae and of blood supply to the epithelium (Brownlee 1956). Pre-weaning and post-weaning diets can affect the thickness and musculature of the epithelium, and the length, width and density of the papillae. Diets that stimulate butyrate and propionate production stimulate ruminal blood flow rate and mitotic activity of the epithelial cells, resulting in denser and larger papillae (Table 3.2). Feeding grain during rumen development stimulates papillae development, but the musculature of the rumen wall is increased by feeding hay. It is generally recommended that pre-ruminant animals be given access to mixed rations of grain and hay to stimulate papillae development in addition to the musculature of the reticulo-rumen.

Areas of the rumen in contact with the ruminal fluid (the ventral sacs) have a greater papillae surface area because this is where the volatile fatty acid absorption

Table 3.2. Effects of fermentation end products and solid feed introduction to pre-ruminant calves on ruminal epithelium development and function

Substrate	Papilla development	Rumen musculature
Acetate	+	?
Propionate	+++	?
Butyrate	+++++	?
Hay	+	+++++
Grain	++++	+
Milk	0	0

Source: Adapted from Mirzaei *et al.* (2015)

is occurring. A large proportion of the dorsal sac is not in contact with fluid and hence has reduced, and often no, papillae. The pillars are devoid of papillae.

> Access to both grain and hay to young ruminants during ruminal development stimulates the development of the papillae and the musculature of the reticulo-rumen.

Anatomy of the ruminal papillae

The epithelium of the rumen comprises typical keratinised, stratified squamous, non-glandular epithelium, like skin, with a *stratum basale, stratum spinosum, stratum granulosum, stratum lucidum* and *stratum corneum* (Fig. 3.8). The basal cells of the epithelium are columnar and attached to the basement membrane by hemidesmosomes. The spinous cells are oval shaped and desmosomes connect them with tonofibrils for nutrient transport. The granular cells are flattened cells connected by tight junctions. The cornified layer is thin and sloughs into the lumen of the rumen, reticulum or omasum. It is unusual for a stratified squamous keratinised epithelium to be an absorptive epithelium, but the

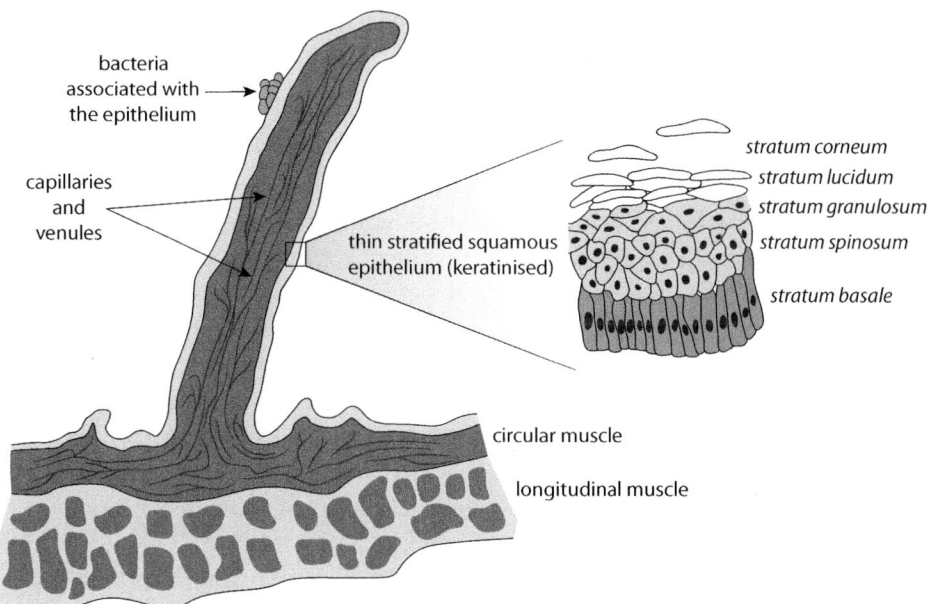

Fig. 3.8. A ruminal papilla (LS) showing the high density of blood vessels in the lamina propria and the thin, stratified squamous epithelium, which allows rapid absorption of the volatile fatty acids. Note the close association of bacteria with the ruminal epithelium. Many of these bacteria are facultative anaerobes that can use the higher oxygen tension that exists close to the blood vessels, thereby maintaining the low oxygen tension in the rumen.

rumen epithelium contains a high density of mitochondria, which provides the energy for active transport of nutrients, and the papillae have a dense network of blood vessels (Fig. 3.8). The mechanisms of absorption of VFA are shown in Fig. 3.21.

The basal cells of this epithelium carry out the vital function of absorbing the VFA produced by ruminal microbes (see Fig. 3.21). There are very tight, 'tight junctions' in the ruminal epithelium to prevent the entry of microbes into the portal circulation. During high-grain feeding, when the ruminal pH can drop below 5.0, the epithelium can become damaged (parakeratosis) and if the integrity of the epithelium is sufficiently damaged, bacteria can enter the portal circulation and cause hepatic abscesses (Nagaraja and Chengappa 1998): a common pathology in feedlot cattle.

The blood supply to the rumen is via a single artery: the coeliac-cranial mesenteric trunk arising directly from the abdominal aorta. Importantly, all of the blood draining the rumen leaves via the portal vein, which transports all of the absorbed nutrients to the liver for processing in biochemical pathways.

> Short-chain volatile fatty acids (acetate, propionate and butyrate), which provide most of the energy for ruminants, are absorbed across the ruminal epithelium by 'papillae' (finger-like, or fungi-like projections of the epithelium).

Anatomy of the reticulum, omasum, abomasum and the reticulo-omasal orifice

The reticulum lies cranial to the rumen at the base of the oesophagus. The reticulum has a keratinised stratified squamous epithelium but it is divided into hexagonal structures in a honeycomb arrangement comprising deep primary 'ridges' producing four-, five- or six-sided structures (Fig. 3.6). Additionally there are secondary ridges that are shallower and that subdivide the deeper units. The base of the units have numerous short horny papillae. It is thought these ridges and papillae play a role in the retention of larger particles in the reticulo-rumen (see 'Reticulo-rumen movements'). Cattle are non-selective feeders and often ingest foreign bodies such as wire or nails, which lodge in the reticulum and may be pushed through the reticular wall by contractions, producing 'hardware disease'.

The omasum lies between the reticulum and the abomasum. In cattle it is a bilaterally flattened sphere and in goats and sheep it is bean-shaped. The distal end is divided internally into numerous longitudinal laminae or leaves (Fig. 3.6), hence the common name the 'bible' or 'book', while the proximal end contains a canal leading directly from the reticulo-omasal orifice to the abomasum (Fig. 3.9). The number of leaves in cattle is highly variable (Becker *et al.* 1963) and

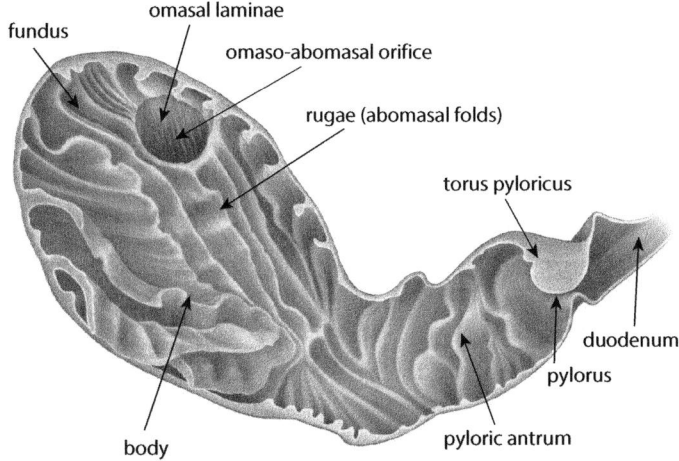

Fig. 3.9. Gross anatomy of the abomasum (internal) of ruminants. Digesta enters at the omaso-abomasal opening and moves towards the pylorus, which contains a thick smooth muscle sphincter controlling the egress of digesta to the duodenum. The wall of the abomasum is a glandular mucosa containing mucus cells, chief cells and parietal cells.

numbers in sheep have yet to be quantified. Like the reticulum and rumen, the omasal epithelium is keratinised, stratified and squamous. The high density of laminae and the digesta held between the laminae makes the omasum a very dense organ. Water absorption makes the mass of omasal digesta a hard, dry mat of material.

The abomasum is a concave tube lying to the right of the midline and extending from the omasum to the pyloric sphincter (the muscular sphincter between the abomasum and the duodenum). The folds in the mucosal wall differ from the rugae of simple-stomached animals in that they are permanent and run longitudinally in a spiral arrangement (Fig. 3.9).

The epithelium of the abomasum is a typical glandular gastric mucosa, comprising columnar cells lining gastric pits. The epithelium contains mucoid cells (producing mucus), chief cells (pepsin), parietal cells (HCl), G cells (gastrin) and enterochromaffin cells (endocrine secretions).

Epithelia of the forestomachs

The epithelium of the reticulum, rumen, and omasum is a stratified squamous epithelium divided into honeycomb ridges (reticulum), filiform or fungiform papillae (rumen) or page-like laminae (omasum). The abomasum has a simple, columnar, glandular epithelium with gastric pits, chief cells and parietal cells.

Reticulo-rumen motility, eructation, rumination and digesta movements

Reticulo-rumen movements

Motility of the rumen is essential for key functions such as: eructation (belching of the gases being continually produced by fermentation); mixing of incoming feed with the microbial inoculum; presentation of the end products of fermentation to the absorptive sites (the ruminal epithelium); sorting of the particles into size and density groupings to allow retention of large particles for further digestion and onward passage of small particles to the omasum (digesta flow); and rumination (presentation of the large particles for re-chewing). Failure of ruminal motility (ruminal stasis) is evident when acidosis reduces the ruminal pH to <5.0 and the rumen movements cease. Bloat (failure to eructate) and compaction of the rumen with undigested contents (failure of digesta flow) results in death of the animal.

The primary (or A) wave (Fig. 3.10) involves a coordinated sequence of contractions of the reticulum, the dorsal sac and the ventral sac, which force the digesta backwards, downwards and then forwards and upwards. This produces regular mixing of the digesta with swallowed saliva and the microbial populations, and brings the fluid containing the VFA, fluids, minerals and ammonia into

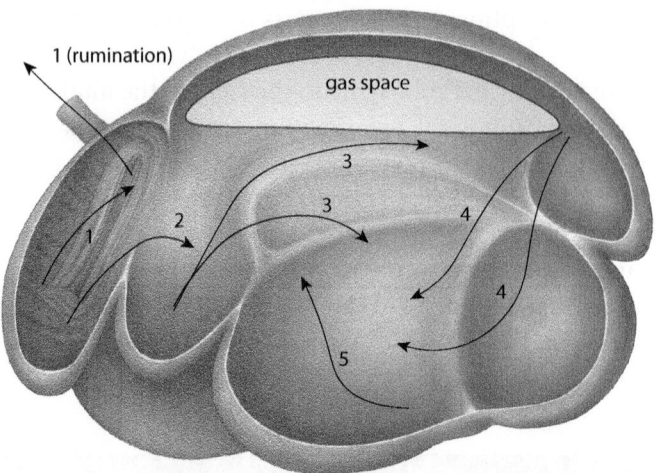

Fig. 3.10. Primary (or A) wave of contraction of the reticulo-rumen. The sequence is: (1) a weak contraction of the reticulum forcing digesta upwards towards the oesophagus (rumination may occur at this point); (2) a strong reticular contraction forcing material over the rumino-reticular fold and into the cranial sac; (3) cranial sac and cranial pillar contract forcing material upwards and backwards into the dorsal and caudal blind sacs; (4) the dorsal sac contracts forcing material forwards and downwards; (5) the ventral sac contracts forcing material forwards and upwards to complete the cycle. The roughage material most likely to present to the cardia for rumination is at the cranial end of the roughage 'raft' floating on the ruminal fluid.

contact with the absorptive epithelium. Primary contractions occur at a frequency of ~1/minute and they last ~30–50 s (Freer *et al.* 1962). An initial weak contraction of the reticulum is followed by a stronger reticular contraction that forces material out of the reticulum into the cranial sac. The cranial sac contracts and the cranial pillar contracts forcing material into the ventral sac. The ventral sac then contracts propelling the material upwards and backwards. The caudo-dorsal sac contracts forcing material forwards and downwards. A ventral contraction then propels material upwards and forwards and over the cranial pillar and over the rumino-reticular fold and into the reticulum. If rumination is to occur, it happens at the start of this primary cycle when the first reticular contraction forces material to the base of the oesophagus (the cardia). The pressure pushes the bolus into the oesophagus where reverse peristalsis takes the bolus to the mouth for re-chewing. At the end of a primary wave, the relaxation of the cranial pillar reduces pressure in the reticulum and the bolus re-enters the reticulum and joins the digesta in the rumen in the next mixing cycle. Figure 3.10 shows the sequence of events involved in a primary cycle.

The secondary movements of the reticulo-rumen are forward-moving and propel the digesta towards the cranial sac and the reticulum (Fig. 3.11). Importantly, these secondary movements produce eructation, the expelling of gas from the oesophagus. The amount of gas released by eructation in cattle can reach 60 L/h = 1400 L/day! The gases are a combination of carbon dioxide and methane plus volatile organic compounds. The breath of ruminants is therefore a

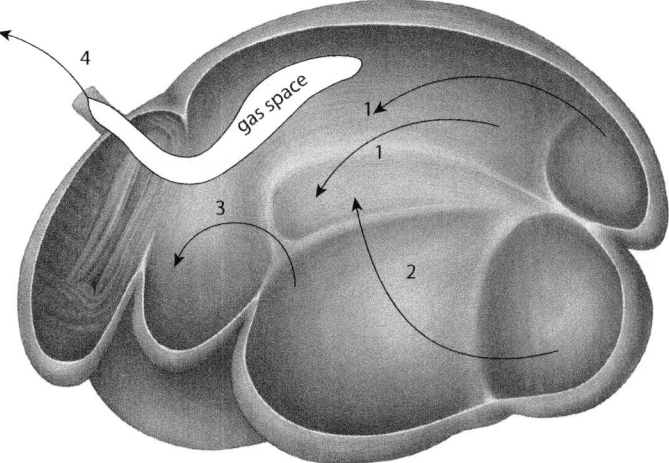

Fig. 3.11. Secondary (or B) wave of contraction of the reticulo-rumen including an eructation cycle. The sequence is (1) contraction of the dorsal sac forcing ingesta forwards and downwards; (2) contraction of the ventral sac forcing material upwards and forwards; (3) ingesta forced over relaxed cranial pillar into cranial sac; (4) forward-moving dorsal sac contraction forces gas space to oesophagus for eructation.

combination of fermentation gases combined with respiratory gases. The secondary movements begin with a contraction of the dorsal sac forcing digesta forwards and downwards. The ventral sac contracts in a forward motion, forcing ingesta forwards and upwards. The gas space is squeezed forwards and downwards towards the cardiac region and the base of the oesophagus. By combining the arrival of the gas at the cardia with breathing, the gas enters the thoracic cavity and is belched.

Together, the primary cycle (mixing and rumination) and the secondary cycle (eructation) produce a circular mixing motion of ingesta, presentation of larger undigested particles to the oesophagus for rumination, and presentation of the gas space to the oesophagus for eructation.

Reticulo-rumen contractions

The reticulo-rumen undergoes two main contractions: primary movements are backward-moving and then forward-moving, producing a circular mixing motion and rumination. Secondary movements are forward- and downward-moving producing mixing and eructation.

Neural control of reticulo-ruminal motility

The reticulo-rumen is innervated by two major neural networks: extrinsic fibres and intrinsic fibres. The extrinsic nerve is the vagus nerve, which contains both afferent (sensory) and efferent (motor) fibres. The vagal afferent fibres are stimulated by ruminal receptors, which are either tension receptors or epithelial receptors. The tension receptors are the most common and are found in the muscle layers of the reticulum, reticulo-ruminal fold, the cranial sac and the reticulo-omasal orifice. These receptors respond to changes in tension or stretch in the reticulo-ruminal wall and influence the frequency and amplitude of contraction and also saliva secretion. The epithelial receptors are found in the basement membrane of the reticulum, cranial sac of the rumen and the longitudinal pillars. The epithelial receptors are either mechanoreceptors or chemoreceptors. Chemoreceptors respond to the VFA in the fluid. The intrinsic nerves are embedded as plexuses below the epithelium of the reticulo-rumen. If the vagus nerve is severed, normal reticulo-ruminal contractions are depressed and the smooth muscle of the reticulo-rumen reverts to low-amplitude contractions innervated by the intrinsic nervous system. The onward passage of feed through the reticulo-omasal orifice is prevented, indicating the importance of vagal innervation to digesta flow from the reticulo-rumen (Okine *et al.* 1998). Feed and gas accumulate in the reticulo-rumen and the animal dies. The splanchnic nerve also plays a role in reticulo-rumen innervation, but its role in contractions appears to be minor.

Factors affecting ruminal contractions

The frequency and strength (amplitude) of ruminal contractions are affected by feeding, diet, VFA concentrations and metabolic disorders, but rates are similar for cattle and sheep (Waghorn and Reid 1983). Over a 24-h period the average frequency of primary contractions is 1.00/minute and secondary contractions 1.00/1.8 min. During feeding the frequencies of both contraction types are approximately doubled. During rumination and resting the frequency of primary contractions drops by ~30% and secondary contractions reduce by ~50%. Fasting has the same effects on contraction frequency as resting and ruminating.

> The frequency of contractions of the reticulo-rumen in sheep and cattle is ~1 every minute (primary mixing contractions), and 1 every 2 min (secondary eructation contractions). Both double in frequency during feeding.

Fluid and particle dynamics in the reticulo-rumen

The rate of digesta outflow from the reticulo-rumen is a key determinant of feed intake in ruminants because it is generally agreed that ruminants consume feed until they reach a maximum level of rumen 'fill', or point of satiation, signalled by the stretch receptors in the epithelium. However, the rumen is never 'full' *per se* and the level of stretch required for satiation varies with physiological state. The kinetics of particle flow in the rumen are complex and fascinating. A model describing particle dynamics in the reticulo-rumen is shown in Fig. 3.12 (after Seo *et al.* 2009).

In this model it is assumed that there are two major pools of particles: those in inescapable pools and those in an escapable pool. Recently consumed forage particles are relatively large and low density and as such associate with a pool of similar particles in the dorsal rumen (Fig. 3.12). Many of these particles are trapped in a floating raft of long particles. Small, light particles are likewise trapped in this raft and cannot sink below the reticulo-omasal orifice. The large particles trapped in the raft stimulate rumination by stimulating the vagal mechanoreceptors in the wall of the cranial rumen and reticulum. Presumably the particles most likely to be ruminated are those trapped in the floating raft and that are presented to the base of the oesophagus (cardia) during reticulo-rumen movements. These are therefore generally the larger particles and those of low density. Rumination then reduces their size, adds saliva, releases trapped gases and

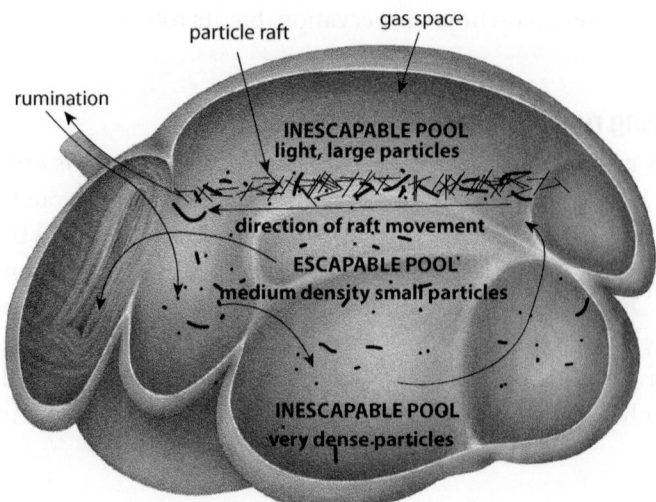

Fig. 3.12. A model of particle dynamics in the reticulo-rumen. Two major pools, defined as inescapable and escapable from the reticulo-rumen, are produced by a combination of particle density (specific gravity) and particle size. Light particles remain in the dorsal sac associated with the fibrous 'raft' and cannot leave the rumen until they are sufficiently small and dense to sink to the level of the reticulo-omasal orifice. Large particles are retained in the raft of inescapable particles and are presented to the base of the oesophagus for rumination. Source: Based on descriptions by Seo *et al.* (2009).

thereby makes them smaller and denser, and therefore more likely to enter the reticulum. The first, and weaker, reticular contraction pushes the fluid/particle pool, which has entered the dorsal reticulum, back into the rumen for further digestion. The second, stronger reticular contraction pushes the fluid/particle pool in the base of the reticulum towards the cardia and through the relaxed, open, reticulo-omasal orifice.

Rumination

'Chewing the cud', or rumination, is obviously a key feature of ruminant animals and is a good indicator of normal, healthy rumen function. Rumination is associated with active reticulo-rumen motility and is triggered by large fibrous particles that stimulate the vagal epithelial receptors and stimulate saliva production. The latter is important for buffering ruminal acids, lubricating the fibrous material, recycling minerals and nitrogen to the rumen, and increasing the density of the ruminal particles. The process of rumination is as follows:

1. An extra reticular contraction occurs a few seconds before the usual biphasic reticular contraction, and this contraction increases the fluid pressures in the cardia region of the reticulum.

2. At the same time, the animal takes a large inspiration (expansion of the thoracic cavity creates negative pressure in the thorax).
3. A bolus of food at the cardia at this time enters the oesophagus in response to the high reticular pressure and low thoracic pressure.
4. The bolus moves up the oesophagus by antiperistaltic contractions.
5. The rate of chewing during rumination appears to be slower than during normal eating.
6. In sheep, the regurgitated bolus is chewed for ~50 s before swallowing again.

Sheep spend ~8–9 h/day ruminating although this is greatly affected by diet. Animals fed grain only or finely ground roughages (including pelleted rations) will not ruminate or will pseudo-ruminate (regurgitation, brief period of chewing then swallowing) because there is little stimulation of receptors in the ruminal epithelium ('scratch' factor).

'Scratch factor'

Rumination (cud chewing) is stimulated by 'scratch factor' – the presence of particles sufficiently large (>2 cm) to stimulate the mechanoreceptors in the reticulo-ruminal wall. Rumination, in turn, stimulates saliva secretion, hence ruminal pH buffering, so providing 'scratch factor', which reduces acidosis and subacute ruminal acidosis.

Eructation

Eructation, or belching, is essential for ruminants to expel the large volumes of gas produced by active fermentation. Sheep produce ~5 L of gas/h and cattle ~60 L/h, although the rate of gas production is greatly affected by the diet, frequency of feeding, and time after feeding. Eructation takes place associated with the secondary ruminal contractions (forward-moving contraction across the dorsal sac). The gas space in the dorsal sac is forced forwards and then downwards towards the cardia. It seems the normal biphasic reticular contractions clear the reticulum of ingesta. As the dorsal gas space moves towards the oesophagus, the reticulo-ruminal fold remains contracted thereby preventing ingesta from refilling the reticulum until the gas has been expelled from the dorsal reticulum. Eructation is part of the normal secondary rumen contractions, whereas rumination requires a special extra-reticular contraction.

Functions of the omasum and abomasum

The role of the omasum is not entirely clear. The organ fills with very fluid digesta leaving the reticulo-rumen, but omasal contents are very dry and tightly compacted. Clearly, fluid absorption takes place in the omasum and in cattle, water, electrolytes and VFA are absorbed. In one study of sheep, 33–64% of the water entering the omasum was absorbed and 40–69% of the VFA were absorbed (Gray *et al.* 1954). There is some thought that the omasum may further retain larger particles for further digestion (mechanical and fermentation) within the laminae. The omasum contracts rhythmically and slowly in a biphasic manner. The canal contractions are thought to move fluid from the omasum to the intra-laminal spaces, while the more forceful contractions of the omasal body forces digesta from the omasum to the abomasum (Stevens *et al.* 1960).

The glandular, acidic stomach of sheep (the abomasum) plays a similar role to the simple stomach of mono-gastric species, namely: acid production (microbial killing); pepsin production (protein digestion); rennin production (coagulation of milk); and synthesis of intrinsic factor (for cobalamin (vitamin B12) absorption). Given the reliance of ruminants on microbial products, the role of acid in microbial death may be particularly important. The fundic region of the abomasum is relatively immotile, the body displays peristaltic activity and the pyloric region shows strong peristalsis.

Microbiology of the reticulo-rumen

Detailed coverage of rumen microbiology is beyond the scope of this text but a summary of the complex diversity and activities of the rumen microbes is necessary to appreciate the strengths and weaknesses of this pre-gastric fermentation system, and the aetiology of many nutrition-induced pathologies in ruminants. More detail on rumen microbiology can be found in Hungate (1966) and Ogimoto and Imai (1981). The symbiotic relationship between the ruminal microbial population and the host animal extracts nutrients efficiently from low-quality roughages. Moderate-quality microbial protein is produced from low-quality, low-protein and high-fibre feeds. The symbiotic relationship allows the host animal to obtain most of its energy needs from the short-chain VFA (acetic, propionic and butyric acids), its amino acids from microbial protein, some of its long-chain fatty acids from microbial lipid and its vitamins (B and K) from microbial synthesis. The microorganisms benefit in turn from being provided with an environment that allows their growth and reproduction (Table 3.3).

Rumen microbiology

The rumen microbial population is complex and dynamic. There are more than 200 species of ruminal bacteria (flora), 20 species of ciliated protozoans (fauna), and at least three species of anaerobic fungi. There are ~10^{10}–10^{11} bacteria/mL of rumen fluid, 10^5–10^6 ciliates/mL and 10^4 fungi/mL. The majority of these

Table 3.3. Characteristics of the reticulo-ruminal environment

Variable	Value/comment
Temperature	38–41°C (slightly warmer than core temperature due to heat of fermentation but at times colder due to cold water and feed inputs)
pH	5.0–7.4 (typically 6.2–7.0); buffered by saliva, and absorption of undissociated acids (see Fig. 3.21)
Anaerobic	−350 mV (Eh) (but with pockets of oxygen inside feed particles and near the ruminal wall where blood oxygen tension is high)
Osmolarity	200–400 mOsm/kg
Gas phase	CO_2 65% CH_4 27% H_2 0.2% O_2 0.6% N_2 7.9%
Diverse microenvironments for microbes	• Wide substrate range (starches, cellulose, hemicelluloses, pectins, proteins, non-protein nitrogen, lipids, soluble sugars) • Microbes attached to plant surfaces, free-floating, attached to reticulo-rumen epithelia, in consortia with other microbes, attached to other microbes, motile microbes follow substrates (e.g. ciliates) • Heterogeneous contents (small, medium, large particles, variable chemical composition dorso-ventral and cranio-caudal)
Other	• Constant substrate mixing • Constant removal of end products and gases • Rapid removal of oxygen (e.g. in swallowed feeds or diffusing through the rumen epithelium) by facultative anaerobes

organisms are strict anaerobes, but a proportion of facultatively anaerobic bacteria can preferentially utilise small amounts of oxygen for their metabolism. These organisms are often associated with the rumen wall where the oxygen tension is higher than deeper in the rumen. They play an important role in utilising small amounts of oxygen that enter the rumen through the ruminal epithelium or from air trapped in forage particles, thereby maintaining the anaerobic environment.

Ruminal bacteria

The ruminal bacteria are classified on the basis of morphology (shape, size), function (mainly via the substrates they utilise predominantly) (Table 3.4) and more recently on the basis of molecular signatures obtained by sequencing DNA (see review by Deng *et al.* 2008). In future, it is likely that gene expression will be used to reflect the microbiome, rather than microbial species *per se.*

Rumen ciliated protozoons

There are ~10^5–10^6 protozoa/mL rumen fluid. Despite being a fraction of the numbers relative to rumen bacteria, the weight of protozoa can be equal to that of

Table 3.4. Classification of rumen bacteria according to predominant substrate used

Group	Major species	Substrates	Products
Cellulolytic	Ruminococcus flavefaciens Ruminococcus albus Butyrivibrio fibrisolvens Fibrobacter succinogenes	Cellulose Hemicellulose Pectin Ammonia Amino acids Branched-chain amino acids	Acetate
Hemicellulolytic	Butyrivibrio fibrisolvens Prevotella ruminicola Ruminococcus flavefaciens Ruminococcus albus	Hemicellulose Cellulose	Acetate
Proteolytic	Rumenobacter amylophilus Streptococcus bovis Prevotella ruminicola Butyrivibrio fibrisolvens	Protein Peptides	
Amylolytic	Prevotella ruminicola Rumenobacter amylophilus Selenomonas ruminantium Streptococcus bovis Succinomonas amylolytica	Sugar Starch Peptides Amino acids Ammonia?	Propionate Lactate Acetate Hydrogen CO_2
Lactate-utilisers	Megasphaera elsdenii Selenomonas ruminantium	Lactate	
Methane producers	Methanobrevibacter ruminantium Methanobrevibacter formicicum Methanosarcina barkeri	Formate CO_2	CH_4 (methane)
Sugar utilisers	Treponema bryantii Lactobacillus vitulinus Lactobacillus ruminus	Soluble sugars Glucose from cellulolysis	

the bacteria due to their larger size (20–200 µm) relative to bacteria (0.4–1.5 µm). An adult cow can have 2 kg of ciliates in her rumen. The impressive motility and grazing behaviours of ciliates could suggest they are the predominant player in ruminal fermentation. However, the total surface area/volume ratio of the bacteria is much greater than that of the ciliates. The bacteria provide more than 99% of the total metabolically active surface area. The rumen ciliates belong to two main families: the *Holotricha* (holotrichs) and the *Ophyroscoloidae* (commonly called the oligotrichs or entodiniomorphs because the major genus is *Entodinia*). The former have cilia covering the entire body surface and the latter have tufts of cilia usually at the anterior end of the cell (Table 3.5).

Photomicrographs of mixed populations of rumen bacteria and protozoa are shown in Fig. 3.13.

The rumen ciliates carry out important functions in the rumen in that they consume large quantities of bacteria. The rate of bacterial engulfment by ciliates ranges from 20 to 3000 bacteria/protozoon/h depending on bacterial species, ciliate species and bacterial density (Coleman and Sandford 1979). Up to 50% of the

Table 3.5. Classification of the rumen ciliated protozoans

Group	Main genera/species	Features
Holotricha	Isotricha prostoma Isotricha intestinalis Dasytrichia ruminantium	• Fast moving • Large (100–200 µm) • Oval shaped • Sugar utilisers
Oligotricha (entodiniomorphs)	Entodinium longinucleatum Diplodinium spp. Epidinium ecaudatum Polyplastron multivesiculatum Ophyroscolex spp.	• Highly variable size (10–50 µm) • Particulate engulfers (including starch)

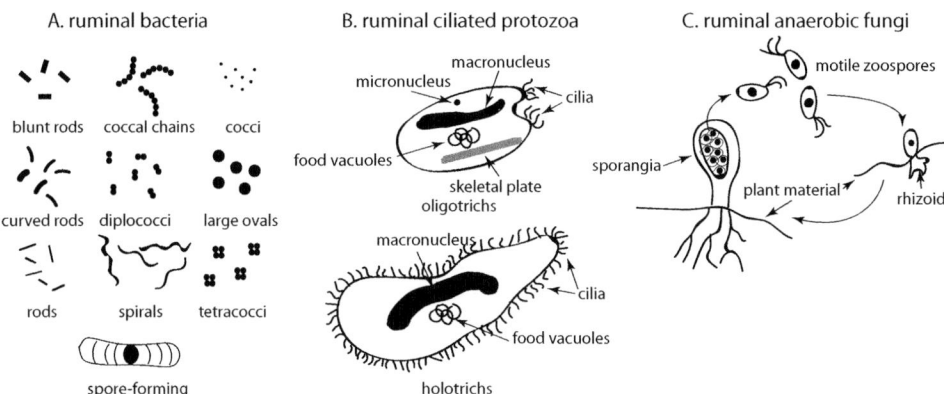

Fig. 3.13. Ruminal microorganisms include bacteria (A), ciliated protozoans (B) and anaerobic fungi (C). The bacterial morphology is highly variable but dominated by cocci and rods. Protozoa belong to two major families, the holotrichs (with cilia covering the entire surface of the organism) and the oligotrichs or entodiniomorphs (with cilia at one end of the organism only). The fungi are present as sporangia attached to plant particles and as motile zoospores seen moving to colonise new plant surfaces.

bacterial population can be consumed by the ciliates, constraining the bacterial population at times when substrates for the bacteria are in plentiful supply. Ciliates also engulf starch particles. Together, these activities buffer the rumen against overly rapid fermentation. Ciliates increase the digestion of feeds in the rumen and thereby increase energy availability, but they also reduce protein availability by reducing bacterial yields (by predation), and by being retained preferentially within the rumen rather than exiting via the reticulo-omasal orifice. They achieve the latter presumably by moving away from the reticulo-omasal orifice and by associating with the inescapable pools (see Fig. 3.12). Ultimately, many ciliates are recycled within the rumen and their contents are reused by other microbes. Recycling is always associated with losses (e.g. of ammonia) and means the ciliates reduce the protein/energy ratio of absorbed nutrients. This has led researchers to consider ciliated protozoons parasites rather than symbionts and some have attempted to reduce their numbers or eliminate them completely from the rumen (Bird *et al.*

1979). Removal of ciliates from the rumen (defaunation) results in faster growth rate, reduced feed intake (hence greater growth efficiency), greater microbial protein flow from the rumen and reduced digestibility of feeds (Eugène *et al.* 2004). The nitrogen efficiency of the ruminal processes is improved by removal of ciliates, but the energy efficiency is reduced, so it is more likely to be beneficial when animals are on low-protein, high-energy diets (Bird *et al.* 1979). The long-term consequences of defaunation (removal of ciliates) on the stability of the rumen ecosystem are not known, but highly productive ecosystems (e.g. ciliate-free) are often unstable.

Ruminal anaerobic fungi

The anaerobic fungi were for many years not identified as rumen inhabitants because most studies were carried out on coarsely filtered rumen fluid and the fungal sporangia reside on the surfaces of the large particles of plant material. Their motile zoospores were often seen scurrying about in the fluid, but they were presumed to be flagellated protozoans. Examination of the forage particles in the ruminal digesta revealed the sporangia of the fungi attached to the surface of solid feed particles. The rumen fungi are the anaerobic fungi *Neocallimastix frontalis, Sphaeromonas communis* and *Piromonas communis* and they can comprise up to 8% of the microbial biomass (Bauchop 1979). The fungi may play an important role in breaking the plant cuticles and allowing access for the bacteria and protozoa (Gordon and Phillips 1998).

Nutrient supply from the gastrointestinal tract of ruminants

The nutrients available for metabolism by ruminants are dramatically different from those consumed, due to the microbial fermentation that takes place in the rumen. Complex carbohydrates are converted to simple mixtures of short-chain VFA. Feed proteins and non-protein nitrogen sources are converted to microbial proteins, and significant losses of nitrogen as ammonia can occur. Unsaturated lipids in pasture forages are saturated in the reducing environment of the rumen. Vitamins B and K are synthesised by the microbes and these become available for the host animal to absorb from the small intestines. Minerals are transformed by redox reactions, are incorporated into microbial cells and complexed with, or split from, organic compounds. Figure 3.14 summarises the major transformations of nutrients on transition through the gastrointestinal tract of ruminants. More details are provided for each nutrient class in the next sections.

Carbohydrate digestion and absorption in ruminants

Typical forage carbohydrates comprise a mixture of soluble sugars, hemicelluloses, cellulose, pectins, starches and lignin in varying combinations and levels. These enter the rumen and are exposed to the extracellular enzymes secreted by the

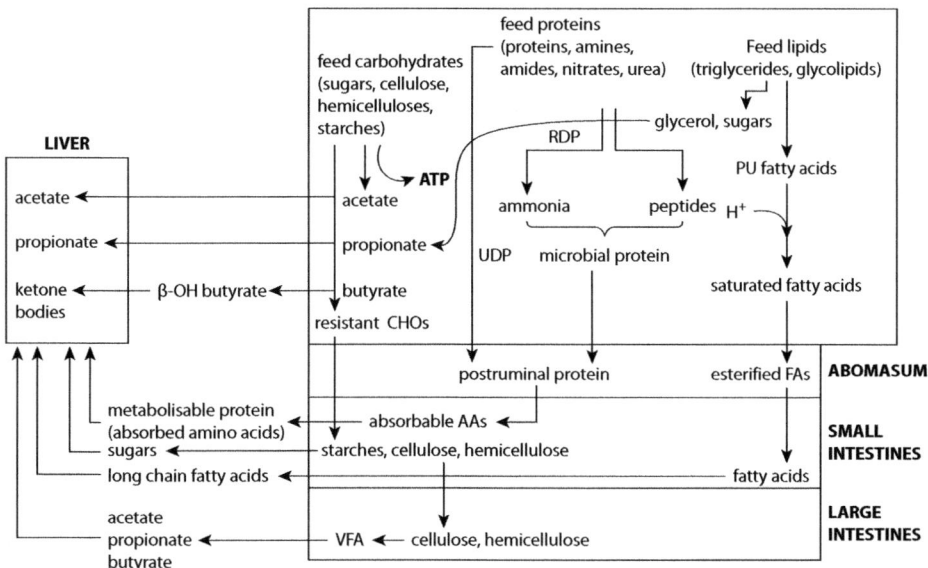

Fig. 3.14. Nutrient supply from the gastrointestinal tract of ruminants. Adenosine triphosphate = ATP, FAs = fatty acids; VFA = volatile fatty acids; RDP = Rumen Degradable Protein; UDP = Undegraded Dietary Protein; AAs = amino acids. Adenosine triphosphate attained from fermentation is used by the microbes to synthesise microbial proteins from simple nitrogen sources like ammonia.

microbes, which occupy various niches within the heterogeneous environment. The complex carbohydrates are broken down to simple sugars, which are then utilised by the microbes as energy substrates, producing the metabolic end products acetate (C2), propionate (C3) and butyrate (C4) and small quantities of higher chain fatty acids. Carbon dioxide, hydrogen gas and methane are also produced as end products. The microbes capture the ATP generated by these transformations and use it for microbial protein synthesis and growth.

Effect of diet and time after feeding on volatile fatty acids concentration and VFA proportions in ruminal fluid

The mixture of VFA produced in the rumen and available for absorption depends on the relative numbers and activity of the microbes generated by the dietary substrates. For example, if the diet comprises a high concentration of fibre (cellulose), the cellulolytic bacteria will predominate and these microbes produce large quantities of acetic acid. If the diet is high in starch, the amylolytic bacteria will predominate and these produce higher levels of propionate. Typical changes in the relative ratios of VFA and lactate in rumen fluid with an increasing ratio of concentrate to roughage in the diet of sheep are shown in Fig. 3.15.

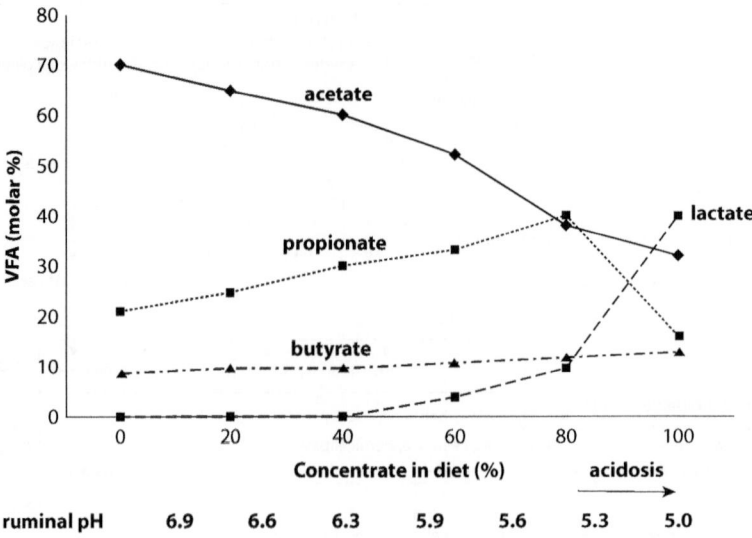

Fig. 3.15. Effect of increasing concentrate to roughage ratio in the diet of sheep on the molar proportions of acetate, propionate, butyrate and lactic acid. Typical minimal pH values after feeding the different diets are also shown.

As the proportion of concentrate in the diet increases, the proportion of acetate progressively decreases and the proportion of propionate increases, until very high starch levels are present, when the propionate is replaced by high levels of lactic acid. The contribution of butyrate to the total acids is relatively low but it contributes to ketone synthesis after transformation in the ruminal epithelium to β-hydroxybutyrate. These relative changes in VFA ratios are important to the host animal, particularly if rapid growth or high milk yields are required.

Effect of diet composition and frequency of feeding on ruminal pH changes

The pH of ruminal fluid depends on the concentration of short-chain VFA (rate of production minus rate of removal), the concentration of lactic acid (rate of production minus rate of removal) and the buffering capacity of the ruminal fluid (rate and composition of saliva production (Fig. 3.16)).

In most feeding situations, the main determinant of ruminal pH is the concentration of VFA in the fluid (Fig. 3.17).

Effect of diet composition on the composition of the short-chain volatile fatty acids in the ruminal fluid

The composition and physical nature of the diet (particle size, pelleting, grinding) influence the relative proportions of the VFA (Fig. 3.15), the rate of

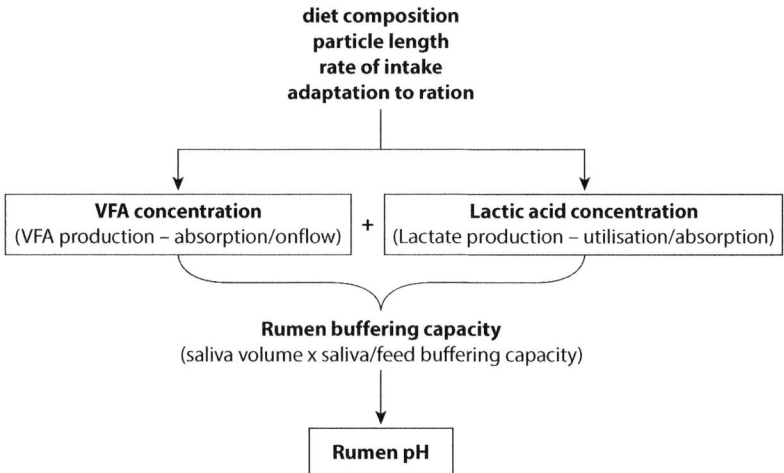

Fig. 3.16. Ruminal pH is determined by the concentration of the volatile fatty acids (VFA) and lactic acid in the fluid and the buffering capacity of the ruminal fluid. VFA concentration depends on rate of production minus removal of VFA by absorption or onflow to the omasum. Lactic acid concentration depends on rate of production minus rate of absorption and utilisation by lactic-utilising bacteria. The buffering capacity of ruminal fluid depends on the rate of saliva production and composition of salivary inorganic buffers. Diet composition, physical form, rate of intake and degree of adaptation to the ration affect all of the above.

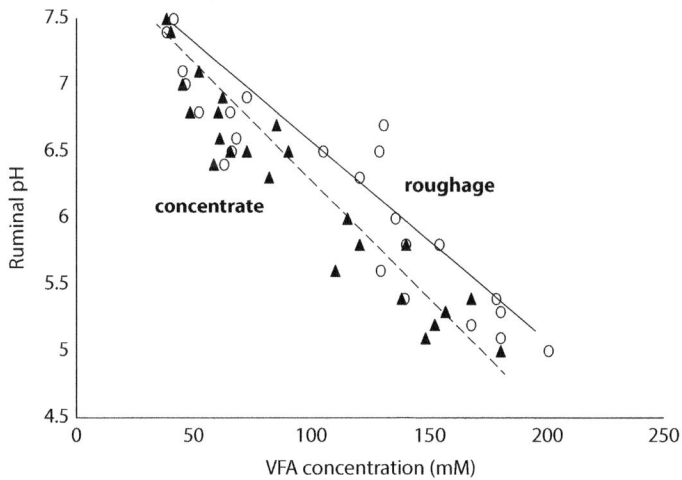

Fig. 3.17. Relationship between ruminal volatile fatty acids concentration (mM) and ruminal pH. Values are from sheep on two diets (roughage = circles and concentrate = triangles) at varying times after feeding. The tendency for deviation of diet values at high VFA is due to accumulation of lactic acid on the concentrate diet. Source: Hynd (1982).

fermentation, the ruminal pH, rate of saliva production, lactate production, rate of outflow of fluid from the rumen and rate of gas production. Highly digestible feeds (high soluble carbohydrates, starches, finely ground, young forages) are rapidly fermented to VFA causing a rapid decline in ruminal pH given the strong relationship between VFA concentration in rumen fluid and fluid pH (Fig. 3.17).

Such feeds support a microbial population characterised by production of high levels of propionic and low levels of acetic acids. If highly fermentable feeds are introduced too rapidly, lactic acid accumulates in the fluid because the lactate-utilising bacteria have not had time to develop and, as a consequence of its low pK_a (it is a stronger acid), the ruminal pH drops rapidly and can fall below 5.0 (see Fig. 7.3).

Figure 3.18 shows daily fluctuations in ruminal pH in a sheep fed a high-grain diet once-daily. The fluctuations reflect rapid VFA production post-feeding followed by an increase in pH as the VFA are absorbed.

Figure 3.19 shows the pH profile of a sheep that did not adapt to the high-grain ration. Note the sudden decline in pH on days 11–12 followed by elimination of pH variations and a failure of pH to return to normal levels (6.0–7.0).

Absorption of volatile fatty acids from the rumen

To understand how VFA are absorbed from the rumen, an understanding of the chemistry of acid dissociation is necessary. Acids exist in a dynamic equilibrium between their undissociated state (in which the H^+ is attached to the conjugate base, such as acetic acid) and their dissociated state (in which the hydrogen ion detaches from the acid leaving the conjugate base to form salts, such as acetate). The relative amounts of dissociated to undissociated acid depends on the pH of the solution. As the pH declines, the proportion of the undissociated form

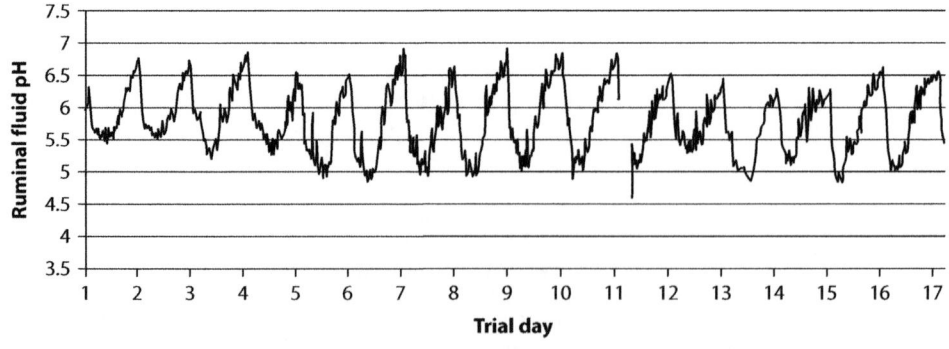

Fig. 3.18.   Fluctuations in ruminal pH of a sheep fed a high-grain diet once-daily for 17 days. pH was logged at 15 minute intervals using an indwelling intraruminal pH logger. Source: Fanning (2015).

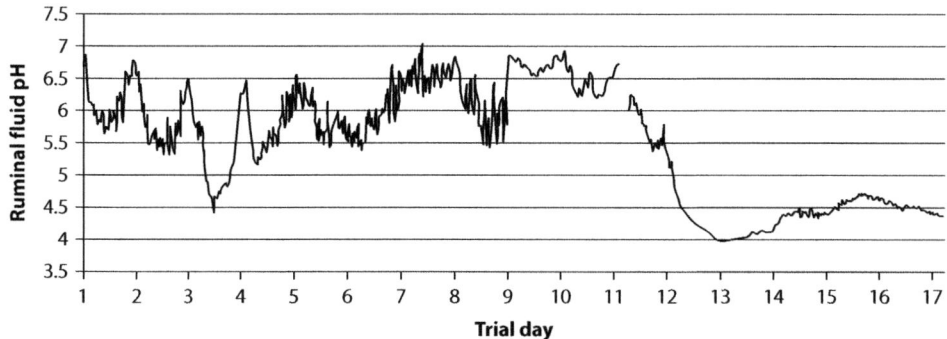

Fig. 3.19. Fluctuations in ruminal pH in a sheep on a high-grain diet fed once-daily for 17 days. pH values below 4.5 are usually associated with ruminal stasis (no movements) and subsequent ruminal impaction, bloat and death. Source: Fanning (2015).

increases. The pH at which half the acid is dissociated and half is undissociated is called the pK_a and for weak acids, such as the VFA, the pK_a is high relative to strong acids such as lactic acid. The dissociation curve for acetic acid is given in Fig. 3.20 and shows a pK_a for acetic acid of ~4.76. The pK_a for propionic acid is 4.9 and for butyric acid it is 4.8. For lactic acid the pK_a is 3.86. This also explains why lactic acid causes acidosis readily in ruminants: it is a stronger acid than the VFA.

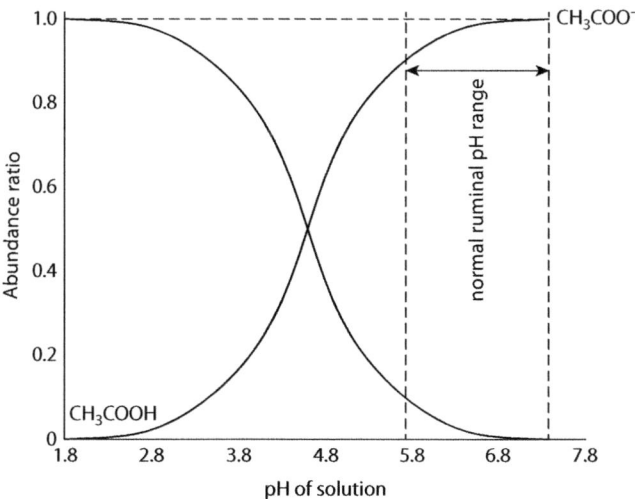

Fig. 3.20. Dissociation curve for acetic acid showing the pK_a of ~4.8, the pH at which half of the acid is in the dissociated form. At typical rumen pH (5.5–7.0) more than 90% of acetic acid is in the dissociated form.

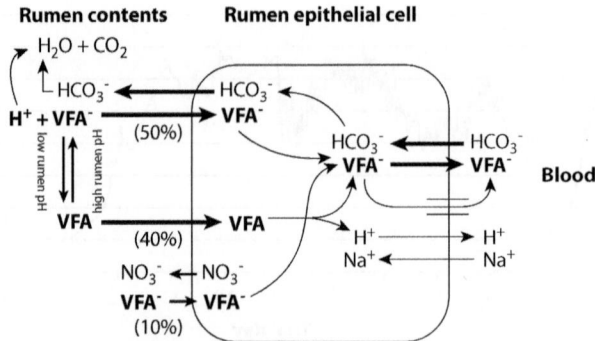

Fig. 3.21. The three main mechanisms of volatile fatty acid absorption from the rumen. The major route is through anion exchange of the dissociated VFA anions with bicarbonate anions in the apical membrane of the rumen epithelial cells. Corresponding exchange at the basolateral membrane provides the bicarbonate for this process as the VFA enter the blood stream. Passive absorption of undissociated VFA also contributes significantly to VFA absorption. These VFA dissociate in the cytosol and the dissociated VFA join the exchangeable anion pool with bicarbonate. A third and minor exchange with nitrate occurs. Sodium/hydrogen exchange pumps maintain cytosolic pH. Dissociated VFA also leave the epithelial cells through voltage-gated channels in the basolateral membrane. (based on descriptions by Penner *et al.* 2009).

Recent studies have revealed that there are at least three mechanisms operating for the absorption of VFA from the rumen (Dijkstra *et al.* 1993; Penner *et al.* 2009; Aschenbach *et al.* 2011). These are summarised in Fig. 3.21.

Approximately half of the VFA are absorbed in the dissociated form in exchange with the anion, bicarbonate. This bicarbonate appears in the epithelial cell in exchange for dissociated VFA at the basolateral membrane (between the cell and the blood). The bicarbonate leaves the cell via the apical membrane (between the cell and the ruminal lumen) and enters the rumen where it can sequester hydrogen ions, thereby buffering the rumen from high acidity. A further 40% or so of the total VFA absorbed is by simple passive diffusion of undissociated VFA. The advantage of absorbing the acid with its hydrogen ion still attached is that this removes the hydrogen ion from the ruminal fluid, again helping to buffer against high acidity. So, regardless of whether the VFA are absorbed in their dissociated or undissociated forms, the ruminal contents are buffered from high acidity. If VFA are absorbed dissociated, bicarbonate buffers the remaining H^+, and if the VFA are undissociated the H^+ ion is removed from the rumen by absorption. The remaining 10% or so of absorbed VFA are drawn into the cytosolic VFA pool by exchange with nitrate ions. The VFA (now all dissociated) leave the epithelial cell and enter the venous blood draining the rumen to the portal system and liver, via either the bicarbonate/VFA anion exchange or through voltage-gated channels. The epithelial cell protects itself from potentially damaging acidity during this process by exchanging the hydrogen ions with sodium ions from the blood.

Protein digestion and absorption in ruminants

The intervention of microbial fermentation in ruminants complicates the supply of amino acids to the absorptive sites in the intestines (Fig. 3.22).

Feed proteins differ in solubility and tertiary structure, and this determines to a large extent their fate in the rumen. Readily soluble proteins are rapidly degraded to peptides, amino acids and ammonia by extracellular microbial proteases. Ammonia, peptides and amino acids are taken up by microbes and converted into microbial protein using the energy they obtain from fermentation (the Fermentable Metabolisable Energy or FME). FME is the energy available from fermentation and is calculated as follows:

$$FME = Feed\ ME - lipids\ ME - UDP\ ME$$

where ME = Metabolisable Energy and UDP = Undegraded Dietary Protein.

FME is a good estimate of the energy available from fermentation because not all the feed energy is available to the microbes. Lipids cannot be used for energy by rumen microbes and the energy locked up in the protein that escapes rumen degradation (UDP) is also unavailable. Some of the microbial crude protein that is

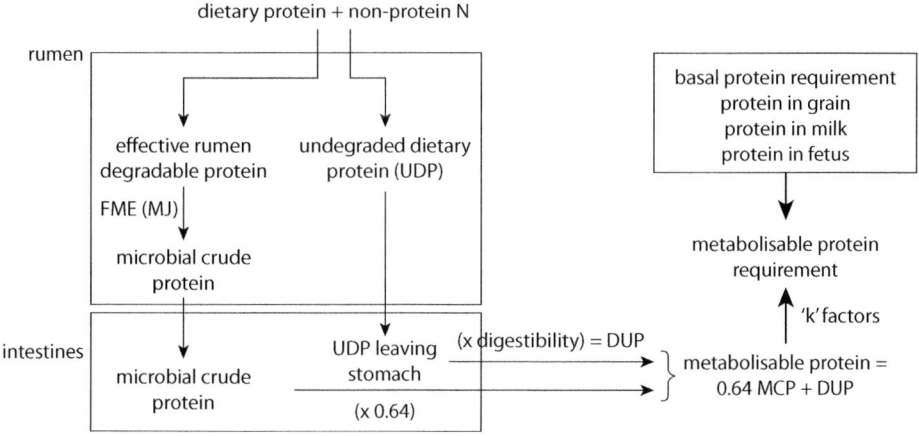

Fig. 3.22. Nitrogen digestion and absorption in ruminants. Feed proteins and non-protein nitrogen are utilised by the microbes to synthesise microbial protein (MCP) using energy available from fermentation (Note: FME = Feed ME – lipids ME – UDP ME). MCP joins with Undegraded Dietary Protein (UDP) to provide the host with metabolisable protein (amino acids).

synthesised is recycled because the microbes die or are consumed by other organisms (particularly ciliates), and this protein re-enters the ammonia pool. MCP that does leave the rumen is not 100% digestible because 25% of it is nucleic acids, which are not digestible. The remaining 75% is also not totally digestible (only 85% of this microbial protein is digestible) so, of all the protein apparently available from the microbes, only 0.64 is absorbable. A variable proportion of the feed proteins are insoluble because of their structure, their interaction with other components of the feed (e.g. tannins or lignin) or because the feed has been treated with heat during preparation (e.g. pelleting). The proportion of UDP in the crude protein of a feedstuff is highly variable, with some feed proteins containing as little as 10% UDP and others as high as 80%. The digestibility of the UDP leaving the rumen is also highly variable (0–90%). The presence of tannins (polyphenolics) in the feed can greatly reduce the intestinal digestion of proteins.

Feeding ruminants is all about managing their microbial population, much like winemaking, but much more complex.

The key to managing microbes is to maintain ruminal pH above 5.5 (ensure 'scratch factor', adapt animals slowly to new diets, and provide buffers in high-risk feeds).

Monitor the microbial population by monitoring rumination, lameness, voluntary intake and faecal consistency.

References

Aschenbach JR, Penner GB, Stumpff F, Gabel G (2011) Ruminant nutrition symposium: role of fermentation acid absorption in the regulation of ruminal pH. *Journal of Animal Science* **89**, 1092–1107. doi:10.2527/jas.2010-3301

Bauchop T (1979) The rumen anaerobic fungi: colonisers of plant fibre. *Annales De Recherches Veterinaire* **10**, 246–248.

Becker RB, Marshall SP, Dix Arnold PT (1963) Anatomy, development, and functions of the bovine omasum. *Journal of Dairy Science* **46**, 835–839. doi:10.3168/jds. S0022-0302(63)89156-0

Bird SH, Hill MK, Leng RA (1979) The effects of defaunation of the rumen on the growth of lambs on low-protein high-energy diets. *British Journal of Nutrition* **42**, 81–87. doi:10.1079/BJN19790091

Brownlee A (1956) The development of rumen papillae of cattle on different diets. *The British Veterinary Journal* **112**, 369–375. doi:10.1016/S0007-1935(17)46456-6

Church DC (Ed.) (1988) *The Ruminant Animal: Digestive Physiology and Nutrition*. Prentice Hall, Upper Saddle River, NJ, USA.

Coleman GS, Sandford DC (1979) The engulfment and digestion of mixed rumen bacteria and individual bacterial species by single and mixed species of rumen ciliate protozoa grown *in vivo*. *Journal Agricultural Science, Cambridge* **92**, 729–742. doi:10.1017/S0021859600053971

Deng W, Xi D, Mao H, Wanapa M (2008) The use of molecular techniques based on ribosomal RNA and DNA for rumen microbial ecosystem system studies. *Molecular Biology Reports* **35**, 265–274. doi:10.1007/s11033-007-9079-1

Dijkstra J, Boer H, van Bruchem J, Bruining M, Tamminga S (1993) Absorption of volatile fatty acids from the rumen of lactating dairy cows as influenced by volatile fatty acid concentration, pH and rumen liquid volume. *British Journal of Nutrition* **69**, 385–396. doi:10.1079/BJN19930041

Eugène M, Archimede H, Sauvant D (2004) Quantitative meta-analysis on the effects of defaunation of the rumen on growth, intake and digestion in ruminants. *Livestock Production Science* **85**, 81–97. doi:10.1016/S0301-6226(03)00117-9

Fanning J (2015) Pathogenesis of subacute ruminal acidosis in sheep. PhD thesis. The University of Adelaide, Australia.

Freer M, Campling RC, Balch CC (1962) Factors affecting the voluntary intake of food by cows. 4. The behaviours and reticular motility of cows receiving diets of hay, oat straw and oat straw with urea. *British Journal of Nutrition* **16**, 279–295. doi:10.1079/BJN19620030

Gordon GLR, Phillips MW (1998) The role of anaerobic gut fungi in ruminants. *Nutrition Research Reviews* **11**, 133–168. doi:10.1079/NRR19980009

Gray FV, Pilgrim AF, Weller RA (1954) Functions of the omasum in the stomach of sheep. *Journal of Experimental Biology* **31**, 49–55.

Hungate RE (1966) *The Rumen and its Microbes*. Academic Press, New York, USA.

Hynd P (1982) Wool growth efficiency: an investigation of the effects of diet and live weight change. PhD thesis. The University of Adelaide, Australia.

Lane MA, Jesse BW (1997) Effect of volatile fatty acid infusion on development of the rumen epithelium in neonatal sheep. *Journal of Dairy Science* **80**, 740–746.

Mirzaei M, Khorvash M, Ghorbani GR, Kazemi-Bonchenari M, Riasi A, Nabipour A, *et al.* (2015) Effects of supplementation level and particle size of alfalfa hay on growth characteristics and rumen development in dairy calves. *Journal of Animal Physiology and Animal Nutrition* **99**, 553–564. doi:10.1111/jpn.12229

Nagaraja TG, Chengappa MM (1998) Liver abscesses in feedlot cattle: a review. *Journal of Animal Science* **76**, 287–298. doi:10.2527/1998.761287x

Ogimoto K, Imai S (1981) *Atlas of Rumen Microbiology*. Japan Scientific Societies Press, Tokyo, Japan.

Okine EK, Mathison GW, Kaske M, Kennelly JJ, Chrisotpherson RJ (1998) Current understanding of the role of the reticulum and reticulo-omasal orifice in the control of digesta passage from the ruminoreticulum of sheep and cattle. *Canadian Journal of Animal Science* **78**, 15–21. doi:10.4141/A97-021

Penner GB, Aschenbach JR, Gotthold G, Rackwitz R, Oba M (2009) Epithelial capacity for apical uptake of short chain fatty acids is a key determinant for intraruminal pH and the susceptibility to subacute ruminal acidosis in sheep. *The Journal of Nutrition* **139**, 1714–1720. doi:10.3945/jn.109.108506

Seo S, Lanzas C, Tedeschi LO, Pell AN, Fox DG (2009) Development of a mechanistic model to represent the dynamics of particle flow out of the rumen and to predict rate of passage of forage particles in dairy cattle. *Journal of Dairy Science* **92**, 3981–4000. doi:10.3168/jds.2006-799

Stevens CE, Sellers AF, Spurrell FA (1960) Function of the bovine omasum in ingesta transfer. *The American Journal of Physiology* **198**, 449–455. doi:10.1152/ajplegacy.1960.198.2.449

van Soest PJ (1994) *Nutritional Ecology of the Ruminant*. 2nd edn. Cornell University Press, Ithaca, NY, USA.

Waghorn GC, Reid CSW (1983) Rumen motility in sheep and cattle given different diets. *New Zealand Journal of Agricultural Research* **26**, 289–295. doi:10.1080/00288233.1983.10427032

Warner ED (1958) The organogenesis and early histogenesis of the bovine stomach. *The American Journal of Anatomy* **102**, 33. doi:10.1002/aja.1001020103

4

Feeding standards for animals

Key points

- Feeding standards are the systems used to align the supply of nutrients from feeds to an animal's nutrient requirements.
- Daily feed intake is a key determinant of animal health, production, production efficiency and profitability of animal enterprises.
- Animals 'eat for energy' and, in general, balance energy intake with energy utilisation using energy-sensing pathways to achieve energy homeostasis.
- Feed intake is controlled in the short term (meal eating) by gut peptides, pancreatic hormones, absorbed nutrients and stretch receptors in the gut. Distal to proximal feedback loops ('brakes') regulate intake, motility and secretions. External factors (environment) and the senses of sight, smell, taste and texture modify the intake response.
- Long-term regulation of energy balance is achieved through hormonal feedback from adipose stores.
- Short-term and long-term signals are integrated centrally in the arcuate nucleus of the hypothalamus.
- Regulation of voluntary intake in ruminants on forage diets is mainly governed by physical mechanisms (gut fill), while regulation on low-fibre (concentrate diets) is mainly by chemical and metabolic mechanisms.
- Regulation of voluntary intake in mono-gastrics is achieved through energy-sensing pathways.
- Diets are formulated in a sequence as follows: water, energy, protein (amino acids), vitamins and minerals, physical form of the ration and feeding program.

- Feeding programs should be evaluated and adjusted iteratively based on frequent assessment of the health, body condition, production and behaviour of the animals.
- Energy systems for ruminants and camelids are based on Metabolisable Energy (ME) systems. Requirements are estimated as Net Energy and then converted to ME based on efficiency of use of ME to Net Energy ('k' factors). The Metabolisable Energy supply from feeds is then aligned to the animal's ME requirement.
- Protein systems for ruminants are based on metabolisable protein. The animal's total requirement for metabolisable protein is calculated and aligned with the supply of metabolisable protein (digestible undegraded protein and digestible microbial protein) from the gastrointestinal tract.
- Energy systems for horses are based on Digestible Energy, and protein requirements are expressed as crude protein, or digestible crude protein and lysine supply.
- Feeding standards for cats and dogs assume an optimum level of energy (or weight of food) is provided to the animal and nutrients are then expressed on an energy density or weight of food basis.
- Pig and poultry diets are formulated on either ME or NE bases. Increasingly NE systems are being developed. Rapidly changing genetics and the increasing use of alternative feed sources and feed additives necessitates continual revision of pig and poultry requirements. Amino acid requirements of pigs and poultry are expressed relative to an 'ideal' protein profile and the dietary supply of amino acids expressed as amino acids absorbed in the small intestine.

Introduction

The term 'feeding standards' refers to the process of relating the supply of nutrients in an animal's feed to the animal's requirements for these nutrients (Fig. 4.1).

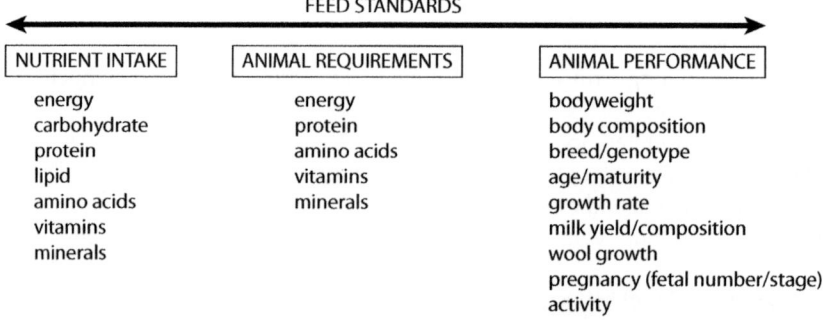

NUTRIENT INTAKE	ANIMAL REQUIREMENTS	ANIMAL PERFORMANCE
energy	energy	bodyweight
carbohydrate	protein	body composition
protein	amino acids	breed/genotype
lipid	vitamins	age/maturity
amino acids	minerals	growth rate
vitamins		milk yield/composition
minerals		wool growth
		pregnancy (fetal number/stage)
		activity

Fig. 4.1. Components of feeding standards. The process can be applied in both directions: that is, you can estimate animal performance from a given intake of nutrients or estimate nutrient requirements from a given level of performance.

They can be applied in both directions: that is, estimate the animal performance from a given intake of specific nutrients (i.e. left to right in Fig. 4.1) or estimate the nutrient requirements for a given level of performance (right to left in Fig. 4.1). Importantly, to relate the animal's nutrient requirement with the supply of that nutrient from feeds requires expression of the nutrient in a common unit (e.g. Metabolisable Energy content of feed (MJ/kg DM), DM intake (kg/day) and Metabolisable Energy requirement of the animal (MJ/day)). For animals that are fed a given quantity of food daily, the requirements are usually expressed as a certain quantity/day (e.g. 1.8 g Ca/day). For other animals (e.g. dog and cat foods), the required levels are expressed on a per kg of feed, or per calorie basis to align the intake of the nutrient with feed or energy intake.

As indicated in Fig. 1.3, the energy required by animals can be expressed as Digestible Energy (DE), Metabolisable Energy (ME) or Net Energy (NE). Protein requirements can be expressed as crude protein (percentage of diet), digestible crude protein or metabolisable protein. For ruminants and camelids, the requirement for protein can be further divided into protein required for microbial synthesis (Rumen Degradable Protein or RDP) and additional protein (Rumen Undegradable Protein or RUP, sometimes called Undegraded Dietary Protein or UDP) required from feed proteins that do not break down in the forestomach. For mono-gastrics, amino acid requirements are now determined as 'standardised ileal digestible amino acids', usually expressed relative to the first-limiting amino acid, lysine.

Different feeding standards systems for animals have been developed in different countries for most livestock species and they are constantly under revision and refinement. The level of precision and accuracy of prediction required in determining the nutrient requirements depends on the productive system and usually the animal species. For pigs and poultry, the level of precision and accuracy required is high because the economic impact of even slightly unbalanced diets is high, given the high proportion of costs incurred by feeding. For grazing animals, the ability to determine the actual nutrient intake from the pasture base as it changes throughout the season, the impact of the grazing environment on the animal's requirements, and the interaction between any supplements fed to the grazing animal and their grazed nutrients are difficult to predict. It is easier to determine the nutrient requirements of animals fed total rations in confinement. Given these differences in feeding systems between countries, between animal species and between feeding situations, the following section explores the principles underpinning the development of computer programs developed for different animal species. Although many programs are very sophisticated, it is essential that graduate nutritionists and health specialists understand the principles on which they are built, to allow critical analysis of animal performance against the expected outcomes from the software.

Feeding standards allow animal performance to be predicted from given nutrient intakes or to predict feed requirements to achieve a desired level of performance/production. For many purposes, however, it is necessary and sufficient for animal nutritionists, veterinarians, consultants and livestock producers to formulate rations 'on-the-run' (i.e. without the use of sophisticated packages or detailed knowledge of the composition of the feeds available). These options are outlined in the chapters on individual animal species. Below is a summary of the principles underpinning the different systems developed in different countries and for different species.

Feed formulation principles

To formulate a diet for a particular animal species or purpose, the sequence outlined in Fig. 4.2 should be followed.

For all animals, diets are formulated using the following general sequence:

1. **Water**: Water is the most important essential nutrient but is often neglected. Suboptimal animal performance will occur if clean, fresh, cool water is not provided.
2. **Energy**: Energy is the primary limiting nutrient for all metabolic processes. Providing feedstuffs that meet the energy needs of the animal also supplies a large portion of the amino acids, vitamins, minerals and lipids. Any deficits in these can then be remedied.
3. **Protein**: Protein, or more correctly, amino acids, are considered next because the supply of essential amino acids can limit the utilisation of energy for maintenance and production. For pre-gastric fermenters, such as

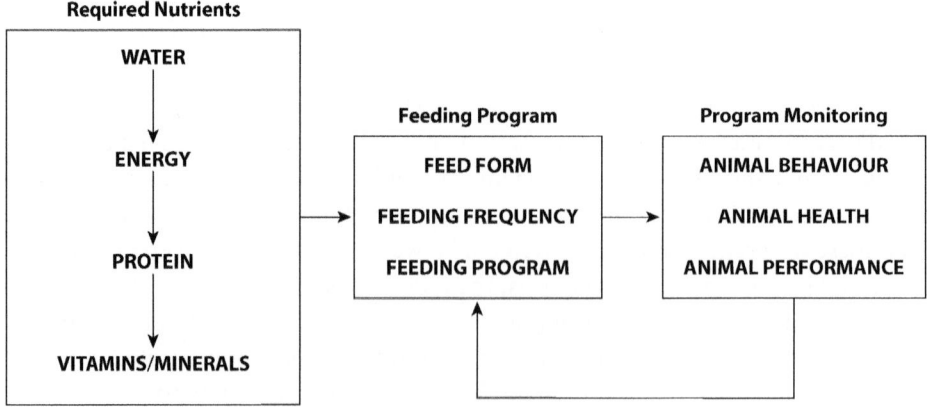

Fig. 4.2. Iterative steps involved in developing a feeding program for animals: establish the nutrient requirements, develop a feeding program, monitor the performance of the animals and revise the program.

ruminants and camelids, protein supply to the host animal is complicated by ruminal fermentation. Part of the requirement for protein is simply to provide the ruminal microbes with simple nitrogen sources (ammonia and peptides) for their protein synthesis. At higher levels of production, there is an additional requirement for non-microbial, dietary amino acids leaving the rumen.

4. **Vitamins and minerals**: Once the energy and protein requirements have been met, the next consideration is the supply of vitamins and minerals to meet the animal's requirement. Both dietary and microbial vitamin supplies must be considered.

5. **Feeding program**: The physical nature of the diet and the program of feeding must now be considered to ensure the diet is palatable, conducive to gut health and function (particularly microbial) and enriches the animal's daily life. Knowledge of the natural evolutionary feeding patterns of the animal can provide useful clues as to the most appropriate program. For example, the daily requirement for nutrients for a ruminant animal might be met in a diet that can be consumed in 10 min/day, but ruminants are grazers and naturally spend 8 h/day eating. Taking this into account in devising a feeding program might mean reducing the energy density, increasing particle size or increasing the fibre content to increase feeding time.

6. **Evaluation of the feeding program**: Frequent iterative assessment of the program and diet should be carried out using a combination of bodyweight measurement, body condition scoring, health assessments and production parameters to allow fine-tuning of the program.

The Pearson's Square: a simple way of making simple rations

A simple means of formulating diets made up of only two ingredients, or simple mixes of feeds, is to use a Pearson's Square. An example of a Pearson's Square to formulate a diet to meet the crude protein requirement of an animal is shown in Fig. 4.3.

The method involves the following steps:

1. Place the nutrient requirement in the box (this approach can be used for all nutrients – protein, energy, minerals, vitamins).
2. Place the two feeds available and their respective nutrient content on the left-hand side (Note: the requirement must lie between the two values for the feeds).
3. Subtract the values along the diagonals (Note: ignore sign).
4. Sum the two values and divide the two values by the sum to determine the proportion of each feed to be fed.

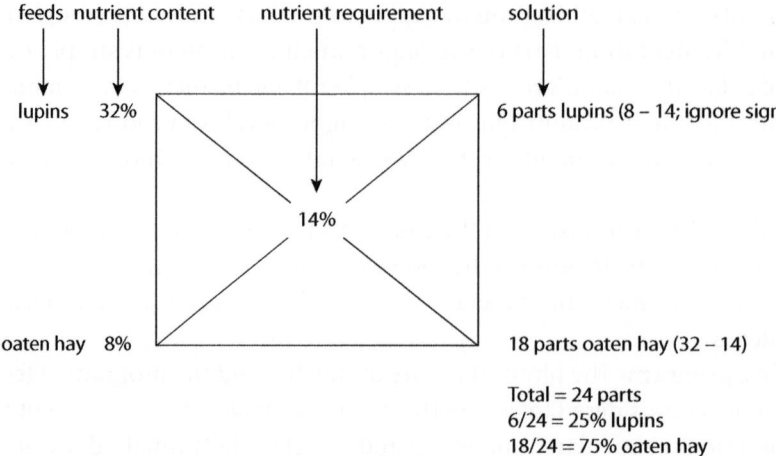

Fig. 4.3. Pearson's Square for formulating simple diets from two ingredients or two mixes of feeds. In this case, the animal's crude protein requirement is 14% and the two feeds available contain 32% CP (lupins) and 8% CP (oaten hay).

For feed mixes, use the same approach but use the combined value for the nutrient content of the mixes. More sophisticated feed formulation systems are described in the next sections and in individual species chapters.

Feed intake regulation and prediction in animals

Understanding the mechanisms operating to regulate the total intake of nutrients by an animal is important because, for companion animals, as for humans, obesity is a major issue with all its attendant pathologies. For production animals, maximising feed intake is often desirable to maximise production. High intakes also increase the efficiency of production because the requirement for nutrients for maintenance of the animal declines proportionately as intake increases. When formulating feeds for animals, it is essential to know that the animal is capable of consuming the entire ration. It is therefore useful to be able to predict the feed intake of an animal. Prediction in mono-gastrics is often based on animal species, genotype and bodyweight. For ruminants, prediction is more difficult, given the impact of fermentation characteristics, interaction between feed components, the effect of maturity (fatness) of the animal and, for grazing ruminants, pasture characteristics and environmental conditions. For both groups, it is generally regarded that animals are always hungry and that intake is regulated largely by physical and chemical constraints that, in the short term regulate meal eating and in the long term regulate energy balance and body fatness. For mono-gastrics, the constraints are mainly chemical and for ruminants on forage diets the constraints are mainly physical. The homeostatic mechanisms regulating feed intake must be efficient because, despite a lifetime of consumption of enormous quantities of food, the variance between adult individuals in bodyweight is relatively low. Indeed obesity and anorexia are really pathologies resulting from failure of the homeostatic mechanisms outlined in the following sections on control of feed intake.

Intake regulation in mono-gastric animals

Internal control of feed intake

Animals have a drive to eat to provide energy for their metabolic pathways. Animals with a high metabolic rate and that are rapidly using energy in metabolism (e.g. for high levels of milk production or high growth rate), have a strong intake drive, while those with a low rate of disposal of energy or in which there are large fat reserves have a lower intake drive. The balance between energy utilisation in metabolic pathways and the supply of energy from feed is detected by 'energy-sensing' pathways that operate in the peripheral tissues and the central nervous system to ensure that energy homeostasis is achieved (Richards and Proszkowiec-Weglarz 2007). In other words 'animals eat to obtain the required amount of energy'. This is illustrated in Fig. 4.4, which shows the relationships between energy density, feed intake and energy intake in pigs (Beaulieu et al. 2009). Similar relationships have been described in poultry (e.g. Classen 2016) and

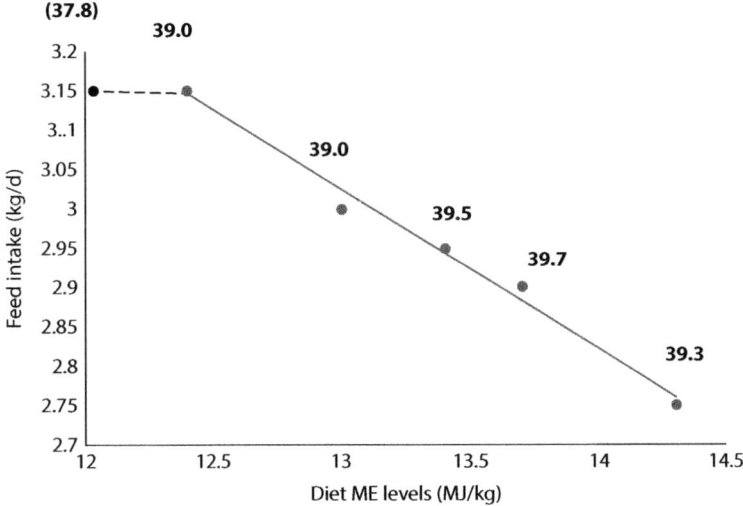

Fig. 4.4. Relationship between feed intake and energy density in the diet of mono-gastric animals. Daily feed intake of pigs (80–115 kg) consuming diets ranging in ME from 12.4 to 14.3 MJ/kg. Values indicated above the data points are total intake of ME at each energy level. The dashed line and bracketed ME intake value have been added to show the theoretical decline in total energy intake as feed intake plateaus due to gut fill. Source: After Beaulieu et al. (2009).

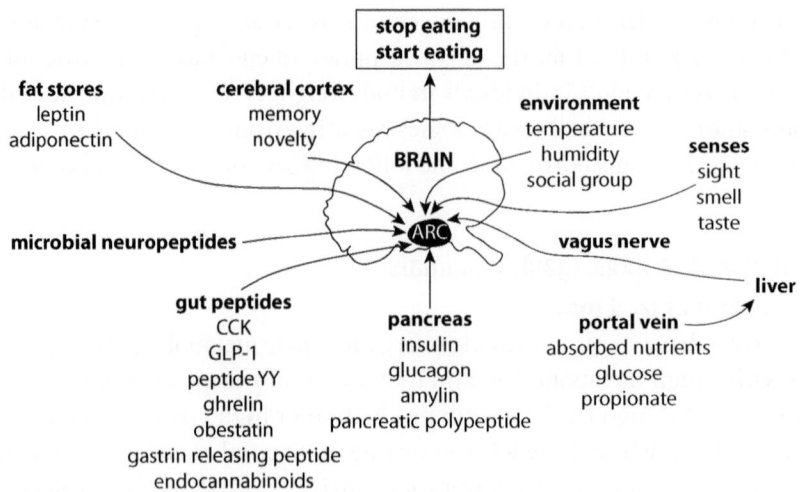

Fig. 4.5. Regulation of feeding in mono-gastrics is controlled by feeding centres in the arcuate nucleus (ARC) of the hypothalamus, which integrates neural, endocrine and nutrient signals from the gut, higher brain, liver and fat stores. Short-term (meal eating) and long-term (energy balance) regulation ensure energy homeostasis is achieved.

probably operate in all mono-gastric animals. As the energy density of the diet of mono-gastrics is decreased, voluntary intake of feed increases. Above an energy density of 13.4 MJ/kg, total energy intake is fairly constant (39.3 to 39.7 MJ/day), but below 13.4 MJ/kg the rate of increase in feed intake does not match the rate of decline in energy density, so total energy intake declines. This failure to compensate completely for a decline in the energy density of the ration is particularly apparent at very low energy densities because the increase in feed intake plateaus and may even decline, as the rate of feed passage limits intake (Fig. 4.4).

The regulation of feeding in the short and long term is controlled by feeding centres in the hypothalamus, which integrate a complex array of signals from the gut, higher brain centres, blood and fat stores (Fig. 4.5).

Short-term and long-term signals regulate the balance between energy demand and energy supply (Fig. 4.5). Short-term signals include the secretion of gut peptides from the entero-endocrine cells in the intestinal epithelium in response to the presence of nutrients in the gut lumen. These gut peptides operate at three levels: the gut itself; the central nervous system (CNS) via the vagus nerve; and the CNS via the blood stream. They influence gut function via mechanisms known as the jejunal, ileal and colonic 'brakes'. These brakes act to control meal size and intake by signalling from the lower gut to the upper gut to change motility, emptying and secretions. The presence of nutrients in the ileum stimulates the release of the peptides Peptide YY and glucagon-like peptide 1 (GLP-1), which affect the motility of the stomach, duodenum and jejunum, gastric and pancreatic secretion, and feed intake (see review by Maljaars *et al.* 2007).

The gut peptides include cholecystokinin (CCK), GLP-1, peptide YY, ghrelin, obestatin, gastrin-releasing peptide and the endocannabinoids. In addition to relaying information from the proximal gut (downstream) to the distal gut (upstream), they activate the vagus nerve, which sends a signal to the brain to signal nutrient status. They also travel to the brain via the blood and activate receptors in the signal satiety (anorexigenic peptides) or increased appetite (orexigenic peptides) (Valassi *et al.* 2008). The presence of food in the gut causes gut distension, which activates the vagus nerve and signals the brain to produce the feeling of 'fullness'.

The presence of absorbed nutrients in the portal vein appear to signal the brain to stop intake. Glucose and volatile fatty acids (particularly propionate) have strong anorexigenic effects on intake. The signals appear to travel through the vagus nerve from the liver to signal incoming absorbed nutrients (Allen *et al.* 2009). This theory, known as hepatic oxidation theory, is particularly important in ruminants (see 'Intake regulation in ruminants' later). Glucose also stimulates receptors in the hypothalamus to reduce intake.

Insulin appears to operate in the CNS to influence feed intake. Increased insulin in the cerebrospinal fluid has been shown to depress intake, while decreased insulin levels in the cerebrospinal fluid increases intake and adiposity (Woods *et al.* 1985).

The longer term regulation of feeding is achieved by satiety signals generated by adipose stores. A suite of adipokines (cytokines from fat cells), including leptin and adiponectin, signal the feeding centres and reduce intake over the longer term. Leptin suppresses neuropeptide Y, which is a potent stimulator of food intake.

Role of the gut microbiome in feed intake regulation

A huge population of bacteria colonises the intestinal system of animals. Most belong to the following phyla: Firmicutes, Bacteroidetes, Verrucomicrobia, Actinobacteria and Proteobacteria. Early estimates were that the total population was a factor of 10–100 times that of the number of cells in the host animal, but more recent estimates are that it is more likely to be about the same number as animal cells (Lee and Mazmanian 2010). Increasing evidence suggests that there is bidirectional communication between the gut microbiome and the brain through the neural, endocrine and immune systems and that dysbiosis (a poorly balanced gut microbiome) can result in hyperphagia (overeating), obesity, Type 2 diabetes and other aspects of the overall pathological state known as the 'metabolic syndrome' (Osadchiy *et al.* 2018). This microbiome/brain interaction is programmed in early life to produce different responses of adult animals to changes in diet composition (Osadchiy *et al.* 2018). The brain can influence the gut microbiome via the autonomic nervous system and hormonal signalling. The microbiome, in turn, can signal the brain to alter feed intake patterns via microbe-derived neuroactive molecules. Different microbiomes extract different levels of nutrients from feeds (Angelakis *et al.* 2013). Nutritionally, then, the response of an animal to a particular dietary regime will depend to some extent (and possibly a large extent) on the microbial population contained in its gut. The importance of this in animal nutrition, health and production is yet to be fully explored and is at the forefront of modern animal nutrition (see Angelakis 2017 for a review). The potential for using probiotics (provision of beneficial microorganisms), or prebiotics (provision of compounds that encourage the development of beneficial organisms) (Markowiak and Slizewska 2018) and reduce reliance on antibiotics is enormous.

External control of feed intake

Factors other than the internal physiological control mechanisms outlined above can influence feed intake. These include characteristics of the feed (texture, smell, temperature, taste and composition), the feeding program (frequency, timing), social factors (presence of other animals and owner behaviour) and the environment (temperature, humidity). The specific effects of these factors are considered for each species in the respective chapters.

Intake regulation in ruminants

Ruminants consume feeds that are generally lower in energy density and higher in fibre than mono-gastrics, so the regulation of feed intake tends to shift away from metabolic regulation as indicated in Fig. 4.5 to regulation based on physical limitations. In contrast to mono-gastrics then, voluntary feed intake tends to increase as digestibility (energy density) increases (Blaxter *et al.* 1961; Blaxter and Wilson 1962). This is thought to reflect the limitations imposed by retention of digesta in the reticulo-rumen. Low digestibility, high-fibre feeds remain in the reticulo-rumen for longer than highly digestible feeds. The high volume of bulky, undigested residue in the rumen triggers stretch receptors in the reticulo-ruminal wall and in turn activate the vagus nerve to signal rumen 'fill'. This relationship, however, has a break point at low levels of fibre (high-energy diets) beyond which regulation of voluntary intake switches back to chemical and metabolic regulation similar to that of mono-gastrics. The relationship between fibre content (NDF) and voluntary intake is shown in Fig. 4.6.

As the NDF content of feeds decreases (digestibility and energy density increases), feed intake increases. As the retention time of material in the reticulo-rumen decreases, so a greater rate of intake can be attained. The major signalling mechanism operating in this zone (from say 35 to 50% NDF) is the triggering of mechanical stretch receptors in the reticulum and cranial sac of the rumen wall,

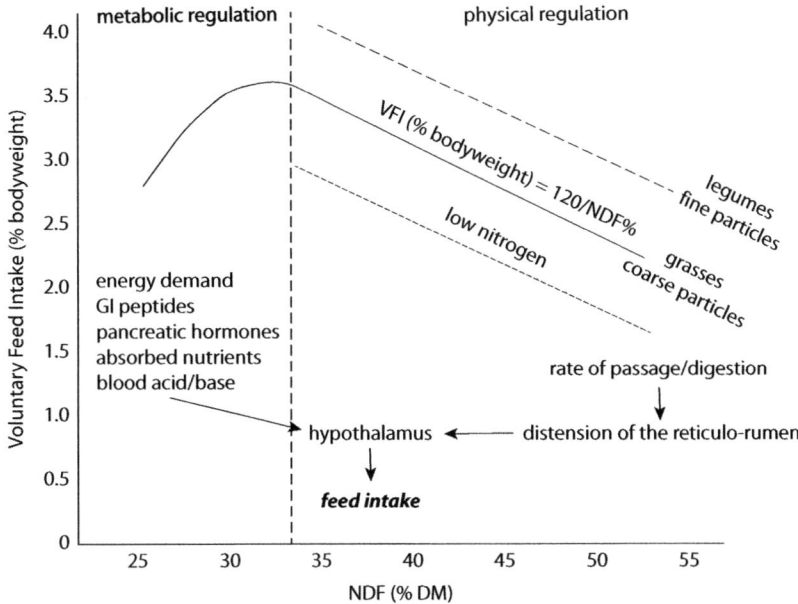

Fig. 4.6. Relationship between neutral detergent fibre content (NDF%) and voluntary intake of feed (as a percentage of bodyweight) in ruminants. The upper dashed line is for feeds that have been finely ground or from legumes and leafy materials, compared with the solid line for grasses and unground feeds. The lower, fine dot line is for feeds that are low in nitrogen or essential nutrients. At any NDF level, legumes and finely ground feeds are consumed at a higher rate than grasses and coarse materials. If essential nutrients for microbes such as nitrogen are limiting, intake is further depressed due to increased retention time of feed in the reticulo-rumen.

which transmit this information to the hypothalamus via the vagus nerve. Although this model broadly holds for moderate- to high-fibre feeds, there is significant variance about the actual intake values at any fibre value. This reflects several factors. First, the important variable is the rate of digesta disappearance from the reticulo-rumen, and this depends on particle size, the characteristics of the fibre (e.g. legumes versus grasses, stems versus leaves, and the availability of other nutrients for microbial activity). This is shown in Fig. 4.6 where intakes are ~20% higher at any NDF level if the material is legume, finely ground or leaf (Freer and Jones 1984). Conversely, for feeds that are low in fibre (high energy density) regulation switches to chemical and metabolic signalling. Now energy homeostasis can be re-established and a more finely tuned balance between energy demands and energy supply achieved. The same suite of gut peptides operates in ruminants as mono-gastrics to regulate digestion and intake, but the absorbed end products of digestion differ. The volatile fatty acids (VFA), acetate and propionate, enter the portal blood supply to the liver and have been shown to reduce feed intake when they are infused into the portal vein of sheep (Anil and Forbes 1988). The hepatic oxidation theory of feed intake regulation in ruminants (see review by Allen *et al.* 2009) suggests that it is these VFA (especially propionate), and not glucose, that activate the hepatic vagal afferent nerves. When ruminant feeds are very high in energy density (e.g. grain concentrates), high levels of acids in the rumen with low levels of saliva input to the rumen can create acute ruminal lactic acidosis and subacute ruminal acidosis. Perturbations to the microbial populations, ruminal environment, ruminal epithelium and blood acid/base balance can induce pathological reductions in feed intake.

Although the generic model in Fig. 4.6 is supported by many experiments, there are many exceptions and absolute values vary widely across ruminant species, diets and animal physiological states. This presumably reflects differences in threshold values for activation of stretch receptors indicating rumen 'fill', changes in hypothalamic thresholds (e.g. during lactation), impacts of silage chemicals, different relative fractional outflow rates of particles in sheep, cattle, goats and deer, and many others. Nevertheless it does provide a general working basis for interpreting ruminant intake behaviour and feed formulation, and the general equation relating NDF (%) and voluntary feed intake (as a percentage of bodyweight) indicated on the 'grasses' line, is a useful field guide for nutritionists.

Nutritional 'wisdom' in animals?

Much of nutritional science is directed towards establishing the nutrient requirements of different animals and formulating diets that meet these specifications. If, however, animals are capable of selecting an optimal ration by free choice from a variety of imbalanced feeds, there would be no need to balance ratio for them. This would have significant advantages and cost savings in ration preparation. This ability to select a balanced ration by free choice from several unbalanced feedstuffs is known as 'nutritional wisdom' and has been the subject of experimentation and debate for many decades. Results are very variable, with a substantial number of trials in laboratory and farm animals clearly demonstrating positive results (Forbes and Kyriazakis 1995), and others suggesting animals have a limited ability to select specific nutrients that are deficient in the diet (Galef 1991). Variable responses in food choices should not be surprising given the following variables involved in diet selection:

- **Animal species**: for some animals it is difficult for them to either detect the different constituents or to select out the desired constituent due to anatomy or physiology (taste, smell, bite size, bite precision).
- **Are the test animals deficient** in a nutrient(s) at the start of the test, in which case we are testing their ability to recognise the nutrient(s) **or are they replete at the start** of the test, in which case we are testing if they can select a diet that maintains essential nutrient balance?

- **Do the different feeds offered differ in ways that the animal can detect** (taste, colour or smell)? This is a function of the feed and the animal species.
- **Have the animals been exposed to the different feeds previously** and have they learned to associate nutritional value with a sensory cue (which may be positive or negative, such as a sweet or bitter taste)?
- **How long has the choice been offered?** Post-ingestive feedback (positive or negative) may take some time to provide the animal with the appropriate clues.
- **Do we know what the true nutrient requirements of the animals are?** Textbook values to which we compare the choices made may not be correct. For example, if the animals are suffering a stress, infection or parasitism the actual requirements may differ from those in the nutrient requirement tables.

There are some examples that strongly suggest animals can select a diet close to optimum for their physiological state. Broiler chickens select from feeds high or low in protein, which are in line with their protein requirements as they change with maturity, sex and lean growth potential (Forbes and Shariatmadari 1994). Lysine- and methionine-deficient birds also selected from the appropriate high-lysine and high-methionine options. Laying birds are capable of selecting diets with an appropriate protein content. Pigs can similarly select foods that match their changing protein requirements with age/fatness and appear to be able to select individual first-limiting amino acids (Bradford and Gous 1991). Cattle appear to be able to select for RDP (Tolkamp *et al.* 1998) and sheep can differentiate foods differing in protein content (Forbes 2007). Layer hens are capable of selecting appropriate calcium levels for egg laying. Given the low solubility of calcium salts, hence lack of taste, the birds must be distinguishing the foods on the basis of other cues. Presumably the feedback that reinforces selection of the higher calcium foods is a post-ingestive feeling of wellbeing (Forbes 2007). Phosphorus-deficient cattle and sheep have been observed to eat bones, suggesting an attempt to obtain P from the bones, but they do not select a P-supplement. This suggests that it is probably just depraved appetite (pica). Sheep, however, are able to correct sodium deficits very accurately when offered NaCl solutions (Denton 1982). Zinc-deficient chickens are able to select a high-Zn diet (Forbes 2007) and thiamine-deficient birds consistently select a thiamine-enriched supplement over a thiamine-deficient supplement (Hughes and Wood-Gush 1971).

Ruminants grazing on high-biodiversity rangelands select from a smorgasbord of feeds varying in toxicity and nutrient density and seem to learn to select diets that meet their nutrient needs, self-medicate to correct maladies and avoid toxicities (Provenza 1996; Provenza *et al.* 2015). This ability arises from flavour-feedback associations, a phytochemically rich landscape and learning *in vivo* to eat 'a nourishing combination of foods' (Provenza *et al.* 2015).

There is only limited evidence that animals can display 'nutritional wisdom' (the ability to select a balanced diet when offered a choice of different feeds), at least in the short term. Over longer periods of time, ruminants do learn to select an optimum ration, provided a phytochemically rich landscape is available to them.

Prediction of voluntary feed intake

Predicting feed intake of animals is important because most animals are fed to appetite (*ad libitum*) and intake is a key determinant of growth, body composition and feed efficiency. Predicting feed intake in mono-gastrics is relatively easy compared with prediction of the voluntary intake of ruminants. For pigs, simple regressions of intake on bodyweight (W), or variants of W (W^2, W^3, and W relative to maximum W) are used. For poultry, breeding companies have established expected daily bodyweight and intake targets for specific genetic lines of broilers. For ruminants, prediction of voluntary intake is complex and might include a large number of key feed and animal variables. For example, the 'Feed into Milk' program (Thomas 2004) estimates forage potential intake (from NIR analysis), concentrate intake, condition score of the cow, live weight, milk energy output, week of lactation, forage starch concentration and crude protein concentration of the concentrate. This equation accounts for more than 90% of the variance in the intake of dairy cows on silages, but only 75% of the variance in forage intakes.

Nutritionists in the field often use 'rules of thumb' in relation to maximum intake of ruminant animals (e.g. sheep and beef cattle typically consume 1.8–2.7% of their bodyweight in dry matter per day; on poor quality feeds = 1.8% of their bodyweight; average quality pastures/hay = 2.0–2.5% and high-quality feeds 2.5–2.7%).

A useful equation for predicting the voluntary feed intake of ruminants on roughage or mixed diets (NDF >30%) is as follows:
Voluntary feed intake (percentage bodyweight) = 120/NDF%

Energy systems for ruminants and camelids

The energy available for maintenance, gestation, lactation, activity and growth and the energy supplied by a particular feed can be expressed as Digestible Energy, Metabolisable Energy or Net Energy. For ruminants, most systems are based on ME and NE. The ME content of the diet is estimated, the NE requirement of the animal is calculated and the ME requirement of the animal is then estimated based on how efficiently the diet is converted from ME to NE. The system used in the UK Metabolisable Energy system (AFRC 1993), the Australian Standing Committee on Agriculture's feeding standards for ruminants (CSIRO 2007), the USA Nutrient Requirements of Beef Cattle (NRC 2016) and many of the systems used in Europe (Van der Honing and Alderman 1988), use variations on the protocols outlined in the following sections.

1. Estimate the Metabolisable Energy content of the diet

The ME content of feeds are derived from tables of average ME values for the particular feedstuffs, from NIRS analysis of the feed or from analysis of the feed composition by 'wet chemistry'. Most equations relate the metabolisable density of a feed (M/D in MJ/kg DM) to diet digestibility (positive), fibre content (negative) or to combinations of feed components (e.g. crude protein, ether extract, modified acid detergent fibre, ash content, DMD). For example, the CSIRO (2007) system derives M/D for roughages, silages and 'energy' feeds as follows:

$$\text{Roughages: M/D (MJ/kg DM)} = 0.172\text{DMD} - 1.71$$

$$\text{Silages: M/D (MJ/kg DM)} = 0.711\text{DOMD} - 1.37$$

$$\text{Grains/protein sources: M/D (MJ/kg DM)} = 0.134\text{DMD} + 0.235\text{EE} + 1.23$$

2. Estimate the maintenance Metabolisable Energy requirement of the animal

The NE required for maintenance of an animal is equivalent to its fasting heat production. To calculate the ME required for maintenance depends on how efficiently the ME in the diet is converted to NE and this depends on the feed 'quality'. Similarly the 'k' factors for efficiency of utilisation of ME for NE are estimated for growth (k_g), pregnancy (k_c) and lactation (k_l).

The 'k' factors vary with diet quality because nutrients such as propionate and glucose are used more efficiently in biochemical pathways. The following equations are used to estimate the effects of diet quality (CSIRO 2007):

$$\text{For maintenance, } k_m = 0.02\text{M/D} + 0.5 \text{ (typical values are 0.60–0.74)}$$

$$\text{For growth, } k_g = 0.043\text{M/D (typical values are 0.22–0.52)}$$

$$\text{For pregnancy, } k_c = 0.133$$

$$\text{For lactation, } k_l = 0.02\text{M/D} + 0.4 \text{ (typical values are 0.56–0.64)}$$

where M/D = the Metabolisable Energy density of the feed (MJ/kg DM).

Similarly the UK system estimates the k values from the metabolisability of the diet (= ME/gross energy).

Now we can estimate the ME requirement of the animal for maintenance because

$$ME_m = NE_m \text{ or } FHP/k_m$$

The fasting heat production relates directly to fasted bodyweight and depends on animal species, breed, sex and age.

Typical values for the maintenance requirement of ruminants (ME_m) are given below for different systems used in different countries. All are based on estimates of the FHP = $cW^{0.75 \text{ or } 0.67}$ and an appropriate k_m factor:

Sheep:

$$ME_m \text{ (MJ/day)} = 0.25W^{0.75}/k_m \text{ where } W = \text{fasted bodyweight}$$
$$= \text{live weight}/1.08 \text{ (AFRC 1993)}$$

$$ME_m \text{ (MJ/day)} = 0.28W^{0.75}/k_m \text{ where } W = \text{fleece-free live weight (CSIRO 2007)}$$

Cattle:

$$ME_m \text{ (MJ/day)} = 0.53W^{0.67} \text{ where } W = \text{fasted live weight (AFRC 1993)}$$

Goats:

$$ME_m \text{ (MJ/day)} = 0.315W^{0.75} \text{ where } W = \text{fasted live weight (AFRC 1993)}$$

$$ME_m \text{ (MJ/day)} = 0.452W^{0.75} \text{ where } W = \text{full bodyweight (NRC 2007)}$$

$$ME_m \text{ (MJ/day)} = 0.28W^{0.75} \text{ where } W = \text{live weight (CSIRO 2007)}$$

Camelids:

$$ME_m \text{ (MJ/day)} = 0.723W^{0.75} \text{ where } W = \text{full bodyweight (NRC 2007)}$$

Note: these are selected values to demonstrate the ranges and variances between species. More detailed values are used in programs designed for each species. For example, effects of breed, age, sex, maturity, previous level of nutrition and body condition score are usually accounted for in the estimation of maintenance energy requirement. Some systems estimate the maintenance energy requirement and then add the energy required for activity (e.g. CSIRO 2007). Others include activity values added to the fasting heat production value before dividing by the k_m value (AFRC 1993).

The Australian system (CSIRO 2007) calculates the maintenance ME requirement of sheep and goats using the following equation:

$$ME_m \text{ (MJ/day)} = K.S.M. (0.28W^{0.75} \exp (-0.03A))/k_m + 0.1ME_p + ME_{graze} + E_{cold}$$

where K = 1.0 for sheep and goats or 1.2 for *B. indicus* or 1.4 for *B. taurus*, S = 1.0 for females and castrates and 1.15 for intact males (rams, goats, bulls), M = 1 + (0.23 × % of DE from milk), W = live weight excluding conceptus and fleece (sheep), ME_p = the amount of ME being used for production, ME_{graze} = additional energy expenditure for grazing activity/k_m, E_{cold} = additional energy expenditure when critical temperature is less than lower critical temperature.

3. Estimate the Metabolisable Energy requirement for growth

Having calculated the ME required to maintain the animal, the additional energy required to achieve growth can be calculated. The basis of this calculation is estimation of the energy content of the live weight gain. This is difficult because the energy content of gain changes with the maturity of the animal. As animals approach their mature weight, the energy content of the gain increases as fat becomes a greater proportion of the weight gain. Breeds differ in mature bodyweight so this is corrected for in some systems by classifying into early, medium and late-maturing types (AFRC 1993). In this system, the changing energy value of the gain is estimated using a quadratic equation based on live weight. CSIRO (2007) uses a similar approach by adopting standard reference weight ranges for different classes of animal (breed and sex). Having estimated the energy value of the gain and the rate of gain, the extra ME required to grow at this rate is calculated from NEgain/k_g.

So the animal's total ME requirement now is its ME_m + ME_g.

4. Estimate the Metabolisable Energy requirement for lactation

The additional energy required for milk production depends on the milk yield and the milk energy content. Simple equations relating milk yield, day of lactation and milk fat content are used to calculate milk energy output. This, of course, is then corrected for the efficiency of utilisation of ME for NE, which for most diets lies between 0.58 and 0.64. In general, for dairy cows this equates to ~5 MJ ME/kg milk (see Chapter 8).

5. Estimate the Metabolisable Energy requirement for gestation

The CSIRO (2007) and AFRC (1993) systems estimate the additional ME required for pregnancy using the energy content of the gravid uterus at particular times of gestation, accounting for the presence of twins or singles, assuming a certain birthweight and an efficiency of ME to NE of a constant 0.133.

6. Estimate the additional Metabolisable Energy requirement for grazing activity

Allowances for grazing activity include energy used for walking in the horizontal plane (2.6 kJ/km/kg bodyweight), 10 kJ/day standing, 0.26 kJ/body movement from standing to lying to standing), 28 kJ/km vertical plane, 2.5 kJ/h eating and 2.0 kJ/h ruminating. Estimates of the distances and times in each activity must be made to estimate the additional energy required. These values in grazing animals are greatly influenced by topography, feed availability and quality, and distance to watering places.

7. Estimate the additional Metabolisable Energy required to cope with cold stress

The CSIRO (2007) model assesses the extra energy needed to cope with temperatures below the lower critical temperature. Lower critical temperatures are estimated depending on rainfall, wind, coat cover depth and skin surface area.

Energy systems for horses

The main energy system used globally for horses is based on Digestible Energy (NRC 2007), but a ME system (Geor *et al.* 2013) and an NE system (INRA 1984) are also used. The NE system is expressed in feed units known as Unite Fouragere Cheval or UFC units, which relate to the NE contained in 1 kg barley (9.414 MJ). The efficiency of conversion of ME to NE (K_m) is estimated from knowledge of the relative energy substrates (glucose, long-chain fatty acids and amino acids) contained in the feeds. The substrates not digested in the small intestines are assumed to be fermented in the large intestine to VFA. However, the site of digestion of feed components in horses is very dependent on the feeding practices, rate of intake and food processing. Also no account is made for differences in the efficiency of utilisation of ME to NE for different physiological states.

Energy systems for pigs and poultry

Globally, energy systems for poultry are based on ME systems, although there is increasing interest in developing Net Energy systems because even small differences in the efficiency of utilisation of Metabolisable Energy for production can have significant economic implications. Nevertheless, at the time of publication of this book, ME systems dominate poultry feeding systems. The relative uniformity of poultry genetics and of rations compared with ruminants, along with the ease of measuring Metabolisable Energy by total collection of excreta, means ME systems will probably remain in place for some time.

For pigs, Digestible Energy (DE) and ME systems are most widely used, but increasingly NE systems are being used (e.g. NRC 2012). NE systems are superior to DE systems if the efficiency of utilisation of ME differs significantly with feed composition or for the productive purpose to which the energy is being used. There is evidence, for example, that the 'k factors' for fattening versus protein deposition differ. As pressures increase to find alternative feeds and by-products, and as the genetics of modern pigs change, it is possible that NE systems will be more efficient at matching energy requirements with energy supply. The 11th edition of the *Nutrient Requirements of Swine* (NRC 2012) uses an NE system, which allows day-to-day dynamic estimates of nutrient requirements to be made. In general, this edition provides higher estimate of nutrient requirements than earlier editions, reflecting increases in pig performance levels. As the pig industry moves towards the inclusion of higher fibre and higher protein diets and away from high-starch diets, the use of NE systems will become more important. The problem with NE systems is that it is not practical to determine the NE content of feeds, so they are estimated indirectly from DE, and from components of the proximate analysis (e.g. CP, starch, EE and ADF content).

Energy systems for cats and dogs

Energy systems for cats and dogs are based on ME. As for other mono-gastrics, nutrient requirements are expressed as minimum and maximum concentrations in the feed (on a dry matter basis or energy density basis). The reason nutrient requirements in dogs and cats are expressed per unit of energy density is that animals adjust their feed intake to meet their energy requirement (broadly), so if other nutrients are expressed per unit energy density, the animal's nutrient requirement will be met if daily energy requirement is achieved. An international scheme administered by the Association of American Feed Control Officials (AAFCO) in the USA established nutrient profiles for cat and dog foods (Dzanis 1994). These are based on minimum and maximum nutrient concentrations in commonly used, non-purified feeds for cats and dogs. The nutrient concentrations are expressed on a dry matter basis (e.g. % or mg/kg) or on an energy density basis (e.g. g/1000 kcalME). If the profiles are expressed on a dry matter basis, it is assumed the caloric density is 4000 kcalME/kg, so in reality the nutrients are on an energy basis. Feeds that are greater than 4000 kcalME/kg need to have their formulations corrected for energy density, otherwise animals fed to a specific energy intake will receive inadequate nutrients. Feeds containing less than 4000 kcalME/kg are not corrected for energy. The National Research Council (NRC 2006) also published minimum nutrient requirements and recommended allowances for cats and dogs.

Given the wide variance in dog size, physiology, metabolism and propensity to obesity, meeting the optimum energy intake for dogs is difficult. Dog and cat feeding programs require frequent review based on target bodyweights, body condition scores and health checking (see Chapter 11).

Protein systems for ruminants

Metabolisable protein system

In 1990 a new system for describing the requirements and supply of dietary protein for ruminants was introduced (ARC 1990). The system, known as the metabolisable protein system has been adopted in various forms and using slightly different terminology in the protein systems for ruminants in different countries (Verité *et al.* 1987; Fox *et al.* 1992; AFRC 1993; Thomas 2004; CSIRO 2007). The underlying basis of the metabolisable protein system is summarised in Fig. 3.22.

In summary, feeds contain crude protein that is either converted into simple compounds (peptides, amino acids and ammonia), which are available to the microbes for microbial protein synthesis or escapes ruminal degradation, providing the animal with intestinal amino acids. The extent to which a feed protein can contribute to the former pool is called the effective RDP, and its actual capture into microbial protein depends on the availability of Fermentable Metabolisable Energy (FME), which is the dietary ME minus the ME derived from fat and Undegraded Dietary Protein. This microbial crude protein contains 0.75 true protein, which is only absorbed at an efficiency of 0.85, producing a digestible microbial protein value of 0.64 (0.75 × 0.85) of the microbial protein leaving the rumen. This joins with the DUP leaving the stomach (estimated from feed composition) to produce the metabolisable protein, which is used with variable efficiency (k factors) to meet the tissue protein requirements for maintenance, growth, lactation, pregnancy and wool growth (AFRC 1993). If the MP requirement is greater than the supply of digestible microbial crude protein from the intestines, then DUP is required.

Several changes have been made to this basic system, particularly for estimating the protein requirements and dietary supply for dairy cows (Fox *et al.* 1992; Thomas 2004). In particular, attempts are now being made to align the intestinal supply of amino acids with the 'ideal' protein requirement, in a similar manner to that used for pigs and poultry.

Estimating the protein requirements of ruminants

Protein requirements are estimated using a factorial method in which all the protein needs are summed as indicated below:

1. Estimate the obligatory nitrogen losses.
   Most systems (AFRC 1993; Thomas 2004; CSIRO 2007; NRC 2016) estimate the basal losses of nitrogen (protein) in urine and faeces, and from skin (hair, scurf, wool) on the basis of live weight or metabolic bodyweight. Some systems (e.g. CSIRO 2007) make allowance for an effect of feed intake on faecal losses. To convert the endogenous basal losses of protein into dietary protein supply, the efficiency with which microbial protein is converted into digestible protein is estimated (between 0.6 and 0.7).
2. Estimate the protein required for live weight gain.
   The protein content of live weight gain changes with maturity, mature bodyweight and sex. To calculate the extra protein needed for growth, the feeding systems account for these differences in varying ways. AFRC (1993) uses weight and weight gain to estimate net protein requirement for growth

and then applies factors to account for breed and sex effects. CSIRO (2007) uses an equation that uses bodyweight relative to a standard reference mature weight.

3. Estimate the protein requirement for lactation.

    The protein content of milk is estimated based on standard values for goats, sheep and cattle and multiplying by the milk yield.

4. Estimate the protein requirement for gestation.

    The total protein requirement for growth of the fetus is estimated on the basis of exponents of days from conception for each species.

Protein systems for pigs and poultry

Meeting the amino acid requirements of animals

Having met the animal's requirements for energy, the nutritionist must then attempt to match the quantity and quality of the amino acids required for body functions and protein accretion in muscle, milk, fibre and fetuses, with the supply of these amino acids from the diet (Fig. 4.7).

Traditionally, the requirements of pigs and poultry for amino acids were based on the concept of first-limiting, second-limiting, and so on, essential amino acids and the use of 'biological value' as a metric for protein quality. Increasingly the requirements for pigs and poultry are expressed in terms of the 'ideal protein' (van Milgen and Dourmad 2015). This system, known as the 'ideal protein' concept, is based on an estimated 'ideal' balance of amino acids, being that which reflects the balance of essential amino acids in the body protein. Usually the amino acids in the 'ideal protein' are expressed relative to the requirement for lysine, because lysine is commonly first-limiting. The 'ideal' protein in the feed is then corrected for digestibility and endogenous losses of amino acids into the intestines to allow matching with the animal requirement. This value is known as 'standardised ileal digestibility'.

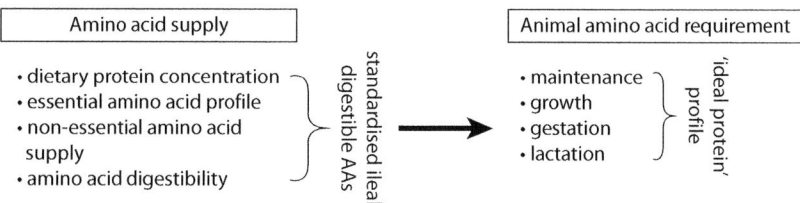

Fig. 4.7. Matching the animal's amino acid requirements with the dietary supply of amino acids. For pigs and poultry there is increasing use of the 'ideal protein' concept and 'standardised ileal digestible' (SID) amino acids to describe the demand and supply respectively. Sources: Based on descriptions by Urbaityte *et al.* (2009) and van Milgen and Dourmad (2015).

Standardised ileal digestible amino acids (SID)

Obviously the first determinant of the quality of a protein is its digestibility. That is, what proportion of the dietary protein is absorbed and available for metabolism? Whole-tract digestibility, however, is inaccurate because hindgut microbial activity can alter the disappearance of protein, which is still not available to the animal. No amino acids are absorbed post-ileum so assuming the ileum is the 'endpoint' is reasonable. Because it is the supply of the essential amino acids that is the important issue for the animal, the ileal digestibility of essential amino acids is used as the benchmark for pig and poultry feed formulation. Some of the amino acids appearing at the ileum, however, are not related to the dietary protein, but are endogenous losses from the gut, saliva, bile and digestive secretions. The term 'standardised' ileal digestible protein, therefore is corrected for these non-feed losses as follows:

$$\text{SID (g / kg)} = \frac{\text{dietary AA flow} - (\text{ileal AA flow} - \text{ileal basal AA flow})}{\text{dietary AA intake}} \times 1000$$

The 'basal amino acid flow' at the ileum is corrected for the effects of feed intake and is therefore 'standardised' to a common intake.

Protein systems for horses

Protein systems for horse are relatively crude compared with the sophisticated systems in place for pigs, poultry and even ruminants. The NRC (2007) system uses crude protein and lysine intake as the key components. A French system, the Matieres Azotees Digestible Cheval (MADC; Martin-Rousset 2001), attempts to account for protein digestibility and site of digestion of amino acids. The system estimates the sites of digestion and hence the extent of absorption of individual amino acids. A major problem with the MADC system is that we have so little information about the digestion of forages and the variables that can change the sites of digestion. Nevertheless, at least this system uses digestible crude protein, and some estimate of the availability of amino acids can be made by using standard factors for different classes of feeds (e.g. green forages, concentrates, straws and silages. In the NRC (2007) system, CP requirements are estimated using a series of standard equations for maintenance, growth, pregnancy, lactation and exercise. The lysine requirement is then estimated as CP requirement × 0.043, assuming that 4.3% of the total protein should be present as the first-limiting amino acid.

Protein systems for cats and dogs

As for other nutrients, AAFCO (Dzanis 1994) and NRC (2006) suggest minimum levels of crude protein and levels of individual amino acids as a percentage of the feed containing 4000 kcals/kg feed for the different productive purposes.

Alternatively, the CP is expressed as a percentage of ME calories. It is assumed that the protein is of high quality, particularly for young animals.

References

AFRC (1993) *Energy and Protein Requirements of Ruminants: an Advisory Manual Prepared by the AFRC Technical Committee on Responses to Nutrients.* Commonwealth Agricultural Bureaux, Slough and CABI, Wallingford, UK.

Allen MS, Bradford BJ, Oba M (2009) Board invited review: the hepatic oxidation theory of the control of feed intake and its application in ruminants. *Journal of Animal Science* **87**, 3317–3334. doi:10.2527/jas.2009-1779

Angelakis E (2017) Weight gain by gut microbiota manipulation in productive animals. *Microbial Pathogenesis* **106**, 162–170. doi:10.1016/j.micpath.2016.11.002

Angelakis E, Merhej V, Raoult D (2013) Related actions of probiotics and antibiotics on gut microbiota and weight modification. *Lancet* **13**, 889–899. doi:10.1016/S1473-3099(13)70179-8

Anil MH, Forbes JM (1988) The role of hepatic nerves in the reduction of food intake as a consequence of intraportal sodium propionate administration in sheep. *Quarterly Journal of Experimental Physiology (Cambridge, England)* **73**, 539–546. doi:10.1113/expphysiol.1988.sp003174

ARC (1990) *Feeding Standards for Australian Livestock: Ruminants.* Standing Committee on Agriculture, Ruminant Subcommittee, Melbourne.

Beaulieu AD, Williams NH, Patience JF (2009) Response to dietary digestible energy concentration in growing pigs fed cereal grain-based diets. *Journal of Animal Science* **87**, 965–976.

Blaxter KL, Wilson RS (1962) The voluntary of roughages by steers. *Animal Science* **3**, 351–358.

Blaxter KL, Wainman FW, Wilson RS (1961) The regulation of food intake by sheep. *Animal Science* **3**, 51–61.

Bradford MMV, Gous RM (1991) The response of growing pigs to a choice of diets differing in protein content. *Animal Production* **52**, 185–192. doi:10.1017/S0003356100005821

Classen HL (2016) Diet energy and feed intake in chickens. *Animal Feed Science and Technology* **233**, 13–21.

CSIRO (2007) *Nutrient Requirements of Domesticated Ruminants.* CSIRO Publishing, Melbourne.

Denton DA (1982) *The Hunger for Salt.* Springer-Verlag, Berlin, Germany.

Dzanis DA (1994) The Association of American Feed Control Officials dog and cat food nutrient profiles: substantiation of nutritional adequacy of complete and balanced pet foods in the United States. *The Journal of Nutrition* **124**, 2535S–2539S.

Forbes JM (2007) A personal view of how ruminants control their intake and choice of food: minimal total discomfort. *Nutrition Research Reviews* **20**, 132–146.

Forbes JM, Kyriazakis I (1995) Food preferences in animals: why don't they always choose wisely? *Proceedings of the Nutrition Society* **54**, 429–440.

Forbes JM, Shariatmadari F (1994) Diet selection for protein by poultry. *World's Poultry Science Journal* **50**, 7–24.

Fox DG, Sniffen CJ, O'Connor JD, Russell JB, van Soest PJ (1992) A net carbohydrate and protein system for evaluating cattle diets: cattle requirements and diet adequacy. *Journal of Animal Science* **70**(11), 3562–3577.

Freer M, Jones DB (1984) Feeding value of subterranean clover, lucerne, phalaris and Wimmera ryegrass for lambs. *Australian Journal of Experimental Agriculture and Animal Husbandry* **24**, 156–164.

Galef BG (1991) A contrarian view of the wisdom of the body as it relates to dietary self-selection. *Psychological Review* **98**, 218–223.

Geor RJ, Harris PA, Coenen M (2013) *Equine Applied and Clinical Nutrition; Health, Welfare and Performance.* Saunders Elsevier, Philadelphia, PA, USA.

Hughes BO, Wood-Gush D (1971) Investigations into specific appetites for sodium and thiamine in domestic fowls. *Physiology & Behavior* **6**, 331–339. doi:10.1016/0031-9384(71)90164-8

INRA (1984) (Eds R Jarrige and W Martin-Rousset) *Le Cheval, Reproduction, Selection, Alimentation, Exploitation.* Institut National de la Recherche Agronomique, Paris, France.

Lee YK, Mazmanian SK (2010) Has the microbiota played a critical role in the evolution of the adaptive immune system? *Science* **330**, 1768–1773. doi:10.1126/science.1195568

Maljaars J, Peters HPF, Masclee AM (2007) Review article: The gastrointestinal tract: neuroen-docrine regulation of satiety and food intake. *Alimentary Pharmacology and Therapeutics* **26**(Suppl 2), 241–250.

Markowiak P, Slizewska K (2018) The role of probiotics, prebiotics and synbiotics in animal nutrition. *Gut Pathology* **10**, 21. doi:10.1186/s13099-018-0250-0.

Martin-Rousset W (2001) Feeding standards for energy and protein for horses in France. In *Advances in Equine Nutrition. Volume 2.* (Ed. JD Pagan) pp. 31–94. Kentucky Equine Research, Versailles, KY, USA.

NRC (2006) *Nutrient Requirements of Dogs and Cats.* National Research Council, The National Academies Press, Washington, DC, USA. doi:10.17226/10668

NRC (2007) *Nutrient Requirements of Horses.* 6th revised edn. National Research Council, The National Academies Press, Washington, DC, USA.

NRC (2012) *National Research Council. Nutrient Requirements of Swine: Eleventh Revised Edition.* The National Academies Press, Washington, DC, USA.

NRC (2016) *Nutrient Requirements of Beef Cattle.* 8th revised edn. National Research Council, The National Academies Press, Washington, DC, USA.

Osadchiy V, Martin CR, Mayer EA (2018) The gut-brain axis and the microbiome: mechanisms and clinical implications. *Clinical Gastrointestinal and Hepatology* **17**, 322–332. doi:10.1016/j.cgh.2018.10.002

Provenza FD (1996) Acquired aversions as the basis for varied diets of ruminants foraging on rangelands. *Journal of Animal Science* **74**, 2010–2020. doi:10.2527/1996.7482010x

Provenza FD, Meuret M, Gregorini P (2015) Our landscapes, our livestock, ourselves: restoring broken linkages among plants, herbivores, and humans with diets that nourish and satiate. *Appetite* **95**, 500–519. Epub 2015 Aug 4. doi:10.1016/j.appet.2015.08.004

Richards MP, Proszkowiec-Weglarz M (2007) Mechanisms regulating feed intake, energy expenditure, and body weight in poultry. *Poultry Science* **86**, 1478–1490. doi:10.1093/ps/86.7.1478

Thomas C (Ed.) (2004) *Feed into Milk: a New Applied Feeding System for Dairy Cows.* Nottingham University Press, Nottingham, UK.

Tolkamp BJ, Kyriazakis I, Oldham JD, Lewis M, Dewhurst RJ, Newbold JR (1998) Diet choice by dairy cows. 2. Selection for metabolisable protein or for ruminally degradable protein? *Journal of Dairy Science* **81**, 2670–2680. doi:10.3168/jds.S0022-0302(98)75824-2

Urbaityte R, Mosenthin R, Eklund M (2009) The concept of standardised ileal amino acid digestibilities: principles and application in feed ingredients for piglets. *Asian-Australasian Journal of Animal Sciences* **22**, 1209–1223. doi:10.5713/ajas.2009.80471

Valassi E, Scacchi M, Cavagnini F (2008) Neuroendocrine control of feed intake. *Nutrition, Metabolism, and Cardiovascular Diseases* **18**, 158–168.

Verité R, Michalot-Doreau B, Chapoutot P, Peyraud JL, Poncet C (1987) Revision du systeme des proteins digestibles dans l'intestine (PDI). *Bulletin Technical CRZV Thiex, INRA* **70**, 19–34.

van der Honing Y, Alderman G (1988) Feed evaluation and nutritional requirements: ruminants. *Livestock Production Science* **19**, 217–278. doi:10.1016/0301-6226(88)90092-9

van Milgen J, Dourmad J-Y (2015) Concept and application of ideal protein for pigs. *Journal of Animal Science and Biotechnology* **6**, 15. doi:10.1186/s40104-015-0016-1

Woods SC, Porte D, Jr, Bobbioni E (1985) Insulin: its relationship to the central nervous system and to the control of feed intake and body weight. *The American Journal of Clinical Nutrition* **42**, 1063–1071. doi:10.1093/ajcn/42.5.1063

5
Grazing animal nutrition

Key points

- Grazing herbivores, mainly sheep, cattle, goats and camelids, are important environmental, social and economic contributors to global food supply by converting low-quality roughages to high-quality animal protein, milk, fibre and work.
- The **nutritive value** of a forage is measured as the animal response per unit of intake of the forage. The **feeding value** of a forage is the animal response when fed the forage *ad libitum*.
- The feeding value of forages in Mediterranean, semi-arid, temperate, subtropical and tropical environments varies widely throughout the year as the native and introduced plants respond to changes in rainfall and temperature. Matching the animal's fluctuating energy requirements with fluctuating forage nutrient supply is a major challenge facing livestock producers.
- Grazing animals are faced with a wide variety of phytochemicals in plants and plant parts differing in chemical composition, taste, smell, toughness, irritants and toxins. Selection of which feeds to consume is complex and is directed towards maximising the pleasure of eating, obtaining a balance of nutrients suited to the metabolic demands of maintenance, growth, pregnancy, lactation and exercise, and minimising the intake of deleterious compounds.
- Herbivores adjust intake rate, bite size (volume) and grazing time depending on the biomass available to them and its quality, in an attempt to meet their energy requirements.

- Supplements are designed to alleviate deficiencies of energy, protein, non-protein nitrogen, and minerals. Choice of supplements should be made on the basis of least-cost per limiting nutrient. On this basis, minerals are often the most cost-effective supplements.
- All forages contain potentially deleterious compounds that reduce animal performance. They include nitrates, tannins, soluble proteins, soluble carbohydrates, essential oils, alkaloids, goitrogens, cyanogens, oxalates, oestrogens and a suite of mycotoxins produced by fungi intimately associated with the forage plant. Many of these deleterious substances play important roles in reducing insect attack and overconsumption by herbivores, but many are also toxic to the grazing ruminant.
- Ryegrass staggers, paspalum staggers, Bermuda grass staggers, tall fescue toxicoses, facial eczema and lupinosis are diseases produced by consumption of toxic fungi growing on forage plants.
- Strategic grazing management is aimed at optimising sustainable production by: establishing high-quality, resilient pastures; maximising biomass production; maximising the efficiency of biomass utilisation; and maintaining the desired pasture species with minimal weed invasion over time. Grazing management systems involving stock management (stocking rates, stock movements) in line with plant growth phases is a key to achieving profitable and sustainable grazing systems.

Introduction

Managed grazing on grasslands, including sown pastures and rangelands, represents the largest form of land use globally, covering some 25% of the world's land area or 33 million km^2 (Asner *et al.* 2004). About 60% of the world's agricultural land is classified as 'grazing land'. More than 1.5 billion animal units (defined in terms of animal size and growth rate relative to cattle = 1), were present in grazing systems on Earth in 1990 (Asner *et al.* 2004). The five countries with the largest grazing land areas are Australia (4.4 million km^2), China (4.0 million km^2), USA (2.4 million km^2), Brazil (1.7 million km^2) and Argentina (1.4 million km^2).

The animals that graze on these pastures, and the crop residues from crop production, support the livelihoods of more than 800 million people. The grazing systems that are in place to manage grazing animals are critical factors in managing natural ecosystems, and conserving water, habitat, energy, genetic resources and carbon. Ruminants, both domesticated and wild, predominate as the major group of grazing animals on grasslands. Many concerns are raised about the use of ruminants in agricultural production on the grounds that they are inefficient at converting food to product, that they produce large amounts of greenhouse gas and that they destroy natural ecosystems. However, well-managed grazing systems are an efficient means of converting low-quality, high-cellulose and low-nitrogen plant material into valuable meat, milk, fibre and work. This is preferable to allowing the forage-carbon to be oxidised (slowly by decomposition or rapidly in fires) to carbon dioxide, with no productive output. Given the economic, environmental and social value of grazing animals, an understanding of their nutritional management is vital to ensure an ongoing contribution of animals to global benefit.

The terms 'grassland' and 'rangeland' are usually used to describe low-input ecosystems based largely on native vegetation, while the term 'pasture' is usually used to describe more-intensively managed systems (using irrigation, seeding, fertiliser). The climatic zones in which the grasslands lie include Mediterranean, temperate, semi-arid temperate, arid, subtropical and tropical, and this greatly influences the forage types, seasonal distribution and nutritive value of food on offer. Typically in regions of high and reliable rainfall (e.g. New Zealand), livestock production is measured on a 'per hectare' rather than 'per animal' basis, because stocking rates are high. As stocking rates increase, production/animal decreases and production per hectare increases. For regions with harsher climates, such as parts of Australia, South Africa, South America and Europe, supplements are used to fill 'feed gaps' when pasture growth is low or zero. In these systems, production is a compromise between production per head and production per hectare. Semi-arid areas of the world are characterised by low and highly variable precipitation and often high rates of evaporation, limiting the moisture available for plant growth. Sparse yearly rainfall is often concentrated in a short rainy season. Animals grazing on rangelands are often competing with native herbivores, feral herbivores, birds, reptiles and insects, and this combined grazing pressure has led to overgrazing and degradation of many rangeland ecosystems.

Some plants evolved to attract grazing herbivores to their vegetative parts as a means of spreading seed, either attached to the animals' coats or after consumption and excretion in faeces. These plants, therefore, produce large amounts of seed and have relatively short vegetative growth periods. Selection and breeding of pasture plants for grazing systems has created plants that produce large amounts of leafy vegetative material and reduced seed production. Supporting this rapid vegetative growth with irrigation and fertiliser can create nutritional and metabolic problems such as nitrate poisoning, milk fever and hypomagnesaemia. Selection of pasture plants for insect resistance, or invasion of the plants with fungi, can also create significant animal health problems such as facial eczema and various 'staggers' syndromes.

This chapter considers the seasonal changes in pasture production, the feeding value of pastures, feeding behaviour of grazing animals, supplementation of grazing animals to meet their nutritional requirements, grazing systems that maximise productivity and common nutritional and metabolic diseases of grazing animals.

Components of the grazing system

Figure 5.1 shows a simplified diagram of the key components of grazing systems. The rate of production of pasture biomass, and the quality of that biomass as an animal feed, depends on the pasture species, the fertility of the soil in which it is grown, and the input of rainfall, temperature and sunlight. The type of grazing management then determines how efficiently the biomass is used and what level of animal production is achieved. Note the important role of grazing management on the production of pastures and also the extent of utilisation of the biomass produced. This makes grazing management decisions critical to the biological and economic viability of grazing systems.

Maximising animal production from pastures in a sustainable manner

The key decisions in optimising animal production from pastures and grasslands are:

1. Choose and manage pasture species that are well adapted to the soils, climate and climate variability of the region.
    A key component of grazing management is the choice of fodder plants that suit the environment (including local conditions down to paddock level). Key issues in this choice are relative value of annuals versus perennials; deep-rooted

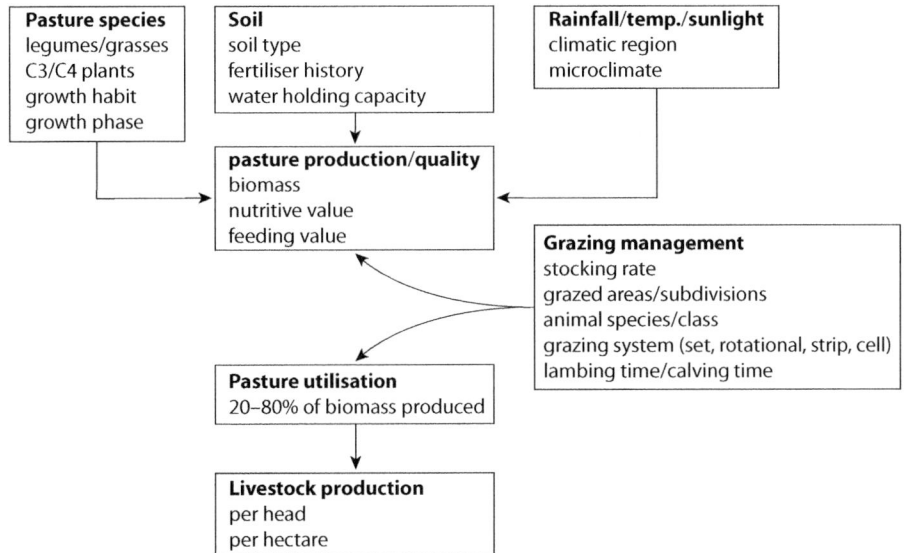

Fig. 5.1. Key components of grazing systems. The production and quality of pastures is determined by the pasture composition (species mix), soil fertility and climatic region, but grazing management also greatly influences not only the production and quality of pastures but also the extent of utilisation of the pasture biomass. The extent of pasture utilisation can vary from as little as 20% of the biomass to 80% of the biomass produced and this is a major determinant of livestock production at pasture.

plants that may be more resilient and responsive to out-of-season rains; C3 versus C4 plants (see 'Feeding value of C3 versus C4 plants' later); the use of fodder trees and shrubs; and the relative level of legumes versus grasses in the sward are key determinants of pasture choice.

2. Choose grazing systems and the timing of animal husbandry practices that impact on the demand for feed at different times of the year in relation to pasture growth.

   The key components of this system are choice of stocking system (set stocking, continuous grazing, rotational grazing, cell grazing, planned grazing, strip grazing), stocking rate, timing of calving/lambing, and the timing of the sale of animals.

The complexity of managing grazing systems in variable climatic and economic environments is high, but is made easier by decision support software such as GrazFeed®, GrassGro® and GrazPlan® developed in Australia by the CSIRO (Donnelly *et al.* 1997). Such packages allow strategic and tactical planning based on historical meteorological data for the property in question. They are based on a detailed understanding of the climate/soil/pasture/animal interactions, which are described below.

Seasonal patterns of forage production in temperate, Mediterranean, tropical and arid regions

Pasture growth in Mediterranean climates

Mediterranean environments (30–42° latitudes, largely coastal) are characterised by cool, wet winters and hot, dry summers, with a period of rapid pasture growth in spring (Fig. 5.2). Countries and regions with Mediterranean climates include: Italy, Spain, Portugal, Chile, South Africa, southern Australia and the East Coast of USA (California). At higher latitudes, pasture species include self-seeding annual grasses such as barley grass (*Hordeum leporinum*), annual ryegrass (*Lolium rigidum*) and brome grasses (*Bromus diandrus, Bromus rigidus*), and annual legumes such as the medics (*Medicago truncatula, M. scutellata, M. polymorpha, M. rugosa*). New, alternative legumes adapted to deep sandy and acid soils such as serradella (*Ornithopus compressus*), biserrula (*Biserrula pelecinus*) and Balansa clover (*Trifolium michelianum*), have also been introduced. The higher latitudes and higher rainfall zones in Mediterranean climates are characterised by perennial grasses such as cocksfoot (*Dactylis glomerata*), phalaris (*Phalaris aquatica*), perennial ryegrass (*Lolium perenne*) and Mediterranean-adapted tall fescue (*Festuca arundinacea*). The perennial legumes in these zones are white clover (*Trifolium repens*), bird's foot trefoil (*Lotus corniculatus*) and lucerne (*Medicago sativa*). Typical growth patterns of pastures, quality and quantity of

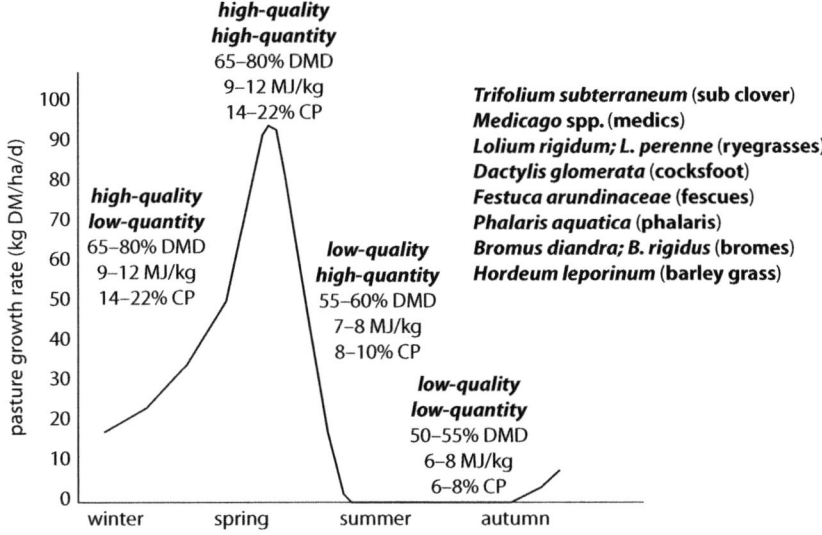

Fig. 5.2. Growth and nutritive value of typical pastures in Mediterranean climates such as southern and western Australia, coastal California, Spain, Chile, south-west South Africa and parts of Italy. Hot, dry summers and cool, wet winters create highly variable pasture growth rates and pasture feeding value. DMD = dry matter digestibility; MJ/kg = megajoules of Metabolisable Energy/kg dry matter; and CP is crude protein %. Some typical pasture species found in this zone are indicated. (Note: actual values will vary greatly with environment, season, location, pasture species and weeds.)

plant growth, and typical pasture plant species in Mediterranean climates are shown in Fig. 5.2.

Typically in Mediterranean environments, annual pastures germinate from seeds that were self-sown from the previous year and that lay dormant in the ground until sufficiently wet to germinate. The resulting plant grows relatively slowly from this 'break in the season' because temperature is limiting growth. At this point, animals have moderate- to high-quality feed but are limited by the quantity of feed on offer (FOO). In spring the temperature increases, nitrogen becomes available from microbial activity, and plants grow rapidly (up to 100 kg DM/ha/day) and the FOO is of high quality. Animal growth rates at this time can exceed 500 g/d and 1.5 kg/d for sheep and cattle, respectively. As summer progresses, annual plants produce seeds, the plant senesces and the seed enters the soil or is spread by wind, birds or attached to animals. The nutritive value of the plant material drops dramatically at this mature stage. Perennial plants during summer can respond to summer rainfall and can provide a green 'pick' for animals, which is higher in energy, protein and minerals. Many animals in Mediterranean and temperate environments are on mixed animal/cropping farms and over the late summer period have access to crop residues (usually cereal residues and residues of canola and grain legume crops). The period from late summer to the break of season in autumn is characterised by low quantity and low quality of available feed and is a period known as the 'autumn feed gap'. This feed gap can be managed in several ways:

1. Provide supplementary feeds to meet the energy, protein and mineral needs of the animals to make up for the pasture shortfall.
2. Conserve fodder (hay, silage) during the period of rapid forage growth (spring/early summer) to feed back in the gap.
3. Remove the animals from the newly germinating pasture to allow good establishment of the young plants without grazing pressure (this is referred to as deferred grazing).
4. Introduce pasture species that 'fill the gap' by having a longer growing season.
5. Graze animals on sown cereal crops during their early growth phase.

Pasture growth in cool temperate climates

Cool temperate climates, characterised by cool summers, cool winters and fairly uniform rainfall across seasons, are found in New Zealand, parts of south-eastern Australia, western Europe, north-western North America and southern South America. Figure 5.3 shows a typical pasture growth curve for such regions. Cool to cold temperatures in winter limit plant growth and moisture can limit plant growth in late summer, depending on the latitude.

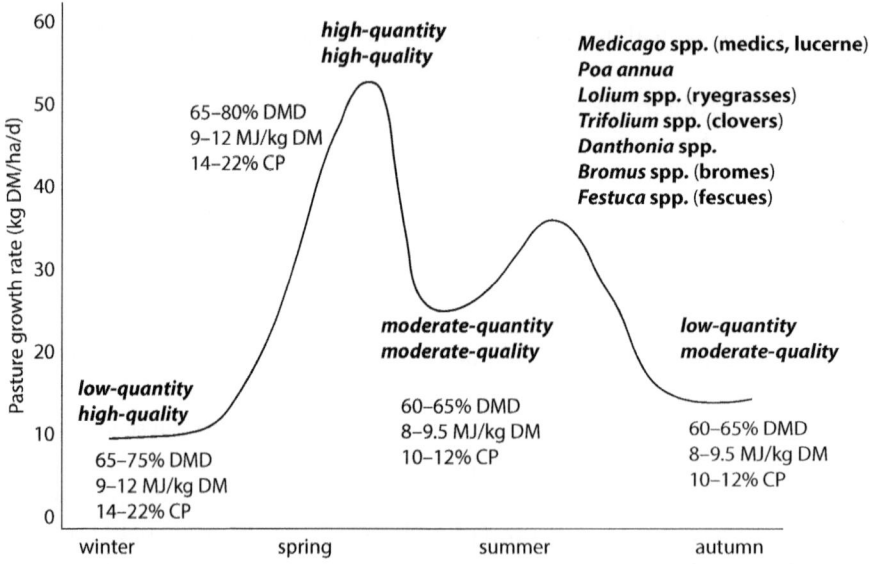

Fig. 5.3. Growth and nutritive value of typical pastures in temperate climates, which span from the subtropics to the polar circles. DMD = dry matter digestibility; MJ/kg = megajoules of Metabolisable Energy/kg dry matter; and CP is crude protein %. (Note: actual values will vary greatly with environment, season, location, pasture species and weeds.)

Typical pasture genera in this region are *Medicago* (medics), *Poa*, *Lolium* (ryegrasses), *Trifolium* (clovers), *Danthonia*, *Bromus* (bromes) and *Festuca* (fescues). The nutritive value of pasture plants in these regions is generally high and stocking rates are relatively high compared with drier or more-variable climatic regions.

Pasture growth in tropical and subtropical climates

The tropics are defined as the region between the northern latitude of the Tropic of Cancer and the southern latitude of the Tropic of Capricorn. The zone is divided into tropical rainforest or Equatorial, Tropical Monsoon, or Tropical Wet and Dry (Savannah). Temperatures in all these zones are above 18°C for 12 months of the year. Rainfall varies from all year to distinct rainy and dry seasons. Typical pasture growth in these zones is depicted in Fig. 5.4.

Tropical plants generally have a lower digestibility and promote a lower voluntary feed intake than temperate plants, largely due to their higher fibre content (Minson 1981; Fig. 4.6). Many tropical plants use the C4 photorespiration pathway (see 'Feeding value of C3 versus C4 plants' later), and grow very rapidly and efficiently. In doing so, they also reach maturity quickly and their digestibility drops rapidly (Fig. 5.4). Tropical plants contain less protein than temperate plants,

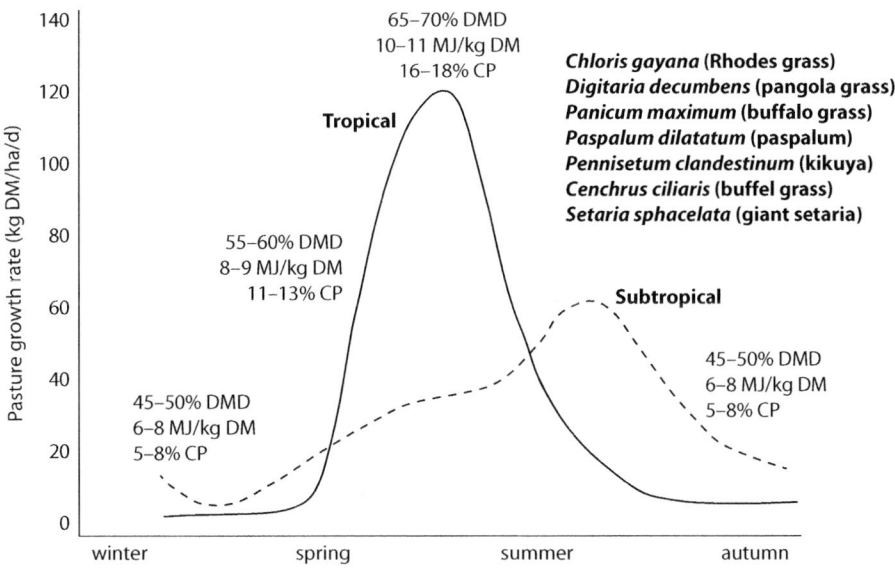

Fig. 5.4. Typical pasture growth characteristics in tropical and subtropical zones found in northern Australia, southern Africa, southern India, Indonesia and other parts of Asia. Pasture plants are commonly C4 plants with very rapid growth and a short period of production of high-quality feed followed by a rapid decline in feeding value. Some typical tropical and subtropical pasture plant species are indicated. (Note: actual values will vary greatly with environment, season, location, pasture species, forage shrubs/trees and weeds.) Source: Adapted from Hill *et al.* (2009).

their fibre content is higher and voluntary intake is lower. Animals on these pastures often respond well to supplements of protein and phosphorus (Ternouth 1990). When tropical plants are grown on nutrient-depleted soils, the digestibility and intake are reduced even further (Minson 1981).

Plant growth in semi-arid and arid environments

Plants growing in arid zones vary greatly in nutritive and feeding value. The fall into three main groups:

1. **Ephemeral annuals**: these germinate after rains and can form dense stands of shallow-rooted, rapidly growing, short flowering plants. Examples are annual lupines (*Lupus* spp.), *Allium* spp., *Trillium* spp. and *Tulipa* spp.
2. **Succulent perennials**: these store water in parenchymal tissue, and can be eaten by livestock in droughts (e.g. cacti).
3. **Non-succulent perennials**: these comprise the majority of arid zone plants. They are hardy plants including grasses, woody herbs, shrubs and trees. Typical plants include halophytic plants such as saltbushes (*Atriplex* spp.), wattles (*Acacia* spp.), *Leucaena*, buffel grass (*Cenchrus ciliaris*), *Sporobolus* spp., *Digitaria* spp. and *Brachiara* spp. Many of these have features that increase resistance to grazing. These include spines, thorns, volatile oils, tannins, high-salt, lignins, glucosinolates, alkaloids, resins and waxes.

Although energy and protein supply to livestock in extensive arid and semi-arid regions are often limiting, phosphorus is also commonly deficient and limiting to animal growth and fertility in particular. In dry seasons vitamin A can also be deficient in plants in these regions.

The feeding value of pasture plants

Nutritive value versus feeding value of pasture plants

The quality of a forage can be described in terms of its 'nutritive value', which is defined as the animal response per unit of feed consumed, or its 'feeding value', which is defined as the animal response when the feed is available *ad libitum*. Feeding value is a function of voluntary feed intake and efficiency of utilisation of feed minus the maintenance requirement of the animals (Ulyatt 1978). Variations in the voluntary intake of feeds often accounts for more than 50% of the variation in feeding value, so 'feeding value' is a better indicator of the quality of forages for animals. The feeding value of a pasture plant depends largely on its digestibility (which reflects its energy content and voluntary intake), palatability (affecting intake), its ruminal retention time (affecting intake), its protein content (which affects the availability of nitrogen for ruminal microbes and its supply of Undegraded Dietary Protein), and its content of essential minerals (for microbial

Table 5.1. Comparative growth rates of sheep fed various grasses and legumes

Plant species	Relative live weight gain	Reference
Perennial ryegrass	100	
Timothy, common	129	Ulyatt (1978)
Lucerne	170	Ulyatt (1978)
Lotus pedunculatus	143	Ulyatt (1978)
White clover	186	Ulyatt (1978)
Sub clover	121	Freer and Jones (1984)
Lucerne	146	Freer and Jones (1984)
Phalaris	102	Freer and Jones (1984)

activity and animal production). These factors change with plant species, plant growth stage and fertiliser use.

Feeding value of legumes versus grasses

Numerous studies have shown that legumes alone, and in combination with grasses, significantly increase the feeding value of ruminant diets (Ulyatt 1973). Even tropical legumes have a higher feeding value than grasses (Milford and Minson 1966). Typical relative growth rates of sheep fed various grasses and legumes are shown in Table 5.1.

Legumes support faster growth than grasses because they support higher voluntary feed intakes at any digestibility of the feed (Fig. 5.5).

Higher voluntary intakes on legumes, in turn, are a consequence of legumes having a shorter retention time in the rumen (Fig. 5.5).

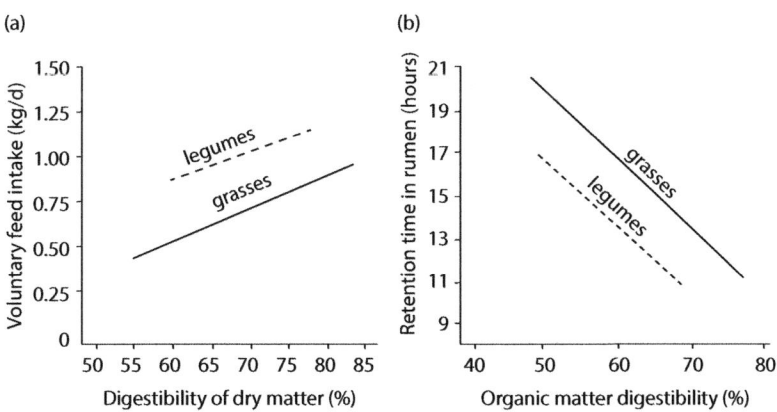

Fig. 5.5. Legumes such as subterranean clover (*T. subterraneum*) and lucerne (*M. sativa*) support higher voluntary feed intakes (a) than grasses such as phalaris (*P. aquatica*) and ryegrass (*L. perenne*) at the same digestibility of dry matter. The higher intake of legumes reflects faster rate of passage (shorter retention time in the rumen (b)). Sources: Based on descriptions by Thornton and Minson (1973) and Freer and Jones (1984).

Table 5.2. Sites of digestion of perennial ryegrass and three legumes in sheep

Parameter	Ryegrass	White clover	Red clover	Sainfoin
N intake (g/day)	38	35	29	34
N digested in stomach (g/100 g N intake)	27	18	4	−1
N digested in intestines (g/100 g N intake)	58	64	78	75
Protein absorbed/energy absorbed	11.5	12.5	15.0	14.8

Source: Ulyatt *et al.* (1977)

Legumes have a higher protein content than grasses and have more readily fermentable carbohydrate (water-soluble sugars and pectin), and less structural carbohydrate (cellulose and hemicellulose) (Fulkerson *et al.* 2007). The majority of the protein in legumes is digested in the intestines rather than the rumen and the ratio of protein absorbed relative to energy absorbed is higher (Table 5.2)

Better performance of animals grazing on legumes than those on grasses is due to higher voluntary feed intake and greater postruminal amino acid absorption.

Feeding value of C3 versus C4 plants

Temperate plant species fix atmospheric carbon using a pathway known as the C3 pathway, named after the first acid produced in the carbon capture pathway (3-phosphoglyceric acid). Eighty-five per cent of plants use this pathway to fix carbon but it is inefficient in that there is a loss of carbon in an oxidation step. C3 plants are adapted to cool, wet and low-light environments. C4 plants bypass the loss of carbon by storing it in a low-oxygen compartment. C4 plants are adapted to warm, dry climates. C3 plants tend to be of higher feeding value than C4 plants.

Seasonal changes in the feeding value of pastures

Early growth of pastures is high in Metabolisable Energy and protein and will support high levels of animal production, late pregnancy and lactation (Fig. 5.6). As the plants mature and begin to flower, the leaf-to-stem ratio declines and the energy and protein levels will support only moderate levels of animal production. At senescence, the dry grass and stems contain low levels of most nutrients (energy, protein and minerals) and are sufficient only for maintenance of live weight in dry stock. As plants approach senescence, the quality and quantity of remaining herbage declines further and live weight losses occur.

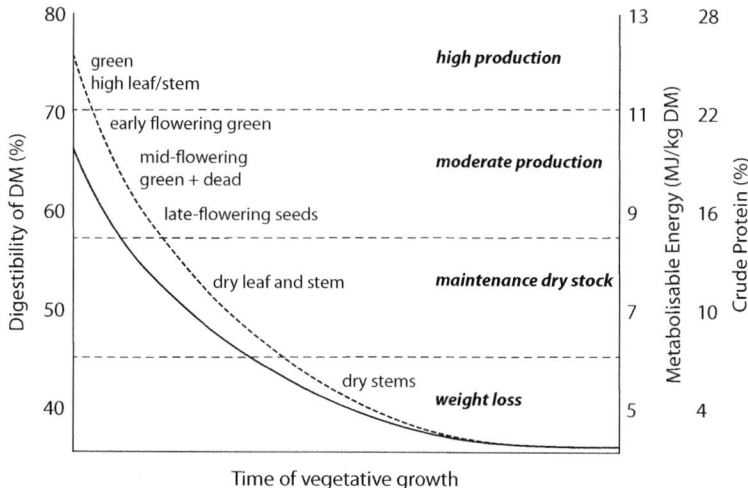

Fig. 5.6. Seasonal changes in the nutritive value of temperate (dashed line) and tropical (solid line) pasture plants with time after germination.

Feeding behaviour of grazing animals

Animals choose diets that maximise the availability of energy to maintain life processes, but they also choose feeds that provide them with positive experiences. Previously it was thought that more-palatable feeds were chosen simply because they tasted good, and unpalatable feeds were rejected because they tasted bad. It is now well established that feeds are chosen to minimise unpleasant experiences (including toxicities and deficiencies), optimise nutrients available for metabolism and even to achieve self-medication (Villalba *et al.* 2006). In other words, animals can identify and select diets that make them 'feel better' (see Fig. 5.7). This so-called 'nutritional wisdom' relies on the availability of biochemically diverse feed sources from which the animal can select, and sufficient previous experience to allow the feedback associations to be formed (see review by Provenza 2018). Monoculture forages and feedlot diets, in contrast, do not provide sufficient variation to allow the herbivore to select a close-to-optimum diet.

This ability to recognise beneficial feedstuffs from a suite of available feeds, depends on the ability to associate positive metabolic feedback to the dietary characteristics. Learning which feeds contain components that make the animal feel 'better' depends on there being a wide diversity of feeds containing varied phytochemicals in the landscape (Provenza 2018). It may also be learned during fetal and early postnatal life from maternal nutrients in amniotic fluid and milk (see Fig. 5.7).

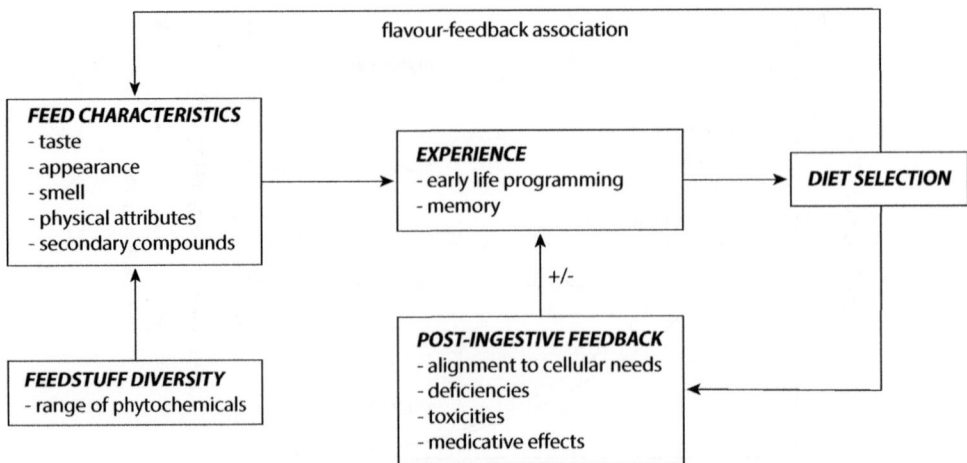

Fig. 5.7. Diet selection by grazing herbivores. The diversity of feedstuffs available for selection dictates the range of components available for selection by the animal. Depending on previous experience of the components (including early life exposure), a diet is selected that provides post-ingestive feedback (positive or negative), which further informs experience/memory. Diet selection depends on the formation of a link between dietary components and post-ingestive feedback.

In grazing environments, the forages available are heterogeneous in distribution, growth habit, nutrient composition and the type and concentration of deleterious compounds. Even within a supposed 'monoculture' pasture there is enormous variation in the distribution of available feeds from which the animal can choose (leaf, stem, seeds and different growth stages of the same plants). The quantity of feed consumed by the grazing herbivore and the nutrient content of that feed is determined by a complex interaction between the animal's physiological state, previous experience of the available feeds, spatial distribution (horizontal and vertical) of the plants and plant parts, and the chemical and physical characteristics of the available plants, as discussed in the next section.

Regulation of dietary choice and intake in grazing herbivores

Herbivores grazing grasslands and pastures have to choose a diet that optimises their survival, health, and ability to reproduce and raise young. On the one hand this means avoiding potentially deleterious and toxic substances, and minimising nutrient deficiencies, and on the other hand maximising energy availability and the pleasure of eating (Villalba *et al.* 2015). Foraging decisions are therefore complex and involve interactions between the chemical and physical components of the feeds on offer, associative effects of combinations of different feeds, and feedback received after ingestion of a feed. The decision is then integrated in the brain against a backdrop of previous experience of the feed stored in memory. The total amount of feed consumed in a meal and the number of meals/day is then

Table 5.3. Diet selection by grazing sheep

Select for:	Select against:
Leaf	Stem
Young, green forages	Older, dry forages
High nitrogen, high digestibility	Low nitrogen, low digestibility
Low fibre content	High fibre content
Low mechanical strength materials	High mechanical strength materials
High cell contents	High cell wall contents
Low physical irritants (prickles)	High physical irritants
Low plant secondary metabolites	Plant secondary metabolites

Source: Kenney and Black (1984)

determined by a balance between energy demands driving appetite and satiety signals signalling meal terminations (Fig. 4.5).

Oro-sensory properties of feeds related to selection by herbivores

Sheep select leaf over stem, young plants and plant parts over mature plants, green material over dead material, and they select against hard, irritating components such as prickles and awns. The selected feeds tend to be higher in protein, higher in digestibility and higher in phosphorus than rejected feeds (Table 5.3).

These selection decisions are based on oro-sensory properties such as tactile stimuli in lips and mouth, sight, smell and taste (Forbes and Mayes 2002) Sight is thought not to play a role in diet selection in sheep; indeed, sheep have a blind spot ~30 cm directly in front of them, and impairing the sight of sheep does not alter the foods chosen (Forbes and Mayes 2002). Touch, taste and smell are used to discriminate foods, but the threshold for rejection may be lowered by hunger or increased energy demands. It is unlikely that animals can identify simple chemical components of feeds as such (e.g. energy content, or nitrogen content *per se*) because these do not exist in the plant as simple compounds. Instead animals discriminate on the basis of receptor-signalling molecules such as sodium (reflecting ash content), 'sweetness' reflecting soluble carbohydrates, or 'ease of harvesting', reflecting fibre content, and so on. The mechanical strength of plants and plant parts reflected in 'softness', 'low physical irritation', 'ease of prehension and harvesting' and high rate of intake, is a significant determinant of selection decisions in sheep (Hodgson *et al.* 1994).

Of all the sensory mechanisms operating to influence diet selection, the role of plant secondary compounds is perhaps the most direct and fastest acting, because they impact directly on taste aversion. Tannins, alkaloids and fungal toxins can be aversive and are found most commonly in legumes and shrubs and less so in grasses and sedges. Condensed tannins can have positive effects in the animal by acting as an anthelminthic and at times of very high soluble dietary protein may be beneficial in reducing ammonia production and reducing the energy costs of urea excretion.

Associative effects of foods

Animals consume a variety of feeds and this provides a better balance of nutrients and secondary plant compounds than is available from any single forage (Villalba *et al.* 2015). Consuming a single feed means satiety will be reached more rapidly when the desired level of the nutrient at highest concentration is reached before the desired level of other nutrients is reached. Consuming a wider variety of feeds also allows dilution of plant secondary compounds, which might be in high concentrations in one plant but low in another. A variety of foods in the selected diet also provides more variation in oro-sensory stimuli (taste, smell, touch, sight) and post-ingestive stimuli (see below). Greater variation in these stimuli increases the total intake and preference (Villalba *et al.* 2015).

Post-ingestive feedback

Grazing animals attempt to maximise energy intake and health by choosing foods that best support these. Post-ingestive feedback is the reward or penalty perceived by the animal after ingestion of a particular feed, which leads to preference for, or rejection of, that feed. The difficulty, however, is that feedback from consumption of a particular forage or forage component is not immediate (with the possible exception of sodium in sodium-depleted animals, because sodium interacts immediately with taste receptors and absorption is rapid). Another difficulty is that consumption of a variety of feeds in a single meal means post-ingestive feedback becomes confounded. For example, when an animal consumes a feed that provides a positive experience post-ingestion together with a feed that does not, the animal equates the positive feedback to both feeds. Similarly, when an animal consumes a feed that produces a negative experience, any other feed consumed at the same time is rejected. Which plant caused the positive or negative feedback? Animals partly overcome this problem by being cautious with new plants. This is called **neophobia**: the fear of the new and unknown. By sampling small amounts of new plants they can trial the toxicity of new plants without consuming too much of the toxin.

These behaviours are deep-seated, as evidenced by the fact that when sheep are given a nutritious feed and are then given a dose of a toxin, the negative feedback from the toxin causes aversion for the nutritious feed. These evolutionary mechanisms for learning avoidance of deleterious plants is probably deeply ingrained in the animal's physiology and is long lasting. The author's experience with sheep that were consuming a balanced ration of oaten hay, barley and minerals, and then infused with a near-toxic dose of methionine, continued to reject these nutritious feed components for the rest of their life (a period of 8 years) presumably because they associated illness with the feeds. This indicates an important role for memory in diet decision making.

Sheep raised on rangelands have different preferences to sheep raised on sown pastures (Arnold and Maller 1977) indicating learned behaviours. Early life experience is also important in generating diet selection. For example, Chadwick *et al.* (2009) showed that dietary exposure of pregnant ewes to salt alters the response off their offspring to salty diets.

Species differences in diet selection

Differences in diet selectivity between herbivore species are a consequence of differences in the anatomy of the mouth, lips, tongue and teeth, differences in their digestive strategies and digestive efficiencies, and differences in their sensitivity to oro-sensorial and post-ingestive receptors. The ability of an herbivore to select nutrient-dense feeds is directly related to bodyweight and digestive efficiency. Large ruminants, which are poor selectors, have wide muzzles compared with selective feeders, which have narrower, longer dental arcades. The cleft upper lip of sheep and camelids allows close grazing and manipulation of plant parts, while the prehensile tongue of cattle is less discriminating with large volumes of material gathered with each bite (see Table 6.1). Larger ruminants, such as cattle, ingest feeds higher in cell wall contents (fibre) and are termed 'bulk roughage eaters'. Sheep and goats, which are much more discriminating than cattle, are called 'nutrient concentrators'. Sheep are extremely capable of selecting preferred material. At certain times of the year, sheep can obtain 80% of their food from plants that are only 1% of the forage available (Leigh and Mulham 1966). Cattle digest fibre better than small ruminants because they have a larger rumen relative to bodyweight and food remains in the rumen for a longer time. The success of the two strategies (the low-selectivity, high-digestibility of fibre in bulk roughage eaters versus the high-selectivity, low-digestibility of fibre in nutrient concentrators, depends on the characteristics of the grazing system. For example, cattle are better suited to mature pastures with little variance in feeds on offer, whereas sheep do better when there is a smaller quantity of more-varied feeds on offer. Indeed this is why co-grazing of herbivores with different grazing strategies can be successful in exploiting different niches within a plant community (Gordon 1988). Even within small herbivores, differences in diet choice allows more extensive utilisation of the available feed resource. For example, in a mixed legume/grass sward, alpacas will favour grasses over legumes, and sheep will favour the legume component. Sheep also select more leaf than alpacas (San Martin 1987). Similarly, goats and sheep select different combinations of feeds in a mixed sward (Gurung *et al.* 1994). Llamas tend to select more of the tall, coarse grasses than alpacas, and llamas eat more stems and less leaves than alpacas. Llamas appear to be bulk roughage eaters, like cattle, and alpacas appear to be opportunistic selective grazers somewhere between sheep and cattle/llamas.

Regulation of total feed intake in herbivores

The feeding value of a pasture is the product of the nutritive value of the feed consumed and the total feed intake. Determinants of the nutritive value of selected feed were discussed earlier. Regulation of feed intake in grazing ruminants involves physical limitations imposed by undigested feed in the gut, changes in circulating metabolite and cell-signalling molecules, fatigue, and factors altering the energy demands of the animal such as pregnancy, lactation, cold stress, exercise and genotype/production level (Fig. 4.5). A major determinant of feed intake in ruminants is the digestibility of the feed. Low-digestibility feeds are retained in the rumen for longer until they reach the critical size and density for onward movement to the omasum (see Fig. 3.12). The generalised relationship between diet digestibility and voluntary intake of feeds is shown in Fig. 5.8.

The following equation describes the relationships between daily intake of feed, bite size, bite rate and grazing time:

$$\text{Intake} = (\text{bite volume} \times \text{bite rate}) \times \text{grazing time}$$

As herbage availability in a grassland decreases, at first there is little impact on the herbage intake because the rate of herbage intake is not affected (Fig. 5.9). However, when the available herbage declines below ~2000 kg dry matter per hectare, the rate of intake starts to decline. To compensate, the animal grazes for longer but the extra time spent grazing does not compensate for the reduction in

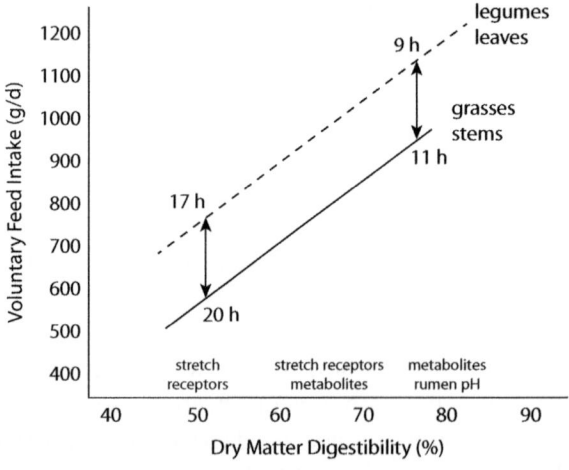

Fig. 5.8. Generalised relationship between diet dry matter digestibility (%), ruminal retention time (h) and voluntary feed intake (g/day). The higher relative intakes of legumes and leaves over grasses and stems is a function of their faster rate of passage through the reticulo-rumen. At low digestibilities, stretch reception is the major satiety signal. At high digestibilities, chemical signals operate to signal satiety.

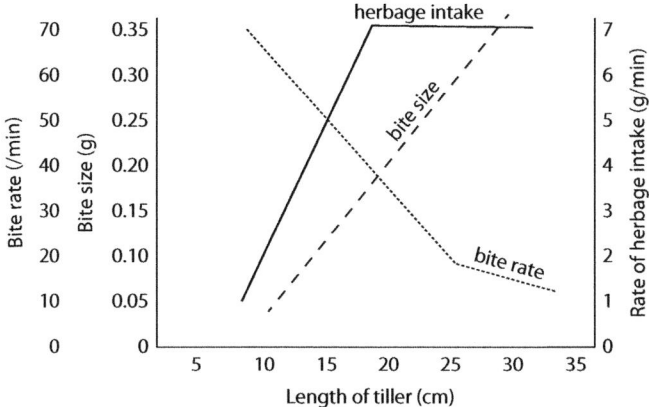

Fig. 5.9. Relationship between bite size, bite rate and total herbage intake as the availability of feed (tiller length) changes. Source: Adapted from Allden and Whittaker (1970).

intake rate, so the total herbage intake declines (Allden and Whittaker 1970). In other words, bite size has the greatest impact on intake, with bite rate and grazing time being the compensatory variables. Bite rate is strongly influenced by the amount of feed-on-offer (height and density).

Pattern of grazing by herbivores

Grazing time and the diurnal pattern of grazing are influenced by the amount of feed available, physiological state (energy demand) and the prevailing climatic conditions. Sheep and cattle graze for between 4.5 and 14.5 h/day, depending largely on feed on offer. Major grazing periods are around dawn and dusk. The breaks between these dawn and dusk grazing bouts decrease as the days get shorter in winter, until grazing in mid-winter is almost continuous. When temperatures are high in summer, the dawn grazing periods start and finish earlier. Heat-adapted animals such as Zebu types of cattle graze for longer in the heat than British breeds of cattle.

Supplementation of grazing herbivores

Supplementary feeding of grazing herbivores is carried out to correct a nutritional deficiency, to maximise productivity or to take pressure off pastures at critical times of plant growth (e.g. during autumn in Mediterranean environments it is common to remove stock from newly germinated pastures to allow good establishment of the new pasture). There are three main types of supplement: energy supplements; protein and non-protein nitrogen supplements; and mineral supplements.

Energy supplements

Energy supplements are usually grains or conserved fodders (hay or silage). In highly seasonal Mediterranean climates, plant growth is largely restricted by moisture in summer, and by temperature in autumn and winter, leaving a period in spring of very high herbage production. This herbage can be conserved as hay or silage and fed back during the autumn/winter feed gap. It is important for producers to calculate the real costs of each available supplement. To do this we need to know the dry matter percentage of the supplement, the Metabolisable Energy content of the supplement (in MJ/kg dry matter) and the cost/tonne of the available feed ($/tonne as fed). Given that most energy supplements are also providing protein, and at times protein may also be limiting, it is useful to know the protein content of the available supplements. To calculate the real energy and protein costs of each supplement do the following calculation:

$$\text{Cost of 1 kg dry matter} = \$/\text{tonne as fed/DM\%}$$

$$\text{Energy content in 1 kg dry matter} = \text{MJ/kg (from table values)}$$

$$\text{Protein content in 1 kg dry matter} = \text{kg/kg dry matter (from table values)}$$

When energy supplements are given to animals at pasture, the impact on production is difficult to quantify because the supplement can cause a reduction in forage intake, no change in forage intake or an increase in forage intake. These so-called substitution effects are due to associative effects of the supplement on digestion of the base feed. Negative associative effects occur when starchy grains are digested in the rumen and the resulting pH decline reduces the activity of cellulolytic bacteria and the digestibility of cellulose. The increased retention time of the fibre in the rumen reduces intake rate (Fig. 5.8). Positive associative effects occur when the energy supplement provides a nutrient that is limiting either microbial activity or animal metabolism. Commonly this will be nitrogen or sulphur, which will stimulate microbial activity in the rumen, or phosphorus or a trace element for animal metabolism. An example of negative associative effects is shown in Fig. 5.10.

As the grain intake increases the intake of the basal straw diet decreases. The total energy intake increases due to the high energy content of the grain supplement. The extent of this substitution effect varies with supplement and basal forage.

Protein and non-protein nitrogen supplements

The value of supplementing grazing animals with nitrogen depends on the level of protein (nitrogen) in the forage, and the protein requirements of the grazing animals (Fig. 5.11).

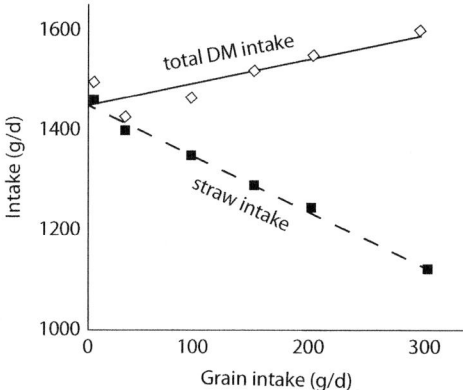

Fig. 5.10. Negative associative (substitution) effects of a grain supplement on straw intake in sheep. Source: Adapted from Dixon and Stockdale (1999).

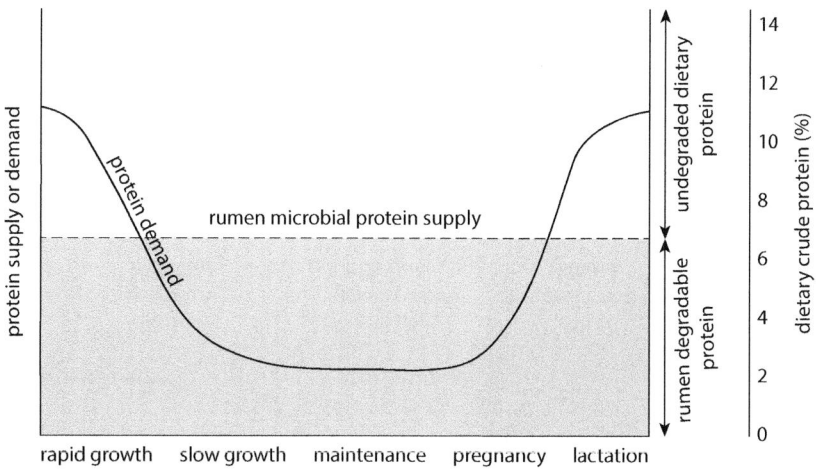

Fig. 5.11. Relationship between protein demand and protein supply in grazing ruminants. Microbial protein is produced from forage protein that has degraded in the rumen to ammonia (Rumen Degradable Protein) and joins with dietary protein that has escaped ruminal degradation (Undegraded Dietary Protein). Microbial protein is adequate for maintenance, slow growth early pregnancy and low milk production. Rapid growth, late pregnancy and high milk yields (early lactation) require dietary protein that escapes ruminal degradation.

Mineral supplements

The most cost-effective supplements for grazing animals are those that correct deficiencies of essential minerals. For a few cents/day, most minerals can be supplied to meet the microbial requirements and the requirements for the animal's metabolism. Correcting either of these will increase food intake and the efficiency of maintenance and production.

Grazing and pasture management for sustainable animal production

Establishing a grazing system that optimises sustainable animal production requires a program of activities involving the following:

1. Establish pasture species that are adapted to the environment and that maximise the feeding value of available herbage throughout the year. Include legumes in the pasture sward and identify grass species with a long growing season.
2. Establish a grazing system that maximises the production and utilisation of herbage biomass. This includes choice of grazing system (set stocking, continuous grazing, rotational grazing, strip grazing, cell grazing or planned grazing) and stocking rate.
3. Choose the timing of husbandry procedures (lambing, calving, shearing, sale of surplus animals) to align pasture production with seasonal energy demands.

Successful grazing management requires an understanding of the phases of plant growth and their impact on plant growth rate.

Plant growth phases

Grasses follow a sigmoidal growth after germination (annuals) or re-shooting (perennials) (Fig. 5.12).

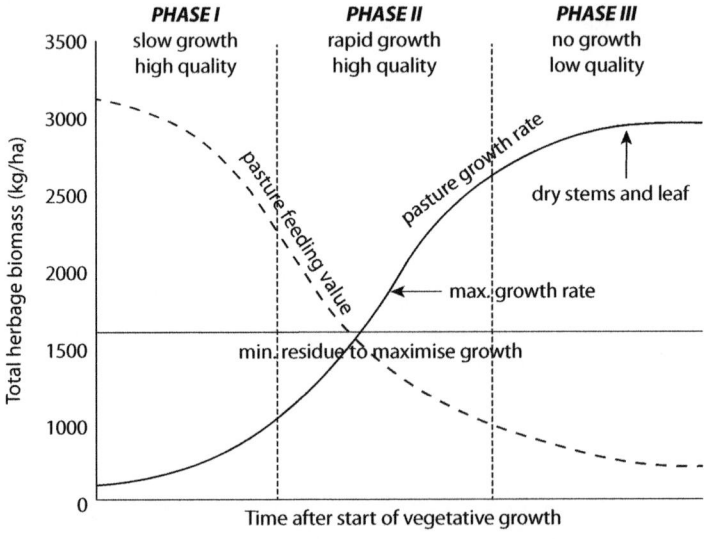

Fig. 5.12. Growth curve of pastures over the growing season showing the three phases of growth and the changes in feeding value as the plants transition through each phase. Maximum growth rate occurs midway through Phase II after which the feeding value declines rapidly as the plant flowers and earlier tillers begin to senesce.

Phase I is characterised by slow growth because the plant has insufficient leaf area to maximise photosynthesis. The nutritive value (digestibility or energy content) of the grass declines slowly as the cell wall structural contents increase. In Phase II the rate of growth increases as the optimum leaf area is approached and photosynthesis is maximised. After this, the rate of growth declines and the feed quality declines rapidly as the plant nutrients are redirected towards flowers rather than tillers.

In the early stages of growth in Fig. 5.12 (Phase I), the plant uses reserves of water-soluble carbohydrates stored in the stems to provide energy for the regrowth of new tillers. As the plant traverses Phase II, photosynthesis increases to a maximum at the optimum leaf area index, and the new photosynthate accumulates in the plant. Management of the rate and severity of defoliation of the plant to ensure rapid recovery of the plant after defoliation is the major objective to grazing management, as discussed in 'Grazing management to maximise animal production per hectare and ensure sustainable grazing systems' below. If the defoliation is too severe the roots are deprived of photosynthate, which is directed towards new leaf growth, and the resilience of the plant is compromised.

Effect of diet selection on pasture growth

Although not as selective in diet choice as sheep and goats, cattle will still select preferred forage species. Low stocking rates, then, allow livestock to continually put pressure on these preferred species and ignore less-preferred species. The result of this is ultimately the loss of preferred species and expansion of less-preferred species. Grazing management systems that increase stocking pressure to the point where the animals have no choice but to eat the less-desired species, or else go hungry, distributes the grazing pressure more evenly (see next section).

Grazing management to maximise animal production per hectare and ensure sustainable grazing systems

Grazing management is aimed at controlling the rate and frequency of plant defoliation to maximise plant dry matter production, maintain high-quality species of plants in the pasture, ensure resilience of the plants to dry periods and optimise the utilisation of pasture (Earl 2014). This is achieved by controlling the following variables:

- stocking density
- recovery period between grazing events
- length of the grazing period
- amount of residual herbage left after grazing.

The objective is to maximise pasture growth rate, pasture quality, the extent of pasture utilisation and persistence of desired pasture species over time.

Continuous grazing

Continuous grazing refers to the practice of grazing animals on the pasture all year, with the only grazing management tool being the actual number of animals, hence stocking rate. The average stocking rate chosen is usually dictated by the most limiting growth period of the year (e.g. winter carrying capacity) or at what is considered by experience to be the 'long-term' carrying capacity. Such systems, in which there is no control over intake and pasture growth phase, are usually well below the biological capacity of the grazing system and inevitably lead to degradation of the land due to removal of desirable species (Tothill and Gillies 1992).

Set stocking

Set stocking refers to stocking large areas of land with small numbers of animals with little or no movement of animals and no assessment of herbage availability. The rate of pasture utilisation (defined as the proportion of herbage mass consumed relative to herbage mass grown) is inevitably low (<30%).

Rotational grazing

Under rotational grazing, animals are moved from one paddock to another to allow pasture time to recover. The paddocks so used are often defined by their physical characteristics (e.g. topography, soil type, vegetation type). Many producers move the animals on the basis of time, assuming time reflects pasture herbage mass, but this will vary with season and site, and a defined rest period for one paddock will be different for another resulting in very varied herbage availabilities.

Strip grazing

Commonly used for dairy cattle, strip grazing is used in high-rainfall areas where the vegetation is relatively uniform (e.g. forage crops or sown pastures). The grazed areas are typically very small and the movements frequent, allowing close management of the biomass remaining after grazing.

Cell grazing

Cell grazing refers to the practice of rotating animals through a large number (usually >10) of small areas at high stocking densities. Movements are based on set times, until herbage mass has been so reduced there is a reduction in animal performance, or until herbage biomass reaches a desired level.

Planned grazing

Planned grazing is based on the principles of plant growth described in Fig. 5.12. The objective is to manipulate grazing intensity and time to achieve maximal pasture growth rate, optimum pasture utilisation, maintenance of desired pasture species over time and maintenance of maximum groundcover (weed control). To achieve this requires knowledge of the herbage mass at any given time, the rate of pasture growth and the requirements of the animals at any time of the year. Generally in high-rainfall areas a set minimum of 1500 kg dry matter/ha is the target, because at this level the plant reserves of carbohydrate are sufficient for rapid recovery and little loss of roots. The quality of the herbage remains high because it stays in the early Phase II stage (Fig. 5.12). The key components of planned grazing are knowledge of the starting quantity of herbage on the paddock, the rate of pasture growth, the number of animals grazing the area and the target residual herbage quantity. Although planned grazing is mainly applicable to higher rainfall, more-intensive production systems, planned grazing can also be implemented in low-intensity, rangelands situations (see 'Grazing management on rangelands' later). Use of smaller paddocks, manipulation of water points to achieve more uniform grazing intensity and more careful attention to stocking rates are key components of increasing cattle productivity and sustainability.

Effect of stocking rate on production per head and production per hectare

As stocking rate increases, the production per animal decreases slowly at first until the optimum stocking rate is reached and after that production per animal falls rapidly (Fig. 5.13). Production per hectare increases as production per animal decreases until an optimum rate is reached. Generally, higher production per hectare can be achieved beyond the optimum stocking rate but this is not sustainable, particularly in relation to parasite burdens, pasture degradation

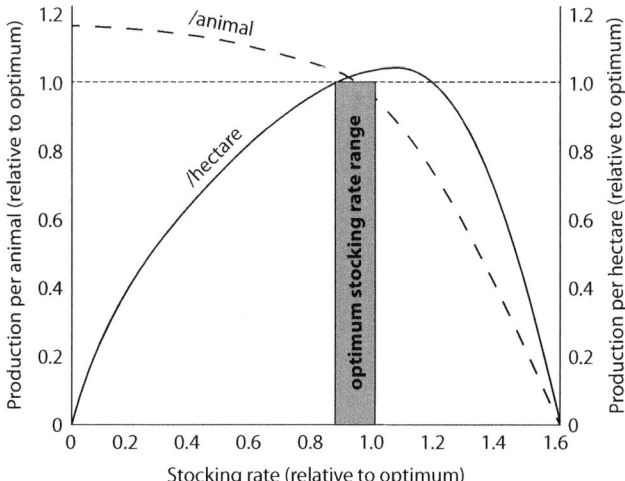

Fig. 5.13. Relationship between stocking rate, production per animal and production per hectare.

(quantity and quality) and soil erosion. In highly variable climates it is advisable to adopt a conservative stocking rate.

Grazing management on rangelands

Managing grazing animals on semi-arid and arid rangelands to achieve healthy ecosystems, profitable enterprises, and efficient and sustainable use of soil, water and plant resources requires a combination of scientific application and local knowledge (Teague *et al.* 2013). To achieve this, good rangelands managers ensure that: animals have enough forage nutrients to meet their needs; a wide variety of plants are consumed; desirable plants are not overgrazed; plants can recover after grazing because sufficient biomass is left after grazing and sufficient time allowed for plants to recover (Teague *et al.* 2013). The following management options can achieve these outcomes:

1. **Grazing deferment**. Removing the animals after grazing to allow recovery of the biomass is critical to the health of rangelands. Recovery times depend on plant species, rainfall and temperature, with drier ecosystems requiring longer recovery periods. To prevent overgrazing of preferred species by the animals, it is essential that periodic removal of the animals is carried out with adequate post-grazing recovery.
2. **Modifying livestock distribution**. The larger a paddock is, the greater the heterogeneity of the landscape and forage available. Livestock develop preferences for certain parts of the landscape, so the 'effective' stocking rates vary from low to high, even when the average stocking rate is moderate. By subdividing the grazed area, more of the landscape is used and previously understocked areas are grazed more heavily and preferred areas grazed less heavily. Livestock distribution in very dry, extensive rangelands can also be manipulated by increasing the provision of watering points. Grazing intensity is greatest close to the watering points. Providing more watering points achieves a similar outcome to paddock subdivision. Increasing stocking density by paddock subdivision or increasing watering sites also forces livestock to learn to select a diet that is more uniform in quality. At low densities, livestock select the most preferred species and gradual deplete these from the ecosystem. At higher densities, they initially deplete the preferred species leaving only less-desired plants, which are usually lower in nutritive value and higher in plant secondary compounds such as tannins. At higher densities, they learn to 'mix the best with the rest' (Villalba *et al.* 2004); that is, they learn about complementarities between plant compounds. The outcome is a wider variety of plants is consumed, reducing the long-term pressure on desirable species. Multi-paddock management also allows producers to move livestock to paddocks with the best chance of meeting their higher nutrient requirements when conceiving, late pregnant or lactating.

Deleterious factors in rangeland and pasture forages

Deleterious factors in pasture plants are those that reduce the capacity for grazing animals to reach their productive potential. They include the following:

- **bloat-inducing compounds**
- **nitrates**
- **plant toxins** such as the glycosides, alkaloids and neurotoxins
- **mycotoxins** such as those causing facial eczema, photosensitisation and other endophyte-related diseases
- **anti-mineral** and **anti-vitamin** compounds
- **lignin, tannins, thorns, awns, burrs** and other physical impediments to voluntary intake and digestion.

As competition for land and for feed resources increases, the role of grazing animals and the forages they consume will become increasingly important. Grazing by ruminants and other forestomach fermenters, which can digest fibre, on land otherwise unusable, produces high-quality food from low-quality resources, thereby adding significantly to the food available for humans. Comparisons of feed conversion efficiencies between ruminants and mono-gastric animals such as pigs and poultry, will always favour the latter if the feed is highly digestible and able to be used directly by humans. However, if the feed is forage containing high fibre and low protein, ruminants will continue to produce when mono-gastrics cannot. Efficient utilisation of forages by ruminants and reducing the impact of deleterious factors that reduce the health and/or efficiency of grazing animals is therefore of prime importance.

Economic losses due to the presence of deleterious factors in pasture and rangeland forages

Economic losses associated with the presence of deleterious factors in forages are difficult to quantify precisely, but some idea of the relative impacts can be gained by making estimates of the impact of different components in the US beef industry (Table 5.4 after Allen and Segarra 2001).

Table 5.4. Relative economic losses associated with various anti-quality components of the forage grazing system in the USA

Component	Losses	US$ million value (per annum)
Decreased forage quality	Decreased FCR	$2000
Hypomagnesaemia	Deaths	$150
Fescue toxicoses	Morbidity	$600
Poisonous plants	Deaths	$340
Total		**>$3000**

Source: Adapted from Allen and Segarra (2001)

Clearly, reducing the negative impacts of poor-quality or toxic forages has enormous potential for improving the efficiency and economic performance of grazing systems.

How do forage chemicals reduce animal performance?

Forage chemicals or physical attributes can reduce forage intake and digestibility, or can have neurological or metabolic/toxic effects (Table 5.5).

Common nutritional and metabolic diseases of grazing animals
Plant poisoning of animals grazing rangelands and forages

Poisoning of animals grazing pastures or rangelands in many countries is common and can result in the sudden onset of clinical signs and deaths in large numbers of animals. Most published estimates of mortality associated with plant poisoning are of the order of 1–3% and the economic losses are in the hundreds of millions of dollars. In addition to deaths, toxic plants can cause weight loss, reproductive losses, photosensitisation, chronic immune compromise and chronic illness.

Table 5.5. Forage chemicals and the performance and health of grazing animals

Compound/characteristic	Intake	Digestibility	Comments
Lignin	-	-	Indigestible
Condensed tannins	-	-	Astringent taste aversion; depressed protein digestibility; inactivate digestive enzymes; anti-microbial effects
Resins	-	-	Indigestible
Silica	-	-	Indigestible
Waxes	-	-	Indigestible
Gossypol		-	Inactivate digestive enzymes
Essential oils		-	Anti-microbial effects
Glycosides (cyanogenic, cardiac, saponins, glucosinolates, diterpenoids, bracken glycosides, calcinogenic glycosides, phenolic)			Haemorrhage; convulsions; death; cyanide poisoning; cardiac pathology
Alkaloids	-		Hepatotoxic; photosensitisation; teratogenic; neurotoxic; reproductive failure
Mycotoxins	-		Neurological; photosensitisation; diarrhoea; jaundice; death
Nitrates			Hypoxia and hypoxaemia; convulsions; death

The most common toxicoses in introduced pasture forages in higher rainfall regions are nitrite poisoning, pyrrolizidine alkaloid poisoning and endophyte-related diseases. Low-rainfall rangelands pose a particular threat to grazing livestock because toxic plants in these environments tend to contain higher levels and more-toxic compounds.

Toxins, or plant secondary metabolites, are present in plants as a defence mechanism against overgrazing by animals. Animals associate the malaise after consuming a toxic plant with that plant and avoid or reduce consuming it in future (Provenza 1996). Problems arise in the following situations:

- Animals unfamiliar with the region are suddenly introduced to new plants in the grazing environment.
- Drought means fewer less-toxic plants available, forcing the animals to consume the less-palatable, more-toxic plants.
- Drought can also increase the concentration of toxins that are largely lipid-soluble.
- Overgrazing places increasing pressure on the less-toxic plants and favours the spread of more-toxic plants.
- Hungry animals will consume otherwise unpalatable plants.

The following section covers the most common toxicoses of animals grazing pasture forages (nitrite poisoning, pyrrolizidine alkaloid poisoning and endophyte-related poisoning) and rangelands (glycosides and various alkaloids).

Nitrite poisoning in grazing animals

Nitrite poisoning is a significant toxicosis in grazing ruminants and can result in sudden and spectacular losses of animals grazing forages and rangeland plants in all continents. Atmospheric nitrogen is 'fixed' into soils and plants by nitrogen-fixing bacteria or rhizobia and this nitrogen joins with nitrogen from fertiliser and manure to form the soil nitrogen pool. This pool of nitrogen is converted to ammonia and then to nitrite (NO_2) and then to nitrate (NO_3) by nitrifying bacteria in the soil (Fig. 5.14). The nitrate is taken up by the plant, with the highest concentrations in the stems and stalks rather than the leaves. Most of the nitrate is converted to plant proteins but excessive nitrate levels remain as unincorporated nitrate in the plant. The nitrates in consumed forage and nitrates in the drinking water, enter the rumen and are then converted to nitrite and then to ammonia by the rumen microbes. The microbes use this ammonia to produce their microbial protein. However, if the rate of nitrite formation from feed and water nitrates exceeds the ability of the ruminal microbes to assimilate the ammonia nitrogen, nitrite accumulates in the ruminal fluid. The excess nitrite is then absorbed across

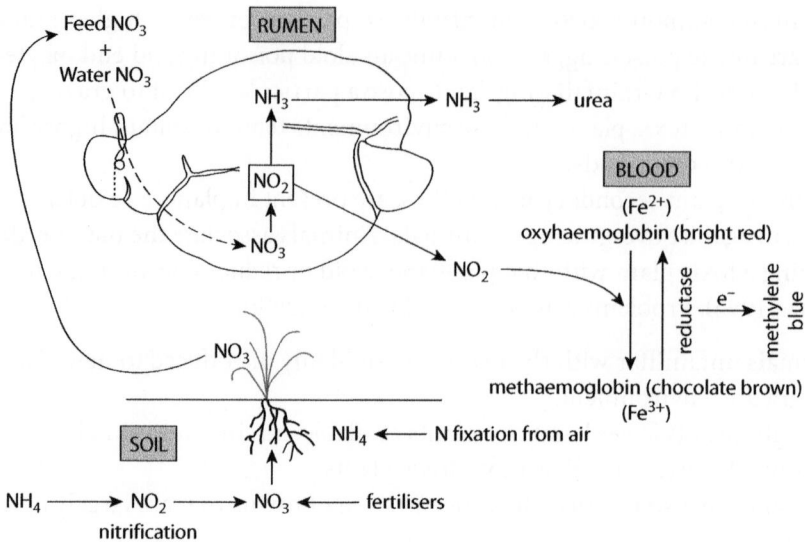

Fig. 5.14. Nitrite poisoning in grazing ruminants. Excessive intake of nitrate (NO_3) is converted to nitrite (NO_2) by microbes in the rumen. Most of the NO_2 is then converted to NH_3 and incorporated into microbial protein. Excessive supply of NO_2, however, is absorbed into the blood stream where it oxidises the Fe^{2+} to Fe^{3+} in haemoglobin, forming methaemoglobin, which cannot carry oxygen.

the ruminal epithelium into the blood circulation where it oxidises the haemoglobin group from the Ferrous (Fe^{2+}) state to the Ferric (Fe^{3+}) state. The ferric methaemoglobin pigment is chocolate brown in colour and cannot hold and transport oxygen. Treatment with methylene blue can reverse the reaction by providing an electron that attaches to the reductase enzyme that oxidases the methaemoglobin back to oxyhaemoglobin (Fig. 5.14).

Plant factors that favour nitrite poisoning

Nitrate concentrations (as KNO_3) in fodder at 1–1.5% of dry matter can cause acute intoxication of ruminants (Parton and Bruere 2002). Legumes are less likely than grasses to take up nitrate because they are actively fixing ammonia into the root nodules (Fig. 5.14). Nitrate-accumulating species are listed in Table 5.6.

The plant parts closest to the ground contain the highest levels of nitrates (stems and stalks). Leaves contain low nitrate levels and the grains and flowers contain very low nitrate concentrations. Nitrates decrease with plant maturity. Climate and growth conditions can greatly influence the level of nitrates in the plants. Decreased light on cloudy days and cool weather increases the concentration of nitrates in plants. Accumulation of nitrates in the soil during droughts can be rapidly taken up by plants when the drought breaks, reaching toxic levels.

Table 5.6. Nitrate-accumulating plants prone to inducing nitrite poisoning in ruminants

Common name	Scientific name
Oats	Avena sativa
Maize	Zea mays
Rye	Secale cereale
Wheat	Triticum aestivum
Barley	Hordeum vulgare
Lucerne (alfalfa)	Medicago sativa
Soybeans	Glycine max
Sorghum	Sorghum bicolor
Johnson grass	Sorghum halepense
Fireweed	Senecio madagascariensis
Docks	Rumex spp.
Pigweed	Amaranthus retroflexus
Ragweed	Ambrosia spp.
Wild mustards	Brassica spp.
Lamb's quarters	Chenopodium album
Field bindweed	Convolvulus arvensis
Ryegrass	Lolium perenne

Clinical signs of nitrite poisoning

The clinical signs of nitrite poisoning relate directly to hypoxaemia and hypoxia: bluish-brown membranes, tachypnoea (rapid breathing), tachycardia (rapid pulse: >150 bpm), salivation, bloat, tremors, staggering, coma, death and chocolate-coloured blood (due to methaemoglobin). Pregnant animals usually abort if they survive a near-fatal episode.

Management to avoid nitrite poisoning

The management strategies to avoid nitrate toxicoses include:

- Avoid drought-stressed forages and accumulator forages (Table 5.6) unless tested for nitrate levels.
- Add energy supplements to high-nitrate feeds to increase microbial activity and reduce the rate of ruminal NH_3 accumulation in the rumen fluid.
- Adjust animals to high-nitrate feeds by providing small amounts frequently during adaptation.
- Do not stock potentially high-nitrate fodders at high rates or strip graze because this will expose the high-nitrate stems and stalks.
- Provide a nitrate-free drinking water supply.
- Do not allow hungry animals access to high-nitrate feeds.
- Harvest potentially high-nitrate fodders at a higher level to reduce the amount of stems and stalks in conserved feed.

- Feed low-nitrate hays before introducing to high-nitrate fodders.
- Do not graze high-nitrate fodders for at least 1 week after a frost.
- Do not feed greenchop if it has become hot after harvest, because the heat generates nitrite.
- Do not feed drought-stressed plants for 3–5 days after rain to reduce the nitrate levels.

Pyrrolizidine alkaloid poisoning

Pyrrolizidine alkaloids (PA) are produced as secondary chemical compounds in many plants throughout the world as a defence mechanism against attack by insects, herbivores and people. More than 350 PAs have been identified in over 6000 species of plants. PA-containing plants are the most common poisonous plants affecting livestock (Stegelmeier *et al.* 1999). The PA-containing plants are usually foreign species, which are invasive in pastures, crops and rangelands. Although most are unpalatable due to their bitter alkaloids, they can be consumed when other feeds are scarce, or when they are harvested in hay or silage. The major PA-containing plants are: *Heliotropium europaeum* (potato weed); more than 30 *Senecio* species including *S. jacobaea* (tansy ragwort), *S. hydrophiloides* (stout meadow groundsel), *S. douglasii* (woolly groundsel), *S. madagascariensis* (fireweed); all members of the *Crotalaria* genus (rattlepods); *Echium plantagineum* (Patterson's curse or Salvation Jane); *Cynoglossum officinale* (houndstongue); and *Amsinckia intermedia* (tarweed). These plants have been variably shown to induce toxicoses in rats, cattle, pigs, poultry, sheep, goats, horses and humans.

The pyrrolizidine alkaloids are a family of alkaloids based on two 5-member rings joined by nitrogen at position 4 (Fig. 5.15).

Variations in the molecules added at R1 and R2 determine the alkaloid name. The PAs include supinine, indicine, echimidine, retrorsine and heliotrine. Some of the PA-containing plants contain a combination of several PAs. For example, *S. jacobaea* contains six different alkaloids in this class.

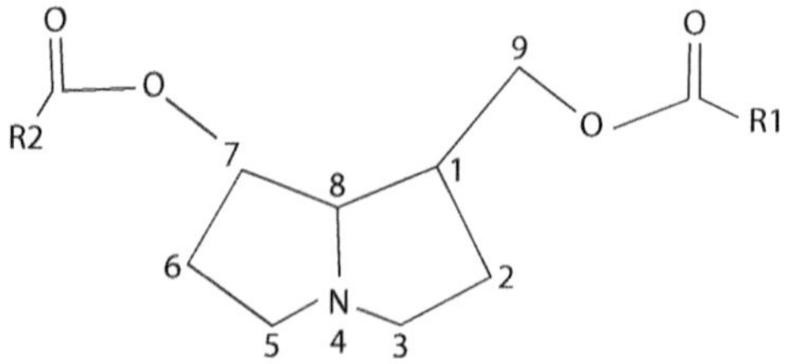

Fig. 5.15. Chemical structure of the pyrrolizidine family of alkaloids. Variations in alkaloids occur at R1 and R2 side chains.

Toxicity of the pyrrolizidine alkaloids

The toxicity of the PAs depends on the chemical nature of the alkaloid, and the species, age, sex, nutrition and physiological state of the animal. Some of the PAs are primarily hepatotoxic, while others produce lesions outside the liver. Monogastrics are particularly susceptible to PA toxicity compared with ruminants, probably due to ruminal microbial detoxification, but also different hepatic detoxification capacity. Sheep and goats are particularly resistant to PA toxicity and merino sheep are much more tolerant of PAs than British breeds of sheep (Stegelmeier 2011). The hepatotoxicity of the alkaloids varies with the copper status of the animal. Sheep consuming high levels of copper in feeds such as subterranean clover (*Trifolium subterraneum*) are more susceptible to hepatotoxicity induced by the alkaloids (Dr D. Salmon *pers. comm.* Australian Sheep Veterinarian's conference, Barossa Valley, South Australia). It seems the ability of the hepatocytes to metabolise pyrroles and to repair pyrrole-damaged hepatocytes is compromised by copper storage in the cells. Young animals are more susceptible to PA poisoning and animals consuming milk from mothers consuming the alkaloid are susceptible.

Pathogenesis of pyrrolizidine alkaloid poisoning

In chronic, long-term intoxication, liver cells damaged by the hepatotoxic alkaloids become enlarged (megalocytosis) as a result of DNA damage and failure of mitosis. The cells then die and are replaced by necrotic lesions and/or fibrotic tissue. In chronic intoxication, liver enzymes and bile salts are slightly elevated in the blood. Sudden high doses of alkaloids cause acute intoxication with necrosis and haemorrhage. The animals become anorexic, depressed, jaundiced and have fluid in the abdomen. The levels of liver enzymes from the damaged liver cells are elevated in the blood, as are the levels of the bile pigments and bile acids (Stegelmeier 2011).

In sheep, a secondary poisoning occurs after long-term exposure to PAs. In the preclinical stage, the liver cells affected by the alkaloids store large amounts of copper. Under certain conditions, the copper is released into the circulation causing profound anaemia due to haemolysis. The animals die rapidly.

Many animals grazing PA-containing pastures live their whole lives with no obvious clinical signs. Others show signs of weight loss, photosensitisation, and neural signs of apparent blindness and aimless wandering. Aimless wandering is a characteristic of intoxication in cattle and horses.

Control of pyrrolizidine alkaloid poisoning

The following integrated strategies can control PA poisoning:

- Graze PA-pastures/rangelands with the least susceptible animals (sheep, goats) and the least susceptible breeds within species (e.g. merinos not British breeds or crossbreeds).

- Supplementary feed animals grazing PA-pastures.
- Use herbicides to control the weeds.
- Dose animals with sodium molybdite/sodium sulphate solution to reduce copper levels.

Endophyte-related poisoning (mycotoxicoses) of grazing animals

Associations between fungi and some pasture species result in the production of mycotoxins, which can have beneficial effects for the pasture, but detrimental effects on animals consuming the plant/fungi combination. The fungi associated with pasture plants are known as endophytes and they produce compounds (mycotoxins) that reduce insect attack, increase drought tolerance and reduce excessive grazing of the plant, hence increasing its persistence in the plant community. Common endophyte-related toxicoses seen in animals grazing improved temperate pastures are listed in Table 5.7.

Facial eczema (Pithomycotoxicosis)

Facial eczema is an important disease of sheep and cattle on perennial ryegrass pastures, but goats, alpacas, deer and llamas are also affected. The dermatitis is caused by the active agent sporodesmin produced by the saprophytic fungus

Table 5.7. Common endophyte-related toxicoses of animals grazing temperate pastures

Disorder	Pasture type/ species	Fungus involved	Toxin(s)	Pathology
Facial eczema	Pasture litter	*Pithomyces chartarum*	Sporodesmin	Photosensitisation; hepatotoxin
Ryegrass staggers	*Lolium perenne*	*Neotyphodium lolii*	Lolitrem B, ergovaline	neurotoxin, tremors
Fescue toxicoses	*Festuca* spp.	*Neotyphodium coenophialum*	Ergovaline, ergopeptides, clavine alkaloids	Vascoconstrictor, hyperthermia
Paspalum staggers	*Paspalum* spp.	*Claviceps purpurea, Claviceps paspali*	Paspalinines, ergotomine, ergometrine, ergotoxine	Tremors; gangrene; haemorrhage
Lupinosis	*Lupinus angustifolius*	*Phomopsis leptostromiformis*	Quinolizidine alkaloids	Hepatotoxin
Bermuda grass staggers	*Cynodon dactylon*	*Claviceps cynodontis*	Paspalitrems A and B	Tremors
Slobbers	*Trifolium pratense; Medicago sativa*	*Rhizoctonia leguminicola*		Excessive salivation

Sources: Waghorn *et al.* (2002); Riet-Correa *et al.* (2013)

Pithomyces chartarum (Smith and Towers 2002). Sporodesmin is concentrated in bile and causes inflammation of the biliary ducts. Complete obstruction of the biliary ducts can occur. This prevents excretion of phylloerythrin, a breakdown product of chlorophyll. Phylloerythrin in sun-exposed skin makes the skin photosensitive, producing severe erythema, swelling and burning of the ears and head in sheep, and dermatitis in other animals. Metabolism of sporodesmin in the liver results in the production of superoxide radicals, which damage the hepatocytes, producing jaundice. In cattle, the clinical signs are diarrhoea, depressed milk production, and dermatitis on unpigmented and hairless skin. All affected animals become photophobic and seek shade. Affected animals should be removed from the pasture immediately and placed in permanent shade with food and water. Reducing grazing pressure reduces the risk because most spores are in the lower plant parts. Supplementation with a slow-release intraruminal bullet of zinc at 20–30 times the nutritional requirement protects animals from facial eczema and is the most effective treatment. As for pyrrolizidine alkaloid poisoning, merino sheep are more resistant to sporodesmin than British breeds. The heritability of resistance is 0.42, suggesting selection for resistance to facial eczema is possible.

Staggers (ryegrass, paspalum, Bermuda grass)

'Staggers' is a neuropathy caused by endophytes infecting three main grass species: paspalum, Bermuda grass (couch) and perennial ryegrass. The clinical signs are tremors, staggering, head shaking and collapse when the animals are excited or made to move. For perennial ryegrass, the endophyte responsible for the mycotoxin is *Neotyphodium lolii* and the mycotoxins are lolitrem B and ergovaline. The endophyte also produces peramine, which inhibits egg-laying in the Argentine stem weevil and thereby protects the ryegrass. Attempts have been made in New Zealand to select strains of *N. lolii* that produce high levels of peramine but low levels of lolitrem B to protect the plant but also prevent ill-health in grazing animals. Sheep grazing perennial ryegrass infected with *N. lolii* have reduced growth rate, staggers, diarrhoea (dags), increased fly strike and heat stress.

Paspalum staggers is caused by infection of *Paspalum* spp. by *Claviceps paspali* or *C. purpurea*, which form sclerotia (fungal bodies) in the seed heads. Clinical signs include breathing difficulties, tremors, excessive salivation, head shaking, diarrhoea and intestinal bleeding, and collapse when excited. Cattle are particularly prone to paspalum staggers because they are less selective grazers than sheep, which tend to avoid the seed heads in which the fungi reside. A gangrenous syndrome is also caused by paspalum endophytes, which produce ergotamine, ergometrine and ergotoxin that constrict smooth muscles of blood vessels causing bleeding and gangrene of the affected parts, similar to fescue toxicoses.

Bermuda grass staggers has been reported in cattle in South America, the USA and in horses in California. Bermuda grass staggers is caused by the endophyte, *Claviceps cynodontis*, which produces the mycotoxins paspalitrems A and B. The syndrome arises mainly in winter after frosts and in pastures heavily invaded by the *Cynodon* weeds.

Tall fescue toxicoses

Tall fescue (*Festuca arundinaceae*) infected with the endophyte, *Neotyphodium coenophialum,* produces a suite of pathologies called gangrenous ergotism (also known as 'Fescue Foot'), dysthermic ergotism, hyperthermia, idiopathic bovine hyperthermia and agalactica (milk suppression). This is an important global disease affecting animals in North America, South America and Australasia. Again, the endophyte protects the grass from pests and makes the plants more drought-tolerant. Peripheral vasoconstriction causes hyperthermia, tissue anoxia, necrosis and gangrene. In cattle, the gangrene is in the limb extremities. The shunting of blood from the periphery to the core produces hyperthermia and this a particular problem for cattle, which, unlike sheep, rely more heavily on cutaneous heat exchange. Agalactica (milk cessation) or hypogalactica (reduced milk production) results from a reduction in circulating prolactin. Even subclinical ergotism reduces weight gain in cattle (Riet-Correa *et al.* 2013).

Lupinosis

Lupinosis is a disease of sheep grazing mature lupin stems or stubbles (*Lupinus angustifolius*) that have been infected by the fungus *Phomopsis leptostromiformis*. The fungus produces the mycotoxin quinolizidine alkaloid. Lupin stubbles after summer rains and warm, humid conditions, become infected with the fungus in the plant stems. The stems remain toxic for months. Lupinosis occurs in Europe, South Africa, the USA, Australia and New Zealand. Animals with lupinosis are lethargic, lose condition, and suffer liver damage with subsequent photosensitivity. Grazing the stubbles immediately after harvest before extensive fungal growth, and grazing at low intensity to reduce stem consumption, are useful strategies to reduce the incidence of lupinosis.

Toxic glycosides in forage plants

Glycosides are naturally occurring compounds found in forage plants and largely based on a sugar attached to a non-carbohydrate group. They include saponins, cyanogenic glycosides, glucosinolates and coumarins (Table 5.8).

Cyanogenic glycosides

Cyanide poisoning occurs in ruminants when they consume plants containing cyanogenic glycosides (Nicholson 2011). These cyanide-containing compounds

Table 5.8. Toxic glycosides found in pasture forages

Group	Plant species	Clinical signs
Nitro-containing glycosides	*Astragalus* spp. *Indigofera* spp. *Lotus* spp.	• Incoordination • Distress • Dyspnoea (difficulty breathing) • Cyanosis • Ataxia
Cyanogenic glycosides	*Trifolium repens* (white clover) *Sorghum* spp.	• Rapid heart rate (tachycardia) • Rapid breathing (tachypnoea) • Recumbency • Convulsion
Cardiac glycosides	*Digitalis* spp. (foxglove) *Liliaceae* spp. (lilies) *Urginea* spp. (sea onion)	• Forceful heart contractions • Restlessness • Laboured breathing • Frequent urination/defecation • Tachycardia
Saponins	*Medicago sativa* (lucerne, alfalfa) *Dioscorea* spp. (yams)	• Inflammation of GI tract • Liver damage • Respiratory failure • Photosensitisation
Glucosinolates	*Brassica* spp. (kale, turnips, swedes, turnips, mustards)	• Goitre
Diterpenoid glycosides	*Asteraceae* spp. (sunflowers) *Xanthium* spp. (cocklebur)	• Depression • Weakness • Convulsions
Bracken glycosides	*Pteridium aquilinum* (bracken fern)	• Polioencephalomalacia • Bone marrow depression • Low leucocyte numbers (immunity depressed)
Calcinogenic glycosides	*Solanum glaucophyllum* (nightshade) *Cestrum diurnum* (jasmine)	• Calcium salts deposited in soft tissues
Phenolic glycosides (phytoestrogens)	*Trifolium subterraneum* (sub clover) *Medicago sativa* (lucerne) *Melilotus* spp. (sweet melilot)	• Abortions • Reproductive disorders • Anaemia
Ranunclin	*Ranunculus* spp. (buttercup)	• Digestive disorders • Weakness • Depression • Anorexia • Diarrhoea

release the cyanide when the plants are chewed by the animals, bringing the glucoside into contact with plant-derived β-glucosidases, which then release the cyanide.

The plants most likely to cause cyanide poisoning in ruminants are all members of the sorghum family, Johnson grass (*Sorghum halepense*), Sudan grass (*Sorghum sudanense*), oats (*Avena* spp.), native fuchsia (*Eremophila* spp.), native couch (*Brachyachne* spp.), red crumbweed (*Dysphania glomulifera*), acacias (*Acacia glaucescens*) and sugar gum (*Eucalyptus cladocalyx*). Absorbed cyanide rapidly depresses cellular oxygen delivery from the oxyhaemoglobin, leaving the venous blood bright cherry red. Cytochrome oxidase enzyme activity is depressed and cellular respiration is depressed. Prevention involves: avoidance of high-risk plants, particularly if they are frost-damaged, drought-affected, immature or wilted; avoidance of hays made from high-risk plants under stress at the time of haymaking; dilution of the high-risk plant material with low-risk plants. Clinical signs of cyanide toxicoses in ruminants are anxiety, excitability, staggering, dyspnoea (difficult breathing), dilated pupils, salivation, muscle tremors, bright red venous blood and convulsions (NRC 2000).

Toxicoses caused by intrinsic plant toxins

Grasses, in general, are low in intrinsic toxic compounds in comparison with other plants, but they nevertheless can produce toxins with varying pathological outcomes in animals. Some arise from direct production of the toxin by the plant, others are produced by fungi or bacteria colonising the grass, and sometimes a combination of plant toxins and microbial toxins produces problems.

Phalaris staggers

Phalaris aquatica and *P. arundinaceae* (reed canary grass) are perennial grasses associated with toxicoses characterised by staggers ('Phalaris staggers syndrome') or cardiac or polioencephalomalacic sudden death ('Phalaris sudden-death syndrome') in sheep and cattle. The neurotoxic compounds causing staggers are methylated tryptamines, tyramine alkaloids and β-carboline alkaloids (Edgar 1994). There is also evidence that other aetiologies such as cyanide poisoning, a rapid-onset polioenchephalomalacia, caused by a thiamine antagonist, and nitrite poisoning, play a role in this spectrum of phalaris-induced neurological diseases (Bourke and Carrigan 1992). The clinical signs of phalaris toxicoses in sheep are tremors, head nodding, twitching, ataxia, incoordination and collapse. In sheep and horses, the methyltyramine toxicoses causes sudden death. Treatment of sheep with slow-release intraruminal cobalt appears to stimulate the activity of detoxifying bacteria (Lee 1956) and is effective in preventing the staggers syndrome, but not the sudden death syndrome. Breeding of low-alkaloid phalaris cultivars has not always resulted in elimination of syndromes because of the complex suite of neurotoxic compounds. The best management options are to minimise disturbance to grazing, not to place hungry sheep on phalaris-dominant pastures, particularly after frost or drought, and to provide either a cobalt bullet to animals or fertilise pastures with cobalt.

Kikuyu poisoning

Kikuyu (*Pennisetum clandestinum*) is a tropical grass that causes poisoning of sheep and cattle. Kikuyu contains oxalates, saponins, nitrates and can be infected by a fungus (*Myrothesium* spp.) that produces a mycotoxin. The plant can also be invaded by army worm, which has been associated with kikuyu poisoning at some times but not others. Mortality in kikuyu poisoning can be high and preceded by anorexia, excessive salivation, dehydration, and cessation of ruminal and intestinal movements.

Annual ryegrass toxicity
Annual ryegrass toxicity or ARGT is a toxicosis caused by an interaction between annual ryegrass (*Lolium rigidum*), a nematode (*Anguina agrostis* or *Anguina funesta*) and a pathogenic bacterium (*Clavibacter toxicus*). Annual ryegrass plants become infected with the nematode larvae, which enter the seed head and form a gall in which they lay their eggs. The eggs hatch into larvae, which can remain dormant for years in the gall. When these larvae become infected with the bacterium, they exude a yellow slime on the seed heads. This slime contains a corynetoxin, related to the antibiotic tunicamycin, which is extremely toxic to animals (Cheeke 1995). The corynetoxin prevents glycosylation of glycoproteins that are essential to enzyme, hormone and receptor function in many metabolic systems.

Brassica toxicity
Brassica species such as kale, turnips, swedes and canola are widely used as fodder crops for grazing animals. They contain two sulphur compounds, S-methylcysteine sulphoxide (SMCO) and glucosinolates, which cause toxicoses. SMCO is also called 'Brassica anaemia factor', and is metabolised in the rumen to dimethyldisulphide, which, at high levels, is absorbed producing haemolytic anaemia, reduced haemoglobin concentrations and the appearance of Heinz bodies in the red blood cells (precipitated oxidised haemoglobin granules). Poor growth, anorexia, diarrhoea, jaundice and ultimately death result from continuous exposure to toxic brassicas. High-nitrogen fertiliser application exacerbates the toxicity. The glucosinolates in brassicas are hydrolysed in the rumen to isothiocyanate which interferes with thyroid function. Signs of this toxicoses therefore are those of thyroid malfunction and goitre.

References

Allden WG, Whittaker IAMcD (1970) The determinants of herbage intake by grazing sheep: the interrelationship of factors influencing herbage intake and availability. *Australian Journal of Agricultural Research* **21**, 755–766. doi:10.1071/AR9700755

Allen VG, Segarra E (2001) Anti-quality components in forage: overview, significance and economic impact. In *Anti-Quality Factors in Rangelands and Pastureland Forages* (Ed. Karen Launchbaugh). pp. 1–4. Bulletin 73. Idaho Forest, Wildlife and Range Experiment Station, University of Idaho, Moscow, ID, USA.

Arnold GW, Maller RA (1977) Effects of nutritional experience in early and adult life on the performance and dietary habits of sheep. *Applied Animal Ethology* **3**, 5–26. doi:10.1016/0304-3762(77)90067-0

Asner GP, Elmore AJ, Loander LP, Martin RE, Harris AT (2004) Grazing systems, ecosystem responses, and global change. *Annual Review of Environment and Resources* **29**, 261–299. doi:10.1146/annurev.energy.29.062403.102142

Bourke CA, Carrigan MJ (1992) Mechanisms underlying *Phalaris aquatica* 'sudden death' syndrome in sheep. *Australian Veterinary Journal* **69**, 165–167. doi:10.1111/j.1751-0813.1992.tb07503.x

Chadwick MA, Vercoe PE, Williams IH, Revell DK (2009) Dietary exposure of pregnant ewes to salt dictates how their offspring respond to salt. *Physiology & Behavior* **97**, 437–445. doi:10.1016/j.physbeh.2009.03.017

Cheeke PR (1995) Endogenous toxins and mycotoxins in forage grasses and their effects on livestock. *Journal of Animal Science* **73**, 909–918. doi:10.2527/1995.733909x

Dixon RM, Stockdale CR (1999) Associative effects between forages and grains: consequences for feed utilisation. *Australian Journal of Agricultural Research* **50**, 757–773. doi:10.1071/AR98165

Donnelly JR, Moore AD, Freer M (1997) GRAZPLAN: decision support systems for Australian grazing enterprises. I. Overview of the GRAZPLAN project and a description of the MetAccess and LambAlive DSS. *Agricultural Systems* **54**, 57–76. doi:10.1016/S0308-521X(96)00046-7

Earl J (2014) Grazing and pasture management and utilisation in Australia. In *Beef Cattle Production and Trade* (Eds L Kahn and DF Cottle) pp. 339–352. CSIRO Publishing, Melbourne.

Edgar JA (1994) Toxins in temperate grasses- implications and solutions. *New Zealand Journal of Agricultural Research* **37**, 341–347. doi:10.1080/00288233.1994.9513072

Forbes JM, Mayes RW (2002) Food choice. In *Sheep Nutrition* (Eds M Freer and H Dove) p. 51. CSIRO Publishing, Melbourne.

Freer M, Jones DB (1984) Feeding value of subterranean clover, lucerne, phalaris and Wimmera ryegrass for lambs. *Australian Journal of Experimental Agriculture and Animal Husbandry* **24**, 156–164. doi:10.1071/EA9840156

Fulkerson WJ, Neal JS, Clark CF, Horagoda A, Nanra KS, Barchia I (2007) Nutritive value of forage species grown in the warm temperate climate of Australia for dairy cows: grasses and legumes. *Livestock Science* **107**, 253–264. doi:10.1016/j.livsci.2006.09.029

Gordon IJ (1988) Facilitation of red deer grazing by cattle and its impact on red deer performance. *Journal of Applied Ecology* **25**, 1–10. doi:10.2307/2403605

Gurung NK, Jallow JA, McGregor BA, Watson MJ, McIlroy BKMH, Holmes JHG (1994) Complementary selection and intake of annual pastures by sheep and goats. *Small Ruminant Research* **14**, 185–192. doi:10.1016/0921-4488(94)90039-6

Hill JO, Coates DB, Whitbread AM, Clem RL, Robertson MJ, Pengelly BC (2009) Seasonal changes in pasture quality and diet selection and their relationship with liveweight gain of steers grazing tropical grass and grass–legume pastures in northern Australia. *Animal Production Science* **49**, 983–993. doi:10.1071/EA06331

Hodgson J, Clark DA, Mitchell RJ (1994) Foraging behavior in grazing animals and its impact on plant communities. In *Forage Quality, Evaluation and Utilisation* (Ed. GC Fahey) pp. 796-827. American Society of Agronomy, Crop Science Society of America, Soil Science Society of America, USA.

Kenney PA, Black JL (1984) Factors affecting diet selection by sheep. 1. Potential intake rate and acceptability of feed. *Australian Journal of Agricultural Research* **35**, 551–563. doi:10.1071/AR9840551

Lee HJ (1956) The toxicity of *Phalaris tuberosa* to sheep and cattle and the preventative role of cobalt. In *Proceedings VII International Grasslands Congress. 6–15 November, Palmerston North, New Zealand*. pp. 387–396. International Grasslands Congress, Christchurch, New Zealand.

Leigh JH, Mulham WE (1966) Selection of diet by sheep grazing saltbush (*Atriplex vesicara*)-cottonbush (*Kochia aphylla*) community. *Australian Journal of Experimental Agriculture and Animal Husbandry* **6**, 460–467. doi:10.1071/EA9660460

Milford R, Minson DJ (1966) Determinants of feeding value of pasture and supplementary feed. *Proceedings of the Australan Society of Animal Production* **5**, 319–329.

Minson DJ (1981) Nutritional differences between tropical and temperate pastures. In *Grazing Animals (World Animal Science B1)*. (Ed. FHW Morley) pp. 143–157. Elsevier, Amsterdam, Netherlands.

Nicholson SS (2011) Southern plants toxic to ruminants. *The Veterinary Clinics of North America. Food Animal Practice* **27**, 447–458. doi:10.1016/j.cvfa.2011.02.008

NRC (2000) *Nutrient Requirements of Beef Cattle. Update 2000*. 7th revised edn. National Research Council, The National Academies Press, Washington, DC, USA.

Parton K, Bruere AN (2002) Plant poisoning of livestock in New Zealand. *New Zealand Veterinary Journal* **50** (Supplement), 22–27.

Provenza FD (1996) Acquired aversions as the basis for varied diets of ruminants foraging on rangelands. *Journal of Animal Science* **74**, 2010–2020. doi:10.2527/1996.7482010x

Provenza FD (2018) Palates link soil and plants with herbivores and humans. *Animal Production Science* **58**, 1432–1437. doi:10.1071/AN17760

Riet-Correa F, Rivero R, Odriozola E, Adrien Md L, Medeiros MT, Schild AL (2013) Mycotoxicoses of ruminants and horses. *Journal of Veterinary Diagnostic Investigation* **25**, 692–708. doi:10.1177/1040638713504572

San Martin F (1987) Comparative forage selectivity and nutrition of South American camelids and sheep. PhD dissertation. Texas Tech University, Lubbock, TX, USA.

Smith BL, Towers NR (2002) Mycotoxicoses of grazing animals in New Zealand. *New Zealand Veterinary Journal* **50**, 28–34. doi:10.1080/00480169.2002.36263

Stegelmeier BL (2011) Pyrrolizidine alkaloid-containing toxic plants (*Senecio, Crotalaria, Cynoglossum, Amsinckia, Heliotropium, Echium* spp.). *The Veterinary Clinics of North America. Food Animal Practice* **27**, 419–428. doi:10.1016/j.cvfa.2011.02.013

Stegelmeier BL, Edgar JA, Colegate SM, Gardner DR, Schoch TK, Coulombe RA, *et al.* (1999) Pyrrolizidine alkaloid plants, metabolism and toxicity. *Journal of Natural Toxins* **8**, 95–116.

Teague R, Provenza F, Kreuter U, Steffens T, Barnes M (2013) Multi-paddock grazing on rangelands: why the perceptual dichotomy between research results and rancher experience? *Journal of Environmental Management* **128**, 699–717. doi:10.1016/j.jenvman.2013.05.064

Ternouth JH (1990) Phosphorus and beef production in northern Australia. 3. Phosphorus in cattle – a review. *Tropical Grasslands* **24**, 159–169.

Thornton RF, Minson DJ (1973) The relationship between apparent retention time in the rumen, voluntary intake and apparent digestibility of legume and grass diets in sheep. *Australian Journal of Agricultural Research* **24**, 889–898. doi:10.1071/AR9730889

Tothill JC, Gillies C (1992) *The Pasture Lands of Northern Australia: their Condition, Productivity, and Sustainability*. Occasional Publication No. 5. Tropical Grassland Society Australia, Brisbane.

Ulyatt MJ (1973) The feeding value of herbage. In *Chemistry and Biochemistry of Herbage. Volume 3*. (Eds GW Butler and RW Bailey) pp. 131–178. Academic Press, London, UK.

Ulyatt MJ (1978) Aspects of the feeding value of pastures. *Proceedings Agronomy Society of New Zealand* **8**, 119–122.

Ulyatt MJ, Lancashire JA, Jones WT (1977) The nutritive value of legumes. *Proceedings New Zealand Grasslands Association* **38**, 107–118.

Villalba JJ, Provenza FD, Han G-D (2004) Experience influences diet mixing by herbivores: implications for plant biochemical diversity. *Oikos* **107**, 100–109.

Villalba JJ, Provenza FD, Shaw R (2006) Sheep self-medicate when challenged with illness-inducing foods. *Animal* **1**, 1360–1370.

Villalba JJ, Provenza FD, Catanese F, Distel RA (2015) Understanding and manipulating diet choice in grazing animals. *Animal Production Science* **55**, 261–271. doi:10.1071/AN14449

Waghorn GC, Adams NR, Woodfield DR (2002) Deleterious substances in grazed pastures. In *Sheep Nutrition* (Eds M Freer and H Dove) pp. 333–356. CABI Publications, Wallingford, UK.

6

Sheep and goat nutrition

Key points

- There are ~1150 million sheep and 1000 million goats worldwide. They contribute significantly to many economies, given their ability to survive on poor-quality, low-protein roughages to produce high-quality meat, milk and fibre.
- Sheep and goats can consume a ration up to three times the protein content of the average feed on offer by selectively grazing between pasture species and within individual plants, using their mobile, cleft upper lips to grasp fine plant parts and seeds.
- The reticulo-rumen of a 50 kg sheep or goat contains from 2–15 L of rumen fluid, which is ~70% of the total capacity of the four stomach compartments.
- Diets can be formulated for sheep and goats from simple equations that determine the maintenance energy requirement, and then use multiples of this value for other functions (e.g. gestation, pregnancy, lactation, growth).
- More sophisticated programs are available to calculate these values 'accurately' and to incorporate them into grazing decision support management packages.
- Protein requirements can likewise be estimated using simple crude protein values for different productive purposes or using software packages based on detailed knowledge of protein digestion and metabolism.
- Wool growth is related positively to dry matter intake but the relationship is one of diminishing returns, so the efficiency of wool growth is higher at lower intakes.
- Wool growth rate is affected by the supply of protein and energy to the follicle but is usually limited by the supply of amino acids, particularly the sulphur amino acids cyst(e)ine and methionine, which are first-limiting for wool growth.
- Wool fibre diameter increases with feed intake and fibre value declines with increasing diameter, so feeding for more wool results in less valuable wool.

- Variable nutrient supply reduces the staple strength of wool because of variations in diameter along the staple.
- Copper deficiency in sheep results in anaemia, weak wool, neurological ataxia in lambs and depigmentation of wool.
- Cobalt deficiency in sheep results in megaloblastic anaemia, anorexia, poor growth, increased photosensitivity and increased infection.
- Polioencephalomalacia (PEM) in sheep and goats is associated with a deficiency of thiamine or an excess of sulphur in the diet. Thiamine deficiency occurs if diets contain thiaminases (e.g. bracken fern) or if thiamine supply from the rumen is depressed by reduced microbial activity or increased ruminal thiaminase production. High-grain diets can reduce thiamine supply. PEM is characterised by opisthotonus (stargazing), head twitching, head pressing and staggering, followed by death.
- Sheep and goats that have consumed cyanogenic glucosides in certain plants show signs of acute cyanide poisoning (cherry red blood, anxiety, dyspnoea, staggering, dilated pupils, excessive salivation, convulsions and death).

Introduction

Domesticated sheep (*Ovis aries*) are small ruminants and are thought to have been developed from an ancestral primitive sheep, the mouflon (*Ovis orientalis musimon*) in the fertile crescent of Iran, Turkey, Syria and Iraq. Today sheep number ~1150 million worldwide, have a mature mean bodyweight of ~50 kg (30–130 kg) and are considered either grass/roughage eaters or intermediate feeders between concentrate selectors/browsers and grazers (Hoffman 1985). Sheep occupy a wide distribution range across diverse environments from deserts to the tropics to the high mountains. Global sheep distribution is Asia (44%), Africa (28%), Europe (11%), Oceania (9%) and the Americas (7%). The success of the Bovidae in general, and sheep/goats in particular, stems from their ability to utilise poor-quality, low-protein roughages, which could only be utilised by mono-gastrics with very large post-gastric fermentation chambers such as the horse. However, as discussed in Chapter 1, pre-gastric fermentation also brings disadvantages and these are reflected in many of the common nutritional problems faced by domesticated sheep and goats in production systems. Common diseases related to nutrition and metabolism are polioencephalomalacia, cobalt deficiency, copper deficiency and cyanogenic glucoside poisoning.

Goats, like sheep, are small ruminants and were the first farm animals to be domesticated *c.* 8000 BC in the Middle East (Boyazoglu *et al.* 2005). This early association with humans resulted in goats being involved in many aspects of human life (religion, economy, nutrition and culture). The number of goats worldwide in 2014 was ~1000 million (FAO 2014) with Asia and Africa accounting for more than 800 million head (Abdel Aziz 2010). China and India hold the two highest goat populations (combined ~280 million), with Pakistan and Bangladesh next (combined 115 million). The global goat population has been increasing over the past decade at a rate of 2–3% per annum because they solve many of the problems of developing societies, but also because goat products in developing countries are increasingly being recognised as high quality (Boyazoglu *et al.* 2005). Past reputations of goats as ecological destructive pests are being replaced by considerations that goats are extremely resilient animals, able to cope with harsh climates (and potential negative climate change scenarios), and can be used to manage vegetation cover by selectively grazing bushes, shrubs and thorny vegetation, hence providing weed and fire control. Many of the preconceptions of goats as environmental destroyers and causes of desertification arose from poor animal management and overstocking. Goats have also been associated with low socioeconomic status, which has driven negative attitudes towards them. In reality, goats occupy a unique niche alongside sheep in that they have different grazing habits, have a different ability to digest plant fibres and produce products with different attributes to other ruminants. In this chapter the similarities and differences between sheep and goats are considered particularly in relation to their nutritional requirements and grazing management.

Digestive anatomy and physiology of sheep and goats
Anatomy of the digestive tract of sheep and goats

Sheep and goats are small ruminants and share similar gastrointestinal tracts to all other ruminants (see Chapter 3). There are few anatomical differences in digestive systems between sheep and goats. The rumen of goats is slightly longer and narrower than that of sheep and the rumen papillae are more fungiform in goats and filiform in sheep (Bhattacharya 1980), but most of the differences in digestive ability appear to arise from differences in diet selection, voluntary intake and retention time of feed in the reticulo-rumen.

The anatomy of the mouth of sheep and goats is designed to allow a high degree of selectivity of plant parts and specific plant species among mixed swards. The cleft upper lip and prehensile tongue allow close grazing when forages are in short supply, and also allow very specific selection of a diet higher in nutritive value than the average feed available. Table 6.1 shows the extent to which sheep can select a diet significantly higher in protein and energy content than the average on offer. On the low-protein barley stubble, the sheep, unlike cattle, managed to select a diet containing more than three times the protein content of the average feed on offer. On mature pastures, selective feeding increased the protein content of ingested feed to almost double that of the average feed on offer. Cattle, on the other hand,

Table 6.1. Comparison of selective feeding behaviour of sheep versus cattle

Pasture/crop residues	Crude protein of feed on offer (%)	Crude protein of feed consumed (%) Sheep	Crude protein of feed consumed (%) Cattle
Barley stubble	4.1	14.9	3.8
Mature annual pasture	11.5	20.5	10.5

Note: Oesophageally fistulated sheep and cattle grazed a barley stubble or mature pasture and the crude protein content of the feed on offer compared with the crude protein content of the feed consumed. Sheep consumed feed two to three times higher in protein than feed on offer. Cattle were unable to select higher quality feed.
Source: PI Hynd (unpublished data)

were unable to improve the quality of the feed consumed, because of the anatomy of their mouthparts and mode of eating (cattle use their prehensile tongue to gather material into their mouth, rather than the mobile cleft lips used by sheep and goats).

The chewing action in sheep and goats is designed to achieve maximum grinding and breaking down of the food bolus during both feeding and ruminating. Ridges on the lower molars and premolars move sideways, backwards and forwards as the lower jaw 'swings' in all three planes. The tongue and oral cavity comprise a thick, keratinised, stratified squamous epithelium providing protection from the coarse material ingested. The tongue contains fungiform papilla distributed across the tongue surface and vallate papillae on the back of the tongue.

Ruminants are unique in having no incisors in the upper jaw. These are replaced by a dental pad, comprising a thickened, cornified epithelium. The adult dental equation for sheep and goats is:

	Incisors	Premolars	Molars
Upper	0	3	3
Lower	4	3	3

This formula means there are no incisors or canines in the upper jaw but there are three premolars and three molars on each side of the upper jaw. The lower jaw contains four incisors, three premolars and three molars on each side. The large gap between the front teeth and the premolars/molars is called a diastema. Sheep are also characterised by the loss of 'milk teeth' progressively with age (Fig. 6.1) as follows:

- Eight milk teeth: birth to 12 months.
- Two adult incisors: 12–19 months.
- Four adult incisors: 18–24 months.
- Six adult incisors: 23–36 months.
- Eight adult incisors: 28–48 months.

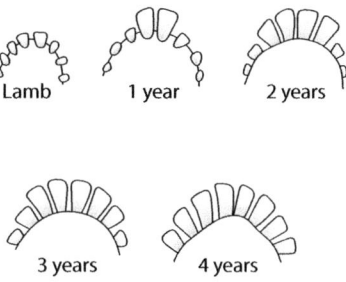

Fig. 6.1. Incisors of the lower jaw in sheep and goats differing in approximate age.

Salivary glands and saliva production in sheep and goats

Saliva plays particularly important roles in ruminants as follows:

- **buffering** the large quantities of volatile fatty acids produced each day in the rumen, hence reducing acidosis
- **lubricating** the hard roughages often consumed
- **maintaining the water content** of the fermenting ruminal digesta
- providing the microbes with **essential nutrients** for fermentation (nitrogen, sodium, potassium, chloride).

There appears to be no salivary amylase in ruminants so there is little early digestion of starches pre-rumen.

Sheep and goats have six pairs of salivary glands (Fig. 6.2).

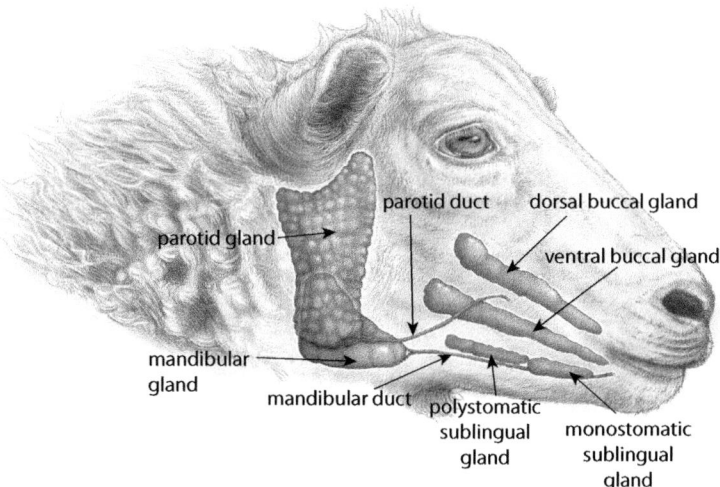

Fig. 6.2. Salivary glands in sheep and goats.

Table 6.2. Salivary glands and their secretions in sheep and goats

Salivary glands	Estimated volume of saliva (L/day)	Saliva type
Parotids	3–8	Serous, isotonic, HCO_3^-, HPO_4^{2-}
Inferior molar	0.7–2.0	Serous, isotonic, HCO_3^-, HPO_4^{2-}
Dorsal and ventral buccal	2–6	Mucous, isotonic. HCO_3^-, HPO_4^{2-}
Mandibular	0.4–0.8	Mixed, hypotonic, little buffer
Sublingual	0.1	Mucous, hypotonic, little buffer
Labial	?	Mucous, hypotonic, little buffer
Total	**6.2–17.0**	

Source: Adapted from Kay (1960)

The volume of saliva depends on the feed consumed, its moisture content, palatability, smell, taste and physical nature (Kay 1960). Adult sheep/goats produce ~6–17 L of mixed saliva per day when consuming roughage diets *ad libitum* (Table 6.2). The parotid glands produce more than half of the total daily saliva secretion. The parotid glands secrete continuously but the submaxillary glands only secrete during feeding (Kay 1960). The saliva from the parotid and inferior molar glands is serous; the submaxillary, sublingual and labial glands are mixed (serous and mucous); and the buccal secretion is mucous only. Sheep and goat saliva is a solution of sodium, potassium, bicarbonate, phosphate and chloride, and contains the buffers, bicarbonates and phosphates. The pH of ruminant saliva is ~8.0 (Counotte and van't Klooster 1979).

The gastrointestinal tract of sheep and goats

Detailed coverage of the anatomy and functioning of the ruminant gastrointestinal system is provided in Chapter 3. Sheep- and goat-specific information is now considered. The four-compartment stomach of small ruminants comprises the reticulo-rumen (compartments 1 and 2), omasum and abomasum. Typical ingesta volumes in each part of the tract are shown in Table 6.3.

The rumen of a 50 kg adult sheep/goat typically holds ~2–15 L of fluid depending on the diet, time after feeding and hydration state of the animal. The

Table 6.3. Range of volumes (L) of digesta in the fermentation compartments of the adult sheep or goat (60 kg) digestive system

	Roughage diet	Concentrate diet
Reticulo-rumen	5–15	2–8
Omasum	0.7–2	0.5–1
Abomasum	1.5–4	0.7–2
Caecum	0.5–1	0.2–0.5
Colon	0.2–0.5	0.1–0.2

Table 6.4. Relative lengths of the intestines of adult (50 kg) sheep and goats

Segment	Relative length (%)	Absolute length (m)
Small intestine	80	26.2
Caecum	1	0.36
Colon	19	6.17
Total	100	32.73

Source: Kararli (1995)

typical forestomach arrangement (reticulo-rumen and omasum) is present in small ruminants. The abomasum ends at the pyloric sphincter: a muscular sphincter regulating the emptying of the abomasum. The small intestine extends from the pylorus to the ileo-caeco-colic junction, a total length in an adult sheep of ~20–25 m (Table 6.4). The caecum is relatively large and contributes significantly to the total energy supply to the animal (~10% of the total Metabolisable Energy). The first 20% of the colon is the proximal colon, which then becomes a spiral colon and finally the descending colon enters the rectum (Fig. 3.1).

Comparative feeding behaviour and digestive physiology of sheep and goats

Although similar in digestive anatomy, sheep and goats differ significantly in feeding behaviour and digestive physiology and hence nutrient supply under different feeding scenarios. Some of these differences are listed in Table 6.5.

Table 6.5. Comparative feeding behaviour and digestive function in sheep versus goats

Characteristic	Goats	Sheep
Grazing behaviour	Bipedal stance Browser Shrub/tree leaf selection Walk long distances	Mainly quadrupedal Grazer Less selective Walk short distances
Digestion of: **high-fibre feeds** **mod-fibre feeds**	Higher Same	Lower Same
Retention time of feed in GIT	Longer	Shorter
Feed variety	Higher	Lower
Recycling of urea to rumen	Higher	Lower
Max. DM intake (percentage bodyweight)	3–6%	3%
Water intake/DMI	Lower	Higher
Tolerance of feed tannins	Higher	Lower

Source: Adapted from Devendra (1986)

The energy, protein, vitamin and mineral requirements of sheep and goats

The detailed algorithms underpinning sheep nutrition are described in Chapter 4. The general equation relating Metabolisable Energy (ME) requirement to bodyweight (W), efficiency of conversion of ME to NE (k_m), age (A), level of production (MEp), grazing activity (MEgraze) and cold stress (Ecold) is as follows:

$$ME_m \text{ (MJ/day)} = \text{S.M.} (0.28W^{0.75} \exp(-0.03A))/k_m + 0.1ME_p + ME_{graze} + E_{cold}$$

However, in many cases, field nutritionists, veterinarians and sheep producers are required to establish feeding programs 'on-the-run' or without the precision and accuracy provided by sophisticated models. If we make a few simple assumptions as follows:

- sex is female or castrate male
- not lactating
- diet is moderate quality so $k_m = 0.72$
- age is adult
- moderate level of production
- grazing on level ground and no cold stress

The equation then simplifies to:

$$ME_m \text{ (MJ/day)} = (0.28/0.72)W^{0.75} = 0.39W^{0.75}$$

This can be further simplified if we assume there is little 'scale' effect of bodyweight over a typical adult animal live weight range. The equation then becomes $ME_m = 0.12W + 1.5$. This equation works well for goats as well. If we compare the ME requirement for maintenance of goats of $0.42W^{0.75}$ (NRC 2007), with the simplified equation, the latter underestimates the ME requirement of goats by only 4%. Given the uncertainties around the impact of goat grazing activity levels, this is a reasonable estimate.

> The maintenance requirement of sheep and goats can therefore be estimated from the following equation:
> ME requirement for maintenance of sheep and goats (ME maint.) $= 0.12W + 1.5$

Now we can add the additional requirements for growth, gestation and lactation. These were calculated assuming average values for k values based on typical feeds that would be used in each scenario, average composition of gain in young, rapidly growing lambs/kids, and typical lactation yields for single and twin-bearing ewes and does.

Table 6.6. Simple values for formulating rations for sheep and goats in different production scenarios

Scenario	ME requirement (MJ/day)	Crude protein requirement (%DM)	Dietary Undegraded Protein content (%)
Maintenance	$ME_m = 0.12W + 1.5$	6–8	0–20
Rapid growth	$ME_m \times 2.0$	14–16	>30
Pregnant (3–4 months) Singles Twins	$ME_m + 3$ $ME_m + 4$	8–10 8–10	20 20
Pregnant (last month) Singles Twins	$ME_m + 4$ $ME_m + 8$	8–10 8–10	20 20
Lactation (1st month) Singles Twins	$ME_m + 8$ $ME_m + 12$	12–14	25–30

Sources: Adapted from NRC (2007); CSIRO (2007)

Table 6.6 shows some simplified equations that can be used to rapidly formulate rations for different classes of sheep and goats.

Although the estimates made using this system may seem to be less accurate than those made using software packages, there are many other inaccuracies and variables that will influence the feeding outcomes (e.g. the accuracy of the feed test, the variance in feed composition throughout the feed batch, impacts of changes in weather and effects on feed intake). Monitoring of the animal performance frequently and adjusting diet is arguably more important than having defined the animals' requirements accurately.

Examples using the rapid feed formulating system outlined in Table 6.6.

Example 1: Formulate a ration and feeding program for 50 kg ewes or does, confined-fed in a drought. Feeds available are barley grain ($250/ tonne), lupin grain ($320/tonne) and meadow hay ($150/tonne).

(a) **Estimate the ME requirement for maintenance (ME_m)**

$$ME_m = (50 \times 0.11) + 1.5 = 7.0 \text{ MJ/day}$$

(b) **Use ME density values for the available feeds (from tables or feed test) to determine the quantity of the feeds required to meet this requirement**

e.g. the ME content of the three feeds are barley = 12.2 MJ/kg DM; lupins = 13.2 MJ/kg DM and hay = 8.0 MJ/kg DM. The maintenance energy requirement can be met by supplying 7.0/12.2 kg barley DM (= 0.57 kg), 7.0/13.2 kg lupins DM (= 0.53 kg) or 7.0/8.0 kg hay DM (= 0.88 kg). Note: convert these DM values to 'as fed' values by dividing by DM% (= 90% for dry feeds).

(c) **Estimate the protein requirements of the sheep**. As for energy, the algorithms for calculating the protein requirements of ruminants (see Chapter 4) are complex. However, in this scenario only sufficient protein to meet the sheep's maintenance requirement is required and this can be met by the microbial supply alone. The feed protein need only be 5–6% to supply sufficient nitrogen for the microbes, and this nitrogen can be in the form of rumen degradable protein. All three feeds contain more than 6% crude protein.

(d) **Select the cheapest feed to meet the requirements.** In this example, the barley is $250/tonne, lupins $320 and hay $150. This means it will cost 16c/day for barley, 19c/day for lupins and 15c/day for hay. Although hay is the cheapest feed, the cost and logistics of feeding hay versus the more dense grains must also be taken into account. In this case, barley grain is the most cost effective.

Example 2: Formulate a ration for feedlot lambs fed for maximal growth

The objective of most feedlots is to minimise the amount of time the lambs spend in the feedlot to reach a goal bodyweight, carcase weight and carcase specifications, with minimal mortalities and adverse health effects (e.g. acidosis, urolithiasis, lameness, mortalities). To achieve these aims, we need a ration that is high in energy and protein (including a high proportion of Undegraded Dietary Protein), balanced for vitamins and minerals and containing sufficient physically effective fibre ('scratch factor' stimulation of rumination).

From a ration formulation point of view, you can use computer programs that calculate the energy and protein requirements of the sheep at different stages of growth and maturity and can estimate a least-cost series of feeds that meet these requirements. In practice, however, feedlot rations are usually formulated from a limited number of grains (cereals, legumes, oilseeds and meals) and forages (hays, chaffs) and it is relatively easy to formulate a cost-effective ration as follows:

(a) **Choose a basic energy source:**

Cereal grains form the basis of most feedlot rations. Barley grain, oats, triticale or wheat are often available but barley and oats are most commonly used because they are less likely to cause acidosis.

Some legume grains are a rich source of both energy and protein (e.g. lupins, beans, peas) and can be used to replace cereals plus protein sources. The advantage of lupin grain is it contains very little (3%) starch (Mohamed and Rayas-Duarte 1995), which allows them to be used in feedlot rations with a short adaptation period and a reduced risk of acidosis.

The M/D of feedlot rations should be in the range **11–13 MJ/kg DM.**

(b) **Choose a protein source:**

Protein can be an expensive part of the ration so it is essential to choose proteins that provide sufficient rumen degradable protein and the minimum level of Undegraded Dietary Protein (UDP) required. Cheaper sources of protein and even non-protein nitrogen (e.g. urea) can supply the RDP.

Typical protein sources are legume grains, oilseed meals, soybean meal, cottonseed meal, canola meal and, sunflower meal. Of these, very useful sources of UDP include lupins (40% UDP), peas (40% UDP), cottonseed and canola meals (35% UDP) and sunflower meal (32% UDP).

The crude protein content of feedlot rations for lambs should be 15–16% of dry matter with a minimum of 30% of this protein as Undegraded Dietary Protein.

(c) **Include a sheep feedlot premix**

To ensure all vitamin and mineral requirements are met in rapidly growing lambs, include a dedicated lamb feedlot premix. The premix may contain rumen-active growth promotants and acid buffers to reduce acidosis.

(d) **Ensure a minimum of 30–35% NDF and a high proportion of roughage particles >2.0 cm ('scratch factor')**

A minimum of fibre (NDF) is essential to ensure healthy rumen function but a proportion of that fibre should be sufficiently long to stimulate rumination, hence high levels of input of salivary buffers into the rumen fluid to prevent large drops in pH.

Lowest cost feed to meet requirements

Given the simplicity of the feedlot ration, it is easy to calculate which grain it is cheapest to include as follows. Compare grains for cost/MJ ME (as for the drought ration earlier); choose the grain that provides that energy at the lowest cost. Repeat for the protein sources. Commonly, barley and lupins are cost-effective sources of energy and protein in feedlot rations. The high UDP content of lupin grain also contributes to its inclusion.

A typical feedlot ration for sheep is shown in Table 6.7.

An iterative (repeat formulation) approach can be taken to minimise the cost/kg feed that still meets the ME, protein and UDP requirement outlined in the table. Try various combinations of grains and roughage until the cheapest feed is found.

Table 6.7. A typical ration for feedlot lambs for maximal growth and maintenance of health

Feedstuff	Percentage as fed
Barley (coarsely ground)	36
Lupins (medium grind)	35
Cereal chaff	25
Vegetable oil	1.00
Molasses	1.5
Salt	0.5
Feedlot premix (including Bovatec, bicarbonate, coccidiostat)	1.0
Nutrient	
ME (MJ/kg DM)	**11.5**
CP (%)	**15.0**
UDP (%)	**35**
NDF (%)	**30**

Pellets versus mixed rations for feedlot lambs

The choice of feeding systems once a ration has been formulated is important in feedlots because each feeding system has advantages and disadvantages as indicated in Table 6.8.

There are few reports comparing the feeding systems for feedlot lambs. One study (Bowen *et al.* 2006) showed that the cheapest option (separately feeding hay and grain) produced comparable growth rates to feeding pellets or a total mixed ration.

Table 6.8. Comparison of feeding systems for feedlot lambs

Feeding system	Advantages	Disadvantages
Total pelleted ration	• No diet selection so certainty of nutrient intake (including premix) • Ease of transport and handling • Easy to use in self-feeders • Low wastage	• High cost • Low 'scratch factor' • Low effective particle size means rapid acid production and low saliva stimulation
Total mixed ration (chopped roughage plus cracked grains) mixed together	• Good effective particle size and 'scratch factor' (saliva and rumination stimulation)	• High cost of mixing and feeding out equipment • High selectivity so uncertain nutrient intake • Separation of feed components (e.g. premix) • Harder to use self-feeders
Grain and roughage fed separately (cafeteria)	• Cheapest system • Good effective particle size and 'scratch factor' (saliva and rumination stimulation)	• Over-consumption of either grain (acidosis) or roughage (poor growth)

Example 3: Formulate a ration for 50 kg pregnant does in the last 4 weeks of pregnancy

The energy requirement of pregnant does increases rapidly in the last 6 weeks of pregnancy. This is particularly the case for twin-bearing does. To determine a ration to meet the requirements of the does, estimate the maintenance energy requirement as follows:

$$ME_m = 0.12W + 1.5$$

That is, $ME_m = 0.12 \times 50 + 1.5 = 7.5$ MJ/day for maintenance.

For single-bearing does ME required $= ME_m + 4 = 11.5$ MJ/day (Table 6.6)

For twin-bearing does, ME required $= ME_m + 8 = 15.5$ MJ/day (Table 6.6)

Feeding the late-pregnant doe adequately to meet these greatly increased energy demands in late pregnancy is critical to ensure pregnancy toxaemia is avoided. The ration crude protein content for pregnant does should gradually increase over the last 6 weeks of pregnancy over the range 9–12% of DM, with an increase in Undegraded Dietary Protein from 20 to 30%.

So typically a 50 kg doe in late pregnancy should be fed a diet containing 10–11 MJ ME/kg DM with 9–12% CP DM and a UDP of ~20–30%. Twin-bearing does should be fed at double the maintenance energy requirement over the last few weeks of pregnancy.

Example 4: Formulate a ration for 50 kg lactating ewes

Additional ME required for lactation amounts to ~8 MJ/day and 12.0 MJ/day for single and twin-rearing ewes (Table 6.6). Together this means the ME requirements of ewes in early lactation on a typical roughage ration (M/D = 10 MJ/kg DM) can be estimated crudely as:

$$ME_m = 0.12W + 1.5$$

The maintenance requirement for a 50 kg ewe would therefore be 7.5 MJ/day. If she is raising single lamb, add the additional 8 MJ/day = 15.5 MJ/day. If she is raising twin lambs, add the extra 12 MJ/day = 19.5 MJ/day.

Lactation requires increased protein intake and an increase in the Undegraded Dietary Protein level (Table 6.6). To achieve the combination of high-energy intake and high intake of protein with a high UDP content will require a concentrate ration containing grains and protein concentrates. To work out a ration, estimate the maximum likely intake of dry matter/day

(= 3% of W = 1.5 kg DM). For the twin-bearing ewes this means having a ration containing at least 19/1.5 = 12.6 MJ/day and for single-bearing ewes the ration would need to contain 15/1.5 = 10 MJ/day. A ration containing 50% barley grain, 20% lupins and 30% medium quality hay would contain 0.5 × 12.2 MJ/kg (barley) + 0.3 × 13.2 MJ/kg (lupins) + 0.2 × 9.0 MJ/kg (hay) = 11.9 MJ/kg DM. The crude protein content would be 0.5 × 12.0% (barley) + 0.3 × 32.0% (lupins) + 0.2 × 9.0% (hay) = 17.4% CP and ~25% UDP. This ration would meet the requirements of the ewes to maximise milk production and lamb growth. Addition of a vitamin and mineral premix to the ration is advised to ensure no deficiencies, which would limit intake and milk production.

Mineral requirements of sheep and goats

Table 6.9 shows the mineral requirements of several classes of sheep (maintenance, growth, gestation and lactation) and is a selection from the NRC (2007) nutrient requirement tables. The requirements are expressed in grams/day for the macrominerals and milligrams/day for the trace elements. It is common to use 'desirable' concentrations of minerals in feeds as a guide to their ability to meet the requirements of the animals and these are indicated in the table. It should be noted that the efficiency of absorption and utilisation of mineral elements is highly variable and influenced by interactions with other minerals (e.g. Cu/Mo/S; Mg/K; Ca/Zn), the concentration of the mineral in the feed, other components of the feed (e.g. iodine uptake is depressed by goitrogens and cyanogenic compounds in brassicas), and the physiological state of the animal. The absolute levels presented in this table therefore should be taken as a guide only and other components of the ration should be considered.

The mineral requirements of goats are less well studied than those of sheep. The values published by NRC (2007) are similar to those for sheep (Table 6.9) so using the sheep values is likely to be adequate for goats.

Vitamin requirements of sheep and goats

Sheep and goats are rarely vitamin deficient because the synthesis of vitamins B and K by the rumen bacteria provide their daily requirement of these micronutrients. Vitamins A, D and E, while present in variable concentrations in typical forages available to sheep, are stored in the liver in sufficient quantities to provide the animal with its requirements for long periods of time (months) when the feeds may be deficient. Vitamin D can be produced by solar ultraviolet radiation, although for sheep with heavy wool cover at high latitudes, solar input may be inadequate. Feeds that have been exposed to the sun (e.g. hay crops) contain sufficient vitamin D for the animal's requirements. Young,

Table 6.9. Mineral requirements of sheep and goats for maintenance, growth, gestation and lactation.[a] Desirable concentrations of the minerals in feeds are also presented.

Note: the desirable values should be taken as guide only as absorption of minerals is highly variable.

Bodyweight (kg)	Bodyweight gain (g/day)	Ca (g/day)	P (g/day)	Na (g/day)	Cl (g/day)	K (g/day)	Mg (g/day)	S (g/day)	Co (mg/day)	Cu (mg/day)	I (mg/day)	Fe (mg/day)	Mn (mg/day)	Se (mg/day)	Zn (mg/day)
40	0	1.8	1.3	0.5	0.4	3.9	0.7	1.2	0.08	2.7	0.4	6	11	0.03	20
40	250	5	3.7	0.8	0.6	6.3	1.3	2.6	0.29	7.1	0.7	78	26	0.45	45
40	400	7	5.1	1	0.7	7.2	1.7	2.9	0.32	9.7	0.8	121	36	0.7	63
60	0	2.2	1.8	0.7	0.6	5.6	1.1	1.7	0.1	4	0.5	8	16	0.05	30
60	250	5.1	3.8	1	0.8	8.3	1.7	3.3	0.37	8.4	0.9	81	32	0.47	53
60	400	7.8	5.9	1.2	0.9	9.6	2	3.9	0.44	11.1	1.1	124	41	0.72	71
40 (late gestation single lamb)	40	5.8	3.4	0.6	0.4	5.3	1.1	2.2	0.12	6.3	0.6	35	23	0.13	35
40 (mid-lactation single lamb)	47	3.5	3.1	0.6	0.7	5.2	1	1.7	0.22	4.8	0.9	11	13	0.16	30
Desirable concentration in feed		1.5 g/kg	1.3 g/kg	0.7 g/kg	1.0 g/kg	5 g/kg	1.2 g/kg	2.0 g/kg	0.11 mg/kg	5 mg/kg	0.5 mg/kg	40 mg/kg	15 mg/kg	0.05 mg/kg	20 mg/kg

[a]From NRC (2007)

green crops over periods of low solar input (e.g. winter) may be low in vitamin D, and animals maintained indoors for periods greater than 6 months may be at risk. Vitamin A content of feeds is reflected by the content of β-carotene, α-carotene and cryptoxanthins in the forage (indicated by the greenness of the plants), so prolonged periods on dry, brown feeds can result in vitamin A deficiencies, but these are rare.

Risk of vitamin A and vitamin D deficiencies

Vitamin D concentrations are low in green, lush feeds in winter (insufficient irradiation).
Vitamin A concentrations are low in dry, brown feeds in summer (insufficient β-carotene from green feed).

Table 6.10 shows the fat-soluble vitamin requirements (A, D and E) for different classes of sheep and goats.

Daily requirements of the B-group vitamins are not available because of the complexity imposed by microbial synthesis of the B vitamins. Although microbial synthesis is thought to provide all of the requirements of the B vitamins, there are significant problems associated with deficiencies of thiamine (polioencephalomalacia) when either microbial synthesis is inadequate (see 'Polioencephalomalacia in sheep' later) or thiaminases are present in the feed. Synthesis of cobalamin (B12) can also be limited by the availability of the trace element cobalt (Lee 1950).

Table 6.10. Vitamins A, D and E requirements of different classes of sheep and goats

Bodyweight (kg)	Weight gain (g/day)	Vitamin A (RE units/day)[a]	Vitamin D (IU/day)	Vitamin E (IU/day)
40	0	1256	224	212
40	250	4000	494	400
40	400	4000	656	400
60	0	1884	336	318
60	250	6000	606	600
60	400	6000	768	600
40 (late gestation)	71	1820	224	224
40 (mid-lactation)	0	2140	604	224

[a] 1 RE unit = 1 µg all-trans retinol or 5.0 µg all-trans β-carotene or 7.6 µg of other carotenoids
Source: NRC (2007)

Nutrition and wool growth

Protein and energy requirements for wool growth

The heritability of wool growth (the proportion of the phenotype that is genetic) in merino sheep is ~0.40 (Safari *et al.* 2007), leaving the remaining 0.60 affected by the environment. Nutrition is the main non-genetic (environmental) component, so wool growth rate and dry matter intake are closely related (Fig. 6.3).

The form of the relationship is one of diminishing returns, which means the efficiency of wool growth (amount of wool per unit intake) declines as intake increases. This has important implications for grazing sheep because increasing stocking rate reduces intake per animal but may increase the total amount of wool per hectare (because if all the harvestable feed is spread over more animals (less per animal), each animal is more efficient at converting the feed to wool. At low intakes of feed, the wool production per animal is lower because the diameter of each fibre is lower and the rate of elongation (fibre length growth) is also lower. Mean fibre diameter is the main determinant of wool price (lower diameter is usually more valuable), so increasing stocking rate not only increases wool production per hectare, but it also increases wool value per kilogram.

> Wool growth rate increases with increased dry matter intake, but the response is one of diminishing returns so the efficiency of wool growth decreases with increased intake.

So does wool growth respond to energy intake or protein intake? This question proves more difficult than first appreciated because, in ruminants, energy in feed is

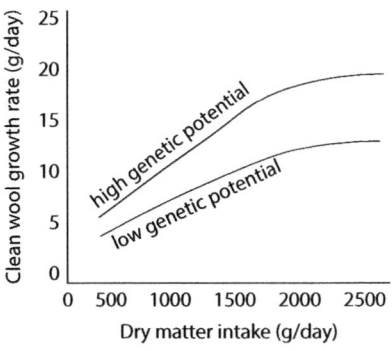

Fig. 6.3. Relationship between clean wool growth rate and dry matter intake for sheep with a high or low genetic wool growth potential. The diminishing returns relationship means wool growth is more efficient at lower dry matter intakes, so for a given amount of feed it is more efficient to allocate the feed to a larger number of sheep, each fed a smaller amount. Sheep or goats with a higher genetic potential for fibre growth respond more to additional feed than low animals with a low genetic potential.

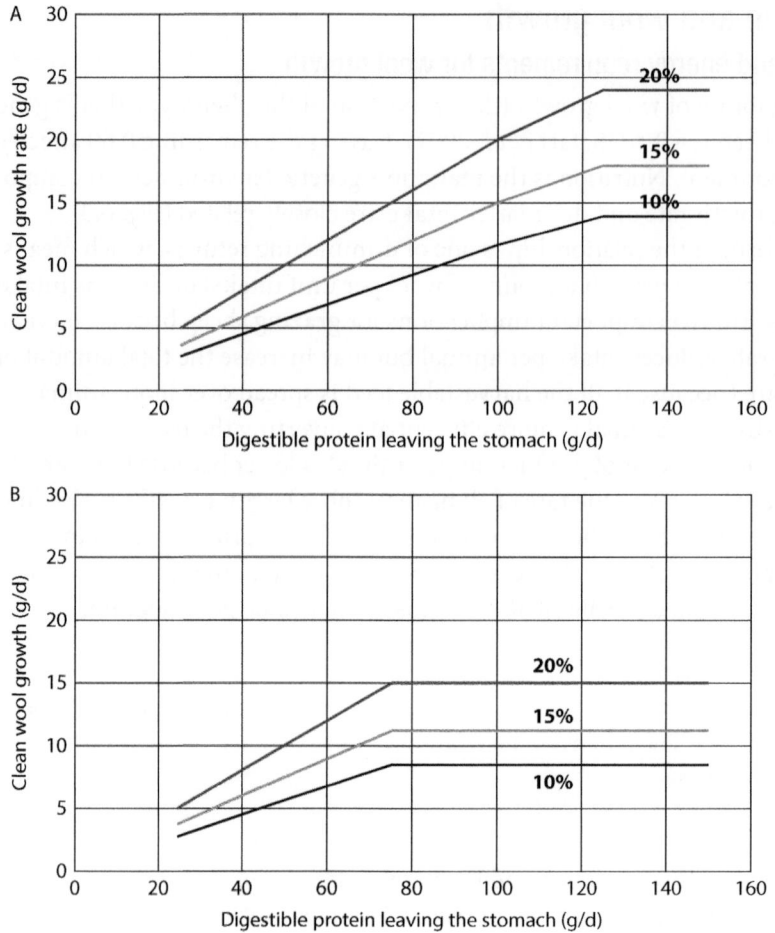

Fig. 6.4. Wool growth responses to increasing digestible protein leaving the stomach (g/day) for sheep differing in efficiency of wool growth and fed a diet containing 10 MJ of Metabolisable Energy/day (A) or 6 MJ/day (B). Efficiencies are defined as clean fleece weights relative to mature bodyweight (e.g. 5 kg clean wool/50 kg mature weight = 10% efficiency). Sources: Adapted from Hynd and Allden (1985); and Hynd and Masters (2002).

converted into microbial protein, which provides amino acids to the host animal. However, it is now clear that wool growth responds to both energy and protein, depending on which is limiting. In other words, there is an optimum ratio of protein to energy to maximise wool growth (Fig. 6.4)

Figure 6.4 shows that when there is sufficient energy being consumed, wool growth responds mainly to the amount of digestible protein leaving the stomachs, and that genetically efficient wool-producing sheep respond to this additional protein more than low-producers. However, if total energy intake is limited (Fig. 6.4B), maximum wool growth is reduced and there is a limited response to the extra protein. Under these conditions, wool growth at high levels of protein leaving the stomachs will respond to additional energy supply. The reason for these responses is that wool growth requires an optimum ratio of protein to energy. The optimum ratio is ~12 g protein leaving the stomachs for each megajoule of ME (Black *et al.* 1973; Kempton 1979). So, in Fig. 6.4B, the lower ME intake means maximum wool growth is met at a significantly lower DPLS value (~80 g/day compared with 120 g/day when energy is not limiting), hence a lower wool growth rate.

DPLS (digestible protein leaving the stomachs) is a combination of microbial protein (limited largely by the fermentable ME available) and UDP. Together these proteins supply the amino acids required to synthesise wool proteins. Given that wool keratin is very high in the sulphur amino acid, cyst(e)ine, it is not surprising that this amino acid, and its precursor sulphur amino acid, methionine, are the first-limiting amino acids for wool growth (see next section).

Wool growth responses to protein composition

The sulphur amino acid, cysteine, is required as major component (14%) of wool keratin, and wool growth responds to additional cysteine entering the intestines up to ~3 g/day. The impact of the amino acid composition of proteins leaving the stomachs is shown in Fig. 6.5.

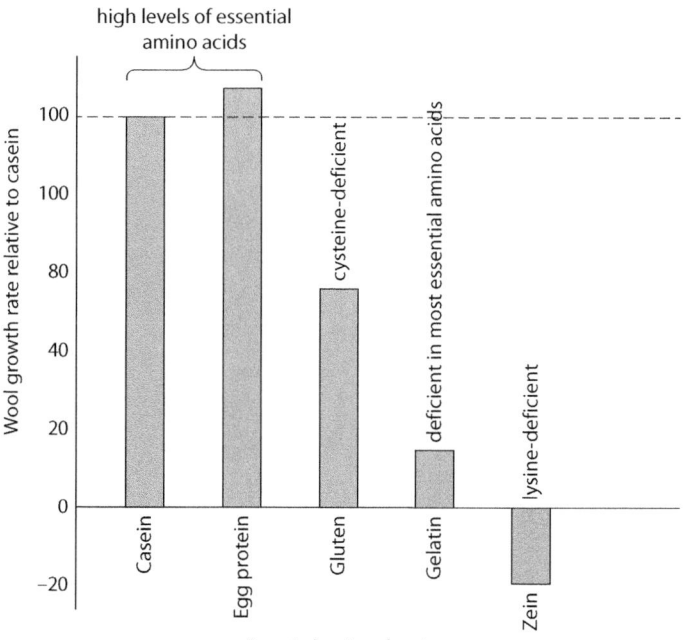

Fig. 6.5. Wool growth responds to the essential amino acid content (and particularly the sulphur amino acid content) of protein leaving the abomasum, relative to the responses to the high-quality milk protein, casein. Source: Based on data from Reis and Colebrook (1972).

Proteins entering the intestines with an amino acid profile similar to casein (high in cysteine and the other essential amino acids), stimulate wool growth. Any deficiency in the essential amino acid profile of a protein results in a lower wool growth response (e.g. gluten is deficient in cysteine and only produces 60% of the wool response to casein). Gelatin, which is deficient in several essential amino acids, produces only 10% of the casein response and zein, which is completely deficient in lysine, actually reduces wool growth. Most of the protein entering the intestines is microbial protein and is reasonably balanced for wool growth. Nevertheless, significant wool growth responses can be achieved using proteins and amino acids that escape ruminal degradation. The author stimulated wool growth rate by 50% in sheep fed a moderate-quality roughage diet (oaten plus lucerne hay) by adding a rumen-protected methionine product. Methionine is a source of cysteine, the first-limiting amino acid and also produces polyamines, which are vital for hair and wool growth (Hynd and Nancarrow 1996).

> Wool and hair growth in animals is limited mainly by the supply of sulphur-containing amino acids (cysteine and methionine) from the intestines.

Minerals and wool growth

Minerals can influence wool growth by affecting feed intake and microbial activity (hence the amount of microbial protein flowing from the abomasum), or by directly affecting the wool growth processes in the wool follicle (Table 6.11).

Table 6.11. Mineral elements affecting wool growth rate and quality

Factor affected	Mineral deficiencies
Feed intake	Sodium Potassium Sulphur Phosphorus Magnesium Cobalt Zinc
Rumen microbial activity	Sulphur Sodium Potassium Cobalt
Metabolism and follicle function	Zinc Copper Selenium Iodine Cobalt

Source: After Hynd and Masters (2002)

Copper deficiencies result in weakened wool, which also lacks crimp ('steely wool'). The weakness is generated by incomplete keratinisation or hardening of the fibre because copper is essential for the oxidation of thiol (S) groups to form the disulphide bonds between cysteine molecules in keratin (see Hynd 2000). Copper deficiency also causes depigmentation of black wool fibres because it is a cofactor in the enzyme tyrosinase required for melanin synthesis. Zinc is involved in wool growth rate and keratinisation, but the mechanisms are not known (White *et al.* 1994). The role of selenium in wool growth is unclear but may relate to protection for oxidative stress and repression of gene expression in the follicle. Iodine is a component of thyroxine and thyroid hormones play an important role in wool and hair growth (Hynd 1994).

Vitamins and wool growth

Although several of the vitamins play important roles in wool growth (Hynd 2000), microbial synthesis of B-group vitamins means vitamin deficiency effects on wool growth are rare.

> Forages that maximise feed dry matter intake, and that have a high content of protein (e.g. legumes), the essential mineral elements and some protein, (containing sulphur amino acids) that escapes ruminal degradation, will maximise wool growth rate and quality.

Common nutritional and metabolic diseases of sheep and goats
Copper deficiency and toxicity in sheep/goats

Copper is a trace element required at very low levels in the diet of sheep (~5 mg/kg DM). Deficiencies are common in areas with sandy, highly leached soils. The widespread use of copper fertilisers has reduced the frequency of copper deficiency in many countries. Deficiencies of copper, however, are complex and involve interactions between soil pH, sulphur levels in the feed and water, molybdenum levels in soil, plant genotype and season. Predisposing factors to copper deficiency are waterlogged soils containing high levels of organic matter, high soil pH (liming), high levels of sulphur and molybdenum, grass-dominant pastures (legumes are higher in copper) and a winter/spring incidence.

Functions of copper in metabolism and clinical signs of copper deficiency

Copper is a component of several key enzymes involved in a multitude of essential metabolic functions (Table 6.12), so a deficiency of copper in the diet of sheep results in many adverse clinical signs and ultimately death.

Table 6.12. Functions of copper in sheep and goat metabolism and clinical signs of copper deficiencies

Function	Enzyme	Signs
Blood formation (haematopoiesis)	Ferroxidase I (caeruloplasmin)	• Anaemia (low haematocrit, low haemoglobin)
Connective tissue synthesis/bone formation	Lysyl oxidase	• Bone fractures • Abnormal development of distal growth plates in bones • Bowing of the forelegs • Osteoporosis • Weak arterial walls
Pigmentation (hair)	Tyrosinase	• Depigmentation of wool
Keratinisation	Enzyme involved in crosslinking of disulphide bonds in keratin	• Poor hoof keratinisation • Low staple strength, wool lacking crimp ('steely' wool)
Immune function	Superoxide dismutase	• Increased infections • Oxidative damage
Nerve function (myelin synthesis)	Cytochrome oxidase	• Hypothermia • Enzootic ataxia (swayback) • Paralysis at birth • Muscle tremors • Teeth grinding • Pupil dilation • Demyelination of the nerve fibres in central and peripheral nerves
Catecholamine production	Dopamine-β-hydroxylase	• Neurological defects, hypothermia, hypotension

Source: Church (1988)

Copper is a key component of the enzyme ferroxidase 1, also known as caeruloplasmin, which contains 95% of the plasma copper. Ferroxidase 1 catalyses the oxidation of the ferrous to the ferric ionic form. The latter is required for transport attached to the protein transferrin, which delivers iron for haemoglobin synthesis in the bone marrow (Fig. 6.6). Depressed haem synthesis results in a reduction in haemoglobin levels in the erythrocytes.

Deficient connective tissue synthesis (poor crosslinking of collagen and elastin) impairs normal bone synthesis, so bones are fragile and easily broken. The forelegs of copper-deficient sheep are bowed because there is an imbalance between osteoblast (bone synthesis) and osteoclast (bone resorption) processes. Failure of myelin synthesis in nerves leads to a condition in lambs known as 'swayback' or enzootic ataxia: a condition characterised by hind limb paresis (weakness).

Copper toxicity
Copper is a toxic element at relatively low levels and, for sheep, levels as low as 10 ppm over extended periods of time can produce toxicity. Poor homeostatic controls

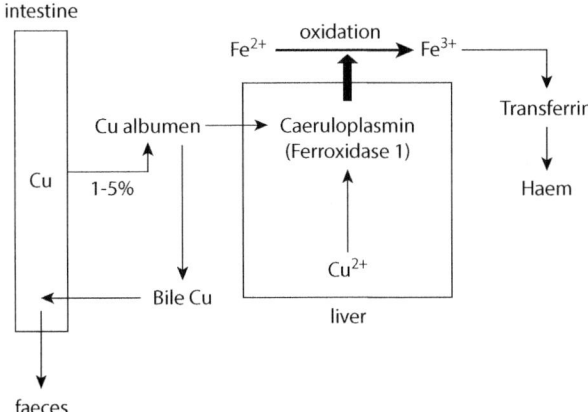

Fig. 6.6. Absorption and excretion routes of copper and its involvement in haemoglobin synthesis.

over copper absorption and excretion, and subsequent storage of excess copper in the liver mean sheep are susceptible to copper toxicity. British breeds of sheep are particularly prone. Consumption of the weed known as potato weed (*Heliotropeum europaeum*) is associated with liver damage caused by hepatotoxic pyrrolizidine alkaloids. These alkaloids cause photosensitisation due to the failure of the liver to breakdown phylloerythrin: a breakdown product of chlorophyll. Phylloerythrin accumulates in the skin and on the exposed extremities such as the ears and face, and makes the skin in these regions particularly sensitive to sunlight. The damaged liver cells also take up copper in abnormally large amounts. Stresses such as transport or malnutrition cause sudden release of the stored copper, which then destroys red blood cells, producing haemolytic anaemia accompanied by jaundice, haemoglobinuria (red urine) and death.

Cobalt deficiency in sheep

Cobalt is a trace element required at a level of only 0.11 mg/kg feed dry matter (Table 6.9). The only known role for cobalt in animal nutrition is as a vital component of vitamin B12 (see Fig. 6.7), although the rumen microorganisms require cobalt for normal fermentation; indeed it is the rumen microbes that synthesis the B12.

Cobalt deficiency is recognised in New Zealand as 'bush sickness', in Great Britain, as 'pining', in South Australia as 'coast disease', in Western Australia as 'wasting disease', in Florida as 'salt disease' and in Michigan as 'Grand Traverse' disease (NRC 2007). Table 6.13 summarises the main factors contributing to cobalt deficiency.

Cobalamin plays several essential roles in animal metabolism and a deficiency of cobalt therefore leads to several adverse clinical signs. B12 is a component of the

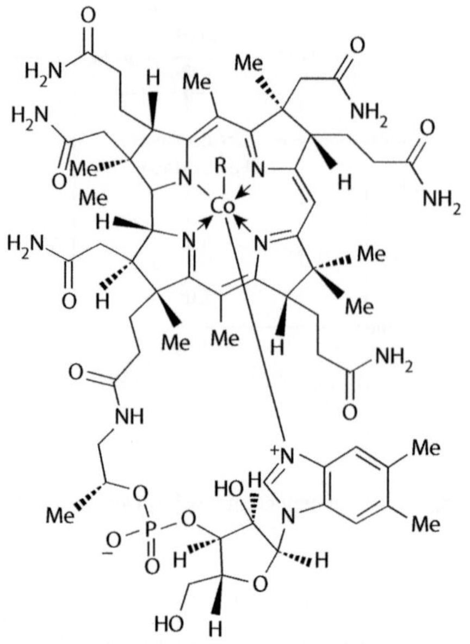

R = CN, OH, Me, 5′-deoxyadenosyl

Fig. 6.7. The vitamin B12 molecule. The R-group determines the form of B12 as cyanocobalamin, adenosylcobalamin, methylcobalamin or hydroxycobalamin. Vitamin B12 contains 4.5% cobalt.

Table 6.13. Predisposing factors to cobalt deficiency

Predisposing factors to Co deficiency
Deep sandy highly leached soils, high limestone content
Grasses not legumes
Alkali soils
Waterlogged soils
Rapid plant growth
Rapidly growing lambs
Grain feeding

enzyme methylmalonyl CoA mutase, the enzyme that catalyses the entry of propionate into the TCA cycle. This is particularly important in ruminant animals given the importance of propionate as an energy source from fermentation and as a major source of glucose (50% of the glucose synthesised by ruminants comes for propionate). Ill-thrift or poor growth of animals occur partly due to anorexia and partly due to poor extraction of energy from the feed. B12 is also a component of the enzyme methionine synthase, which is vital for the synthesis of methionine and the transfer of methyl groups to other compounds (Fig. 6.8).

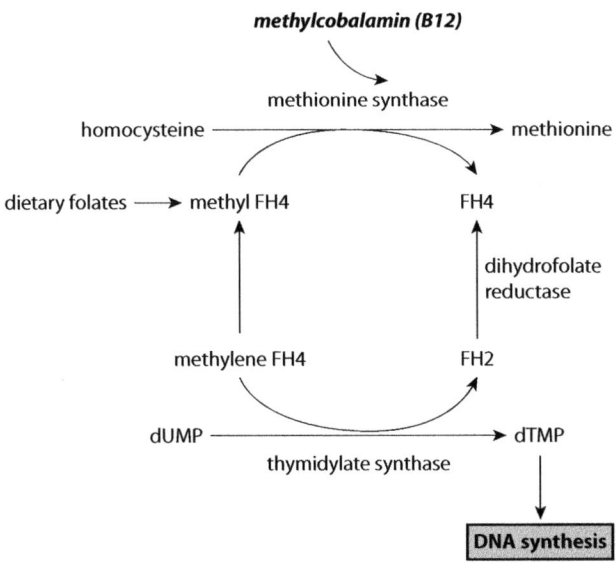

Fig. 6.8. Role of vitamin B12 and folate in DNA synthesis. B12 is a component of the enzyme, methionine synthase, which catalyses the demethylation of methyl tetrahydrofolate methyl FH4. FH2 = dihydrofolate; dUMP = deoxyuridine monophosphate; dTMP = thymidine monophosphate.

The methyl group transfer enabled by B12 activity in methionine synthase is essential for the activity of thymidylate synthase, which is required for DNA synthesis. Failure of DNA synthesis during red blood cell formation leads to the erythrocytes becoming enlarged (megaloblastic) and fewer of them being formed. Methionine synthesis from homocysteine is also critical for normal protein synthesis.

Deficiencies of cobalt/B12 result in anorexia, poor growth, ocular discharge, photosensitivity and increased susceptibility to parasitic and bacterial infections. Megaloblastic anaemia and cardiovascular lesions also occur.

Polioencephalomalacia in sheep

Polioencephalomalacia (PEM), or cerebrocortical necrosis, is a disease of sheep and cattle associated with a deficiency of the B-group vitamin, thiamine, but the aetiology of PEM is complex and seems to involve an interaction between thiamine status, levels of sulphur in the diet, level of thiamine production by ruminal microbes and destruction of thiamine by thiaminases in the rumen (Amat *et al.* 2013). High levels of sulphur in the diet and water are well known to induce clinical signs of PEM (e.g. Gooneratne *et al.* 1989; McAllister *et al.* 1992). There is currently some debate as to whether PEM is entirely a disease of sulphur toxicity or whether there is also sulphur-independent PEM associated only with thiamine deficiency induced by ruminal changes. Sheep given high-sulphur intakes have increased levels of thiamine in the brain, suggesting that high S is inducing an increased metabolic demand for thiamine. Sheep incapable of meeting this elevated thiamine demand may be susceptible to development and progression of the disease. At present, the evidence suggests that high levels of dietary S reduce the thiamine status of sheep by reducing the ruminal synthesis of thiamine (Alves de Oliveira *et al.* 1997), increase the ruminal breakdown of thiamine (Olkowski *et al.* 1992) and increase the metabolic demand for thiamine in the brain. That thiamine is the key factor in S-induced PEM is confirmed by a reduction in the incidence of PEM induced in lambs by high S in the diet, when supplemented with thiamine (Rousseaux *et al.* 1991). High levels of sulphur in plants would result if the protein content of the plant was high and the levels of sulphur-containing amino acids in the proteins were high. Some weeds contain high levels of sulphate and the brassicas (kale, turnips, rape, mustards) and oilseed meals synthesise sulphur-containing compounds in their normal metabolism.

In adult ruminants, thiamine is synthesised by some of the ruminal microbes so it is assumed that a deficiency of thiamine might reflect a depression of the growth or activity of these thiamine-synthesising bacteria. Alternatively, other bacteria that produce thiaminase I, such as *Bacillus thiaminolyticus* and *Clostridium sporogenes*, or thiaminase II by *Bacillus aneurinolyticus*, destroy the ruminal thiamine. It is suggested that a high incidence of PEM in feedlot sheep may reflect an effect of grain feeding on the proliferation and activity of such organisms.

PEM is characterised by necrosis of the neurones of the cerebrocortical grey matter followed by malacia (softening of the neuronal tissue). Clinical signs of the disease are agitation and anxiety, muscle twitching, head pressing, seizures, paddling and a characteristic head position known as stargazing or opisthotonos followed by death. Post mortem of the brain shows a characteristic auto-fluorescence of the necrotic tissue under ultraviolet light.

Urolithiasis (waterbelly) in goats and sheep

The formation of uroliths, also known as urinary stones or urinary calculi, is common in small ruminants. They are most common in bucks or castrate males and in certain breeds, particularly pygmy goats. The uroliths are varying-sized concretions of various mineral salts containing silica, magnesium ammonium phosphate ('struvite'), calcium carbonate and calcium oxalate. The type of stones formed depends on the soil type (e.g. high silicate levels), whether the animals are fed in a feedlot (magnesium ammonium phosphate or struvite) or if the diet contains high levels of oxalates (e.g. *Oxalis pes-caprae*). The mineral concretions form in the kidneys, resulting in obstruction of the urethra. Typically they are associated with feeding unbalanced macrominerals in high-grain, low-roughage rations. Phosphate and magnesium levels are usually high and calcium levels low (producing a Ca:P ratio significantly lower than the recommended ratio of 2:1). Cereal grains are low in Ca and high in P, contributing to the formation of uroliths. The absence of roughage and 'scratch factor' in the diet also contributes to urolith formation because less saliva is produced. Saliva is a major route of excretion of excess phosphorus so excretion of P through the kidneys increases. Excess P in the kidney tubules can precipitate and form the uroliths, which block the renal pelvis and urethra. Breed differences in susceptibility to urolithiasis probably relate to different relative reliance on kidney P excretion versus faecal P excretion. Other risk factors include a reduction in water or salt intake, urinary tract infections and high urinary pH.

Typically the concretions block the end of the urethral process (the 'vermiform' appendage at the end of the penis) or in the sigmoid flexure of the penis. Typical clinical signs of urolithiasis are:

- blood in urine
- straining to urinate
- painful urination and kicking at belly
- dribbling urine
- inappetance
- 'waterbelly' (ruptured bladder and abdominal swelling)
- tachycardia (elevated heart rate)
- tachypnoea (elevated respiration rate)
- elevated blood urea nitrogen
- elevated creatinine (kidney failure).

Although surgery is the most frequent treatment for urolithiasis, prevention is better than cure. Ensuring the diet has a Ca:P ratio of ~2:1, contains roughage sufficient in length to ensure rumination and salivation through 'scratch factor' effect, and ensuring the dietary cation/anion difference is sufficiently low to provide required calcium (see Fig. 8.9; Jones *et al.* 2009). Other preventative measures include acidifying urine by adding sodium chloride to the diet at 1–4% to stimulate thirst.

References

Abdel Aziz M (2010) Present status of the world goat populations and their productivity. *Lohman Information* **45**, 42–52.

Alves de Oliveira L, Jean-Blain C, Komisarczuk-Bony S, Durix A, Durier C (1997) Microbial thiamin metabolism in the rumen simulating fermenter (RUSITEC): the effect of acidogenic conditions, a high sulfur level and added thiamin. *British Journal of Nutrition* **78**, 599–613.

Amat S, McKinnon JJ, Olkowski AA, Penner GB, Simko E, Shand PJ, *et al.* (2013) Understanding the role of sulphur-thiamine interaction in the pathogenesis of sulphur-induced polioencephalomalacia in beef cattle. *Research in Veterinary Science* **95**, 1081–1087. doi:10.1016/j.rvsc.2013.07.024

Bhattacharya AN (1980) Research on goat nutrition and management in Mediterranean Middle East and adjacent Arab countries. *Journal of Dairy Science* **63**, 1681–1700. doi:10.3168/jds.S0022-0302(80)83130-4

Black JL, Robards GE, Thomas R (1973) Effects of protein and energy intakes on the wool growth of merino wethers. *Australian Journal of Agricultural Research* **24**, 399–412. doi:10.1071/AR9730399

Bowen MK, Ryan MP, Jordan DJ, Beretta V, Kirby RM, Stockman C, *et al.* (2006) Improving sheep feedlot management. *International Journal of Sheep and Wool Science* **54**, 134–141.

Boyazoglu J, Hatximinaouglou I, Morand-Fehr P (2005) The role of the goat in society: past, present and perspectives for the future. *Small Ruminant Research* **60**, 13–23.

Church DC (1988) *The Ruminant Animal: Digestive Physiology and Nutrition.* Prentice Hall, Upper Saddle River, NJ, USA.

Counotte GHM, van't Klooster AT (1979) An analysis of the buffer system in the rumen of dairy cattle. *Journal of Animal Science* **49**, 1536–1544. doi:10.2527/jas1979.4961536x

CSIRO (2007) *Nutrient Requirements of Domesticated Ruminants.* CSIRO Publishing, Melbourne.

Devendra C (1986) Feeding systems and nutrition of goats and sheep in the tropics. In *Proceedings of the Workshop on the Improvement of Small Ruminants in Eastern and Southern Africa.* 18–22 August, Nairobi Kenya (Eds KO Adeniji and JA Kategeli) pp. 91–110.

FAO (2014) *FAO Statistical Yearbook Europe and Central Asia Food and Agriculture.* FAO, Budapest, Hungary, <http://www.fao.org/publications/sofa/2013/en/>.

Gooneratne SR, Olkowski AA, Christensen DA (1989) Sulphur-induced polioencephalomalacia in sheep: some biochemical changes. *Canadian Journal of Veterinary Research* **53**, 462–467.

Hoffmann RR (1985) Digestive physiology of deer – their morphophysiological specialisation and adaptation. In *International Conference on the Biology of Deer Production.* 13–18 February, Dunedin. (Ed. PF Fennessy and KR Drew) pp. 393–407. Royal Society of New Zealand, Wellington, New Zealand.

Hynd PI (1994) Determinants of the length and diameter of wool fibres. 2. Comparison of sheep differing in thyroid hormone status. *Australian Journal of Agricultural Research* **45**, 1149–1157. doi:10.1071/AR9941149

Hynd PI (2000) The nutritional biochemistry of wool and hair follicles. *Animal Science* **70**, 181–195. doi:10.1017/S1357729800054655

Hynd PI, Allden WG (1985) Rumen fermentation pattern, postruminal protein flow and wool growth rate of sheep on a high-barley diet. *Australian Journal of Agricultural Research* **36**, 451–460. doi:10.1071/AR9850451

Hynd PI, Nancarrow MJ (1996) Inhibition of polyamine synthesis alters hair follicle function and fiber composition. *The Journal of Investigative Dermatology* **106**, 249–253. doi:10.1111/1523-1747.ep12340634

Hynd PI, Masters DG (2002) Nutrition and wool growth. In *Sheep Nutrition.* (Eds M Freer and H Dove) pp. 165–187. CSIRO Publishing, Melbourne and CABI Publishing, Wallingford, UK.

Jones ML, Streeter RN, Goad CL (2009) Use of dietary cation anion difference for control of urolithiasis risk factors in goats. *American Journal of Veterinary Research* **70**, 149–155. doi:10.2460/ajvr.70.1.149

Kararli TT (1995) Review article: comparison of the gastrointestinal anatomy, physiology and biochemistry of humans and commonly used laboratory animals. *Biopharmaceutics & Drug Disposition* **16**, 351–380. doi:10.1002/bdd.2510160502

Kay RNB (1960) The rate of flow and composition of various salivary secretions in sheep and calves. *The Journal of Physiology* **150**, 515–537. doi:10.1113/jphysiol.1960.sp006402

Kempton T (1979) Protein to energy ratio of absorbed nutrients in relation to wool growth. In *Physiological and Environmental Limitations to Wool Growth* (Eds Black JL and Reis PJ) pp. 208–222. University of New England Publishing Unit, Armidale.

Lee HJ (1950) The occurrence and correction of cobalt and copper deficiency affecting sheep in South Australia. *Australian Veterinary Journal* **26**, 152–159.

McAllister MM, Gould DH, Hamar DW (1992) Sulphide-induced polioencephalomalacia in lambs. *Journal of Comparative Pathology* **106**, 267–278. doi:10.1016/0021-9975(92)90055-Y

Mohamed AA, Rayas-Duarte P (1995) Composition of *Lupinus albus*. *Cereal Chemistry* **72**, 643–647.

NRC (2007) *Nutrient Requirements of Small Ruminants: Sheep, Goats, Cervids and New World Camelids.* National Research Council, The National Academies Press, Washington, DC, USA.

Olkowski AA, Gooneratne SR, Rouseaux CG, Christensen CA (1992) Role of thiamine status in sulphur-induced polioencephalomalacia. *Research in Veterinary Science* **52**, 78–85. doi:10.1016/0034-5288(92)90062-7

Reis PJ, Colebrook WF (1972) The utilisation of abomasal supplements of proteins and amino acids by sheep with a special reference to wool growth. *Australian Journal of Biological Sciences* **25**, 1057–1072. doi:10.1071/BI9721057

Rousseaux CG, Olkowski AA, Chauvet A, Gooneratne SR, Christenson DA (1991) Ovine polioencephalomalacia associated with dietary sulphur intake. *Transboundary and Emerging Diseases* **38**, 229–239.

Safari E, Fogarty NM, Gilmour AR, Atkins FD, Mortimer SI, Swan AA, *et al.* (2007) Genetic correlations among and between wool growth and reproductions in Merino sheep. *Journal of Breeding and Genetics* **124**, 65–72.

White CL, Martin GB, Hynd PI, Chapman R (1994) The effect of zinc deficiency on wool growth and skin and wool follicle histology of male Merino lambs. *British Journal of Nutrition* **71**, 425–435. doi:10.1079/BJN19940149

7

Beef cattle nutrition

Key points

- Beef cattle are large ruminants with a total digestive tract content in adults of more than 250 L, 100–180 L of which are present in the reticulo-rumen.
- Cattle produce large quantities of saliva (50–150 L/day) depending on diet (more on roughage diets). Saliva lubricates ingesta, buffers the rumen contents, recycles nutrients to the ruminal microbes, and provides antibodies, enzymes and anti-foaming agents to the rumen.
- The energy requirements of cattle are expressed as Metabolisable Energy or Net Energy, and the protein requirements are expressed in terms of Undegraded Intake Protein and Degradable Intake Protein.
- Feedlotting of cattle is used to finish the cattle to market specifications when pasture feed cannot consistently meet the nutritional requirements; create a more uniform meat product for the market; access a wider range of markets; or reduce stocking pressure on dry pastures without needing to sell animals in poor condition.
- Acute lactic acidosis and subclinical ruminal acidosis are significant problems in cattle on high-grain diets. Acute acidosis is characterised by accumulation of lactic acid in the rumen and in the systemic circulation due to rapid lactate production by lactate producing microbes, with inadequate development of lactate-utilising organisms. Subacute acidosis occurs when volatile fatty acids accumulate in the ruminal fluid.
- Hypomagnesaemic tetany, or grass tetany, occurs in ~1–3% of cattle grazing pastures in most countries. It is characterised by low blood magnesium, particularly in lactating cows but the disorder is complex, involving interactions between pasture species, ambient temperature, pasture growth rate, potassium fertilisation of pastures, nitrogen intake, ammonia levels in the rumen, ruminal pH and calcium status.

Introduction

Domesticated beef cattle (*Bos taurus* and *Bos indicus*) numbered almost 1000 million head worldwide in 2017, with India (30%), Brazil (22%), China (10%), USA (9%), EU (9%), Argentina (5.4%) and Australia (2.8%) having almost 90% of the total. Per capita annual consumption of beef ranges from 1 kg in Bangladesh to 54 kg in Argentina (Santich 2014). Beef consumption has traditionally been associated with the developed countries where high levels of pig and poultry meat are also consumed. In these countries there has been a downward trend in beef consumption and a concomitant increase in chicken meat consumption from the 1960s to the present, as chicken meat production intensified and became highly efficient, inexpensive and consistent in quality (Santich 2014). Current drivers of beef consumption in many developed economies include concerns over saturated fat intake, animal welfare, greenhouse gas contributions, and deforestation. However, demand for beef in many developing economies is increasing as per capita income increases and the nutritional status of people improves (Santich 2014). Beef is an important source of protein, energy, essential nutrients such as B-group vitamins, and essential minerals, particularly iron and zinc. The iron in beef is present in haem and, as such, is absorbed efficiently. The omega-6 to omega-3 ratio of fats in pasture-fed beef is lower than that of grain-fed beef, and pasture-fed beef contains less total fatty acids. It is generally agreed by nutritionists that beef plays an important role in a balanced diet. The fact that such a nutrient-dense food can be produced from nutrient-sparse plants accounts for the success of these large ruminants worldwide.

Digestive anatomy and physiology of beef cattle
Anatomy of the digestive tract of beef cattle

The dental formula for permanent teeth in cattle is:

	Incisors	Premolars	Molars
Upper	0	3	3
Lower	4	3	3

The age of cattle can be approximated by the number of permanent incisors that have erupted on the lower jaw as shown in Fig. 7.1.

Cattle have a broad muzzle without the cleft upper lip of sheep and goats, and are thus less able to select finely within available feeds. Instead they use their large prehensile tongue to grasp plant parts and draw them into the mouth where the material is shorn off by the incisors against the dental pad. Jerking movements of the head remove the material. The selective ability of cattle is less than that of sheep (see Table 6.1 and Grant *et al.* (1985). Even when grazing the same pastures, sheep and

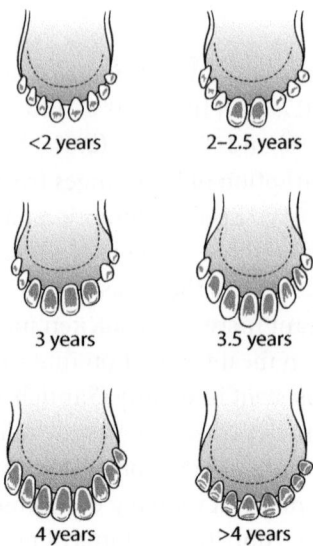

Fig. 7.1. Ageing beef cattle using the number of permanent incisors erupted on the bottom jaw.

cattle differ in the major dietary components (Grant *et al.* 1985). Sheep and cattle graze at different levels in the sward (sheep graze at a deeper level), sheep select finer components and cattle tend to graze taller, more fibrous components (Grant *et al.* 1985). Although it would appear that sheep would therefore have an advantage over cattle in terms of nutrient availability, this does not appear to be the case because cattle digest the lower quality, higher fibre components to a greater extent.

Adult cattle produce between 50 and 150 L of saliva daily. The amount of saliva added to the chewed feed depends on the moisture content of the feed and the feed particle size (Maekawa *et al.* 2002). Concentrates, with small particle size have less saliva added per unit dry matter than forages with large particles because the forages require more chewing to reduce the particle size sufficiently for swallowing. Concentrates are therefore swallowed drier than forages. Table 7.1 shows the amount of saliva/g dry matter produced by cattle eating concentrates, forages or silage.

Table 7.1. Effect of diet type on the rate of saliva addition to the feed, and the rate of saliva production when the cows were not feeding

Diet type	Rate of saliva production during eating (mL/g feed DM)	Resting saliva rate (mL/min)
Silage	4.43	100 mL/min = 144 L/day
Total mixed rations	3.18	
Concentrates	1.19	

Source: Maekawa *et al.* (2002)

Although these estimates are for dairy cows, it is likely that similar values and similar effects of diet occur in beef cattle.

Functions of saliva in beef cattle

Saliva plays several important roles in beef cattle:

1. **Lubrication:** Saliva lubricates the feed bolus to allow swallowing. The more chewing and the tougher and drier the feed material, the more saliva is produced during eating.
2. **Enzyme activity:** Cattle saliva does not contain amylase but does contain lipases, which digest fats in milk-fed calves.
3. **Buffering:** Salivary bicarbonates and phosphates buffer the ruminal pH, so it remains between 5.5 and 7.2.
4. **Recycling of nutrients for microbes:** Proteins entering the rumen are deaminated and the ammonia is either taken up by the rumen microbes or is absorbed into the portal vein. In the liver, the ammonia is converted to urea and either excreted at the kidneys or enters the saliva, which is swallowed and the urea-nitrogen is available again to the microbes as ammonia. Other recycled nutrients include sodium.
5. **Antifoaming agents:** Saliva contains mucus, which has anti-foaming activity thereby reducing the incidence of bloat.
6. **Antibodies:** Saliva contains large quantities of immunoglobulin-A, which may influence microbial activity in some way.

The gastrointestinal tract of a cow is common to other ruminants and shown in Fig. 3.1. The characteristics of the digestive tract of adult cattle are shown in Table 7.2.

Details of digestion and nutrient availability in ruminants are outlined in Chapter 3.

Table 7.2. Typical values for capacities, lengths, digesta pH and epithelium type in the gastrointestinal tract of adult beef cattle (500 kg)

Segment of tract	Capacity (L) or length (× body length)	pH	Epithelium type
Reticulum	16–20 L	5.5–7.2	Stratified, squamous, keratinised
Rumen	100–180 L	5.5–7.2	
Omasum	8 L	5.5–7.2	
Abomasum	27 L	2–3	Simple columnar glandular mucosa
Duodenum	25× body length	5.5–6.0	Simple columnar, villi and crypts, Brunner's glands, mucus cells, endocrine cells
Jejunum		6.5–7.0	Simple columnar, villi and crypts
Ileum		6.5–7.4	Simple columnar, villi and crypts
Caecum	20 L	5.5–6.5	Simple columnar to low cuboidal, crypts, goblet cells
Colon	10 L	6.5–6.7	

The energy, protein, vitamin and mineral requirements of beef cattle

Energy and protein requirements of cattle

The theory underpinning the estimation of the energy and protein requirements of beef cattle is outlined in Chapter 4. These principles are applied in many computer programs designed to estimate the nutrient requirements of beef cattle, such as CNCPS (Fox *et al.* 2004), Tedeschi *et al.* (2016) and GrazFeed (Freer *et al.* 1997). These programs allow the user to specify many classes of cattle, their production levels and the feeds available to match feed supply with the animal's requirements. For many situations, however, the nutritionist in the field often relies on less-sophisticated means of estimating the requirements. In some instances, it is useful to be able to estimate the energy and protein requirements of different animals to broadly align them to the feeds available. A simple method of formulating and checking rations for cattle is outlined as follows:

1. Estimate the maintenance energy requirement (from live weight).
2. Use multipliers of maintenance for pregnant, lactating or rapidly growing animals.
3. From the total ME requirement and the ME density of feed, estimate the required feed intake in DM.
4. Check that the animal can consume this quantity of feed (from the NDF content of the feed; DM intake (% of bodyweight) = 120/NDF%; Fig. 4.6).
5. Check that the protein content of the feed meets the minimum requirement and check if undegraded protein is required.

A simple series of calculations can be used to estimate the energy and protein requirements of beef cattle for maintenance, lactation and growth, as shown in Table 7.3.

The values in Table 7.3 can be readily remembered so ration formulation 'on the run' is relatively easy.

Table 7.3. Simple calculation of the energy and protein requirements of beef cattle

Status	ME (MJ/day)	CP (%DM)	Undegraded Intake Protein (UIP) (% of CP)	Max. DM intake (%W)
Maintenance (M)[a]	0.1W + 8	7–10	<20	1.8 (low-quality) 2.0 (medium-quality) 2.5 (high-quality)
Rapid growth	2 × M	18–20	20–40	2.5–2.7
Pregnancy (last 1/3)	1.3 × M	7–10	20	(as for maintenance)
Lactation (early)	2 × M	10–11	20–30	2.7

[a]The maintenance energy requirement formula was derived from the fasting heat production formulae of AFRC (1993), NRC (2016) and CSIRO (2007) and simplified by plotting the FHP values against live weight in a simple linear regression. Assumptions included a diet M/D of 10 MJ/kg DM, activity allowance (+10%).

Table 7.4. Approximate maximum dry matter intakes of beef cattle on different diets

Diet NDF (%)	Maximum DM intake (% bodyweight)	Typical feeds at this NDF level
20	Intake limited by chemical constraints (e.g. ruminal pH)	Barley grain, sorghum grain, wheat grain
25	Intake limited by chemical constraints (e.g. ruminal pH)	Cottonseed meal, lupins
30	4.0 (unlikely)	Canola meal
35	3.4	Oats
40	3.0	Legume hay, canola silage
45	2.7	Legume hay
50	2.4	Grass silage, cereal hay
55	2.2	Barley hay
60	2.0	Barley straw, oat straw
65	1.8	Grass hay
70	1.7	Wheat straw

Having calculated the daily dry matter intake required to meet the energy and protein requirements of the animal, it is important to make sure the animal is capable of eating this much feed. Table 7.3 shows typical dry matter intakes based on bodyweight and feed quality. Alternatively, a good working estimate of maximum dry matter intake of a beef animal can be determined as follows:

$$\text{Max. DM intake (\% bodyweight)} = 120/\text{NDF\% of feed}$$

Table 7.4 shows typical NDF values of different feeds and the maximum DM intake that feed will allow provided all other constraints are removed (e.g. nutrient deficiencies).

Examples of how you can use this simple, practical method to formulate rations and determine intakes required to meet desired production levels are now discussed.

Example 1: How much of a barley grain/medium-quality hay mixture (50/50) should I feed 200 kg steers to maintain their weight?

$$\text{ME}_{\text{m}} = 0.1 \times 200 + 8 = 28 \text{ MJ/day}$$

A 50/50 barley/hay mix would be approximately (0.5 × 12 = 6 MJ/kg DM from barley) + (0.5 × 9 MJ/kg DM hay) = 10.5 MJ/kg DM. So the steers would require 28/10.5 kg DM/day = 2.7 kg DM/day. The feed DM would be ~90%, so you need to provide 2.7 divided by 0.9 = 3.0 kg/day as fed. It is important to ensure that the animals can consume this amount of feed/day. The feed is medium quality, so cattle should be able to consume ~2.0% of their bodyweight in dry matter/day = 0.02 × 200 = 4 kg/day, well above the required intake. The protein requirement for maintenance is met by microbial supply so no additional dietary protein is required.

Example 2: How much lupin grain should I feed to 500 kg cows in the last third of pregnancy to meet their requirements? The cows are consuming 5.0 kg medium-quality hay/day.

$$\text{ME}_m = (0.1 \times 500) + 8 = 58 \text{ MJ/day}$$

Pregnancy increases their requirements by 30%, so the ME requirement is $1.3 \times 58 = 75$ MJ/day. The hay is providing ~9.0 MJ/kg DM $\times$ 4.5 kg DM (5 kg $\times$ 0.9 DM) = 40.5 MJ/day, leaving a deficit of ~35 MJ/day. Lupins contain ~13 MJ/kg DM, so we need 2.7 kg lupin DM/day = 2.7/0.9 kg DM = 3.0 kg lupins as fed/day plus the 5.0 kg hay (total = 7.2 kg DM/day). Maximum DM intake for 500 kg cows on the hay plus lupins would be 2.7% of bodyweight (= 12.5 kg/day) because the NDF of lupins is 25% and the NDF of the hay is ~60%, so the NDF content of the 3.0 kg lupins plus 5.0 kg hay is ~45% = 120/45 = 2.7% bodyweight DM/day.

   Rapid growth (particularly in young animals), late pregnancy and early lactation increase the requirement for Undegraded Intake Protein (UIP). Lupins contains moderate to high UIP, so this will suffice.

Example 3: How much of a feedlot concentrate ration containing 12.3 MJ/kg DM ME would be required for growing steers (weighing 300 kg and growing at 1.0 kg/day)?

The maintenance energy requirement of the steers is 0.1W + 8 = 38 MJ/day. To grow at 1.0 kg/day requires ~2$\times$M levels = 76 MJ/day. The ration contains 12.2 MJ/kg DM, so they would need to eat 6.22 kg DM/day = 6.9 kg as fed. The NDF content of a feedlot ration would be ~25–30%, so maximum DM intake would be >4.0% of bodyweight = 12 kg DM/day, so this is achievable. The ration should contain at least 18–20% CP with some Undegraded Intake Protein (20–40%).

Mineral requirements of beef cattle

Beef cattle require 17 mineral elements (10 microminerals and seven macrominerals) to thrive. The mineral and vitamin requirements of growing beef cattle are listed in Table 7.5.

   The microminerals or trace elements are required as components of enzymes or as components of special molecules such haem (Fe), thyroxine (I) or vitamins (e.g. Co in B12). The macrominerals are components of structural compounds (bone, teeth and connective tissues) or buffers (bicarbonates, phosphates).

Table 7.5. Mineral element requirements of growing beef cattle

Mineral	Feed content
Microminerals	**(mg/kg DM)**
Cobalt	0.11–0.20
Copper	8–16
Iodine	0.5
Iron	30–50
Manganese	20–25
Selenium	0.10
Zinc	12–35
Macrominerals	**(g/kg DM)**
Phosphorus	1.8–3.2
Sulphur	1.5
Calcium	1.9–4.0
Sodium	0.8–1.2
Magnesium	1.9
Potassium	5.0
Chlorine	2.0

Source: Hynd (2014)

Clinical signs of deficiencies of the essential mineral elements are shown in Table 7.6.

Vitamin requirements of beef cattle

Vitamin deficiencies in cattle are relatively rare given the normal production of the B-group and vitamin K by ruminal microbes. Exceptions are when soils are deficient in cobalt, resulting in vitamin B12 (cobalamin) deficiency, or when thiaminases are present in the feed consumed (e.g. bracken fern poisoning or high-grain rations). Detailed coverage of polioencephalomalacia, the disease caused by thiamine deficiency in ruminants, is found in Chapter 6 (see page 187). Vitamin K can be obtained by cattle from the microbes (in the form of menaquinone) or directly from the feed (in the form of phylloquinone). Vitamin K deficiency in cattle has been reported when cattle graze on mouldy sweet clover hay (*Melilotus albus*) containing the vitamin K antagonist, dicoumarol. The clinical signs are haemorrhage and anaemia resulting from failure of the vitamin-K induced blood clotting mechanism. Vitamins E and A can be deficient when cattle graze poor pastures during the dry season for extended periods of time. The vitamin A content of feeds is directly related to the proportion of green forage available (reflecting the content of the vitamin A precursors, carotene and carotenoids, while vitamin D content is related to the extent of solar input to the feed (high in hay, low in lush green feeds). A summary of the clinical signs of vitamin deficiencies in cattle and factors predisposing to the conditions is shown in Table 7.7.

Table 7.6. Clinical signs of deficiencies of the essential minerals for beef cattle

Mineral element	Clinical signs
MACROMINERALS	
Calcium	• Brittle bones • Muscle weakness (paresis, ataxia)
Phosphorus	• Reduced intake • Depraved appetite (pica) • Reduced fertility • Rickets and osteomalacia • Broken bones (rib fractures)
Magnesium (see section on hypomagnesaemia in 'Common nutritional and metabolic problems in beef cattle')	• Muscle twitching, hyerpexcitability • Tremors • Ataxia • Nystagmus (eye flickering) • Soft tissue calcification • Arrhythmia (ventricular tachycardia)
Sodium	• 'Pica' (depraved appetite) • Reduced feed intake, milk production, growth • Bodyweight loss • Shivering, incoordination • Dehydration • Arrhythmia
Sulphur	• Emaciation • Anorexia • Reduced ruminal digestion • Lactate accumulates in rumen
Chlorine	• Hyponatraemia, hypokalaemia • Poor growth • Anorexia • Emaciation • Dehydration • Pica • Hypertension
Potassium	• Reduced feed intake • Pica • Cellular acidosis • Anorexia • Tetany, cramps, spasms
TRACE ELEMENTS	
Cobalt	• Anorexia • 'Ill-thrift' (poor growth) • Rough coat • Anaemia • Emaciation • Death
Copper	• Depressed growth • Rough coat • Diarrhoea • Skeletal abnormalities • Leg abnormalities and pacing gait • Depigmentation of hair • Anaemia

Table 7.6. (Continued)

Iron	• Unlikely in cattle • Anaemia • Anorexia
Zinc	• Anorexia • Rough coat • Decreased feed intake • Stiff gait • Swollen hocks • Poor growth
Selenium	• White muscle disease (WMD or NMD = nutritional muscular dystrophy) • Reduced fertility • Retained placenta
Iodine	• Unlikely in cattle • Goitre (enlarged thyroid) • Reduced fertility
Manganese	• Unlikely deficiency in cattle • Leg/joint malformations

Source: Underwood and Suttle (1999)

Table 7.7. Clinical signs of, and predisposing factors to, fat-soluble vitamin deficiencies in beef cattle

Vitamin	Clinical signs	Predisposing factors
Vitamin A (retinal, retinol, retinoic acid, β-carotene)	• Poor fertility • Night blindness (xeropthalmia or dry eye; lacrimation; opacity of the cornea; constriction of the optic nerve) • Abortion • Reduced feed intake • Diarrhoea • Scaly skin, dermatitis • Abnormal bone growth • Seizures • Weak calves • Increased susceptibility to infection	• Months on dry feeds • High-grain diets • Silages • Mouldy sweet clover (dicoumarol) • High nitrate feeds
Vitamin D (ergocalciferol = D2 in plants, cholecalciferol = D3 in animals)	• Young animals = rickets (bent forelegs, arched back, swollen joints) • Older animals = osteomalacia • Stiff gait • Irritability • Depressed intake • Laboured breathing	• Inadequate sunlight • No access to sun-cured roughages for months
Vitamin E (α-tocopherol)	• White muscle disease (nutritional muscular dystrophy) • Depressed fertility • Increased susceptibility to disease	• Linked to selenium status • Requirement increased by polyunsaturated fatty acids in diet

Source: Huber (1988)

Feedlot nutrition

Lot feeding of cattle is used for the following reasons:

- to finish the cattle to market specifications when pasture feed cannot meet the nutritional requirements
- to create a more uniform beef product for the market
- to meet a wider range of markets
- to reduce stocking pressure on dry pastures without needing to sell animals in poor condition.

The profitability of feedlots depends on the relative prices of feeds, the price of cattle entering the feedlot (store cattle) and the value of finished cattle. Fixed costs of infrastructure to establish the feedlot and regulatory compliance costs (accreditation, approval) are also very significant. Of the variable costs, feed constitutes 60–70% of the total. Maintaining high growth rates and reaching target live weights and fatness levels are essential. Maximal growth rate can only occur if the animal is consuming close to its maximum capacity (~3–4% of bodyweight) of a ration high in energy and protein, and balanced for minerals and vitamins. The protein must contain sufficient Undegraded Dietary Protein (sometimes called Undegraded Intake Protein or UIP) with a high content of the essential amino acids methionine, lysine, tryptophan and threonine. Typical DM intakes, growth rates, ME densities and crude protein content of feedlot systems are shown in Table 7.8.

Cattle entering a feedlot must be introduced gradually to the high-grain ration to prevent the development of subacute and acute lactic acidosis (see 'Acute ruminal lactic acidosis and subacute ruminal acidosis' later). Most are coming off pastures and have been transported by truck to the feedlot: a stressful event for the animals. The introduction of cattle to high-grain rations, known as 'transition feeding', usually involves gradually increasing the grain or

Table 7.8. Typical feed intakes, weight gains and dietary ME and crude protein contents for feedlot cattle

Cattle	Bodyweight (kg)	DM intake (%W)	DM intake (kg/d)	ME content (MJ/kg)	CP content (%)	Daily weight gain (kg/d)
Weaners	150	2.6	3.9	12.0	15.0	1.0
	200	2.7	5.4	11.5	13.0	1.0
Yearlings	250	2.9	7.3	11.5	12.0	1.3
	300	2.8	8.4	11.0	11.5	1.3
	400	2.6	10.4	11.0	11.0	1.3
Steers	350	2.9	10.2	10.8	11.2	1.4
	400	2.8	11.2	10.8	11.0	1.4
	500	2.6	13.0	10.8	11.0	1.4

Source: NRC (2000)

Table 7.9. Typical transition program to a high-grain feedlot ration

Days in feedlot	Roughage component (%)	Grain component (%)	Observations
0–2	100% hay diet to settle into feedlot		
3–6	80	20	Faecal checking
7–10	60	40	Rumination checking
11–14	40	60	Bunk score checking
15–18	30	70	Lameness checking
19–finish	20	80	

concentrate proportion of the ration while decreasing the roughage component. The starting ratios of roughage to grain and the rate of the step changes vary from feedlot to feedlot. Many start at a ratio of 60:40 grain: roughage and ramp up to 80:20 over a period of 10 days. Others start at higher ratios of roughage to grain to reduce risk. Table 7.9 shows a typical 'risk-averse' transition feeding program.

Regular observation of faeces (see next section), monitoring of feed intake by bunk scoring and lameness checking are critical to ensure early detection of developing acidosis. The proportion of cattle ruminating should also be noted. If fewer than half the cattle lying down at any one time are ruminating, acidosis may be developing. Particular care is needed when adverse weather events occur, because these can disrupt normal feeding behaviour resulting in sudden changes to ruminal grain loads and lactate (acute acidosis) or VFA (SARA) accumulation.

> Rumination checking is vital in feedlots. More than 50% of cattle lying down should be ruminating at any time. If not, acidosis may be developing.

Faecal evaluations in feedlot cattle

Regular evaluation of faeces provides valuable information on the site and extent of digestion of feed and can allow early diagnosis of fermentation problems (see Stallings 1998). The three 'C's' of manure observations are described in the next sections: colour, consistency and content.

Colour

The colour of cattle manure is very variable and reflects the fed type, bile content and passage rate through the GI tract. Manure is dark green when cattle are on fresh green forages and olive brown when on hay rations. Typical total mixed rations (TMR) are yellow-olive colour. The variations in colour between individual animals can be an indicator of selective feeding, which can produce digestive and nutritional problems. If such variations are noted, the ration may require better mixing, or changes made to the particle length. Cattle can select against long particles so ensure the TMR contains particles less than 5 cm. Grey manure usually indicates diarrhoea. Dark manure may indicate haemorrhage in the GI tract (e.g. coccidiosis). Bacterial infections often produce a watery diarrhoea with yellowish manure. The presence of whitish material in the faeces indicates undigested starch, which indicates poor rumen function.

Consistency
The consistency of faeces is a reflection of the water content, which is altered by osmotic imbalances drawing water into the gut. Imbalances can be caused by excessive water intake, which can occur during heat stress or if high levels of rumen degradable protein are consumed. Animals on high RDP diets consume more water to excrete the high levels of urea formed from the ruminal ammonia. Diarrhoea also occurs in response to bacterial infections, which cause osmotic imbalances in the epithelial cells lining the gut. High levels of acid production also damage the epithelia allowing water to enter the lumen.

Content
The contents of manure should be relatively uniform in size without discernible feed components (e.g. whole grains). Large particles of undigested grain or roughage indicate poor or absent rumination and poor rumen function. Rumination is triggered by the 'scratch factor' of feed particles (usually roughage) greater than 1–2 cm. Poor rumination, in turn, results in less salivary buffer entering the rumen and the potential for acidosis developing. Sometimes mucus appears in the faeces, which is indicative of mucosal damage in the large intestine caused by excessive hindgut fermentation. Mucin 'casts' can sometimes be found in the faeces. These are mucin plugs that are produced to heal areas of the intestine damaged by high levels of acid produced by large intestinal fermentation. They appear after washing to be like sausage casings. Finally, foamy or 'bubbly' manure indicates ruminal lactic acidosis or high levels of gas production in the large intestine.

A scoring system for evaluating manure is shown in Table 7.10.

Faeces are the window to gut health!

The three Cs of faecal observations are **colour**, **consistency** and **content**.

Table 7.10. Scoring system for evaluating manure from feedlot beef cattle

Score	Manure appearance	Manure contents	Potential situation
1	Faeces effectively water; may be grey	Water, mucin casts	Animal is very ill; acute acidosis, bacterial infection
2	Thin, custard-like but not water. Splatters on hitting ground	Mucin casts; undigested roughage particles, undigested whole grains	Rumen function poor; carbohydrate spillover to large intestine
3	Thick, custard-like forming a solid pat 2–3 cm high	Some recognisable particles but largely uniform and small particles	Ideal
4	Thick faeces, well formed, stacks in concentric rings up to 10 cm high	Uniform small particles	Ration may be unbalanced
5	Stiff, dry balls of faeces like horse manure	Long retention time may indicate nutrient imbalance in rumen	Animal is dehydrated; ration may be unbalanced

Source: Stallings (1998)

Feeds and feed additives used in feedlot nutrition

Feedlot diets typically contain a high level of cereal grains (corn, wheat, barley, rye, triticale, oats), a roughage source (oat husks, wheat husks, cereal straw, cotton hull waste), a protein source (soybean meal, canola meal, cottonseed meal, grain legumes), a mineral/vitamin premix and additives designed to improve rumen function and animal health (Table 7.11).

Common nutritional and metabolic diseases of beef cattle

Grass tetany (hypomagnesaemia)

Hypomagnesaemic tetany or grass tetany is a common disorder of grazing beef cattle, affecting 1–3% of all cattle at pasture. It can affect all classes of cattle but is most common in adult lactating cows and, like hypocalcaemia, is most common in older cows. Grass tetany has been the most common cause of death in beef cattle in southern Australia for the past 40 years. Loss of magnesium in milk is a major loss and the labile reserve of Mg, which is mainly found in bone, is lower in adult cows. The disease is complex and involves interactions between pasture species, temperature, pasture growth rate, potassium fertilisation of pastures, nitrogen intake, ammonia levels in the rumen, ruminal pH and calcium status (Martin-Tereso and Martens 2014).

Clinical signs of hypomagnesaemic tetany

Grass tetany is characterised by hypomagnesaemia (plasma Mg <1.5 mg/dL or 0.65 mM), and reduced Mg in cerebrospinal fluid (<1.0 mg/dL). There are two

Table 7.11. Typical supplements and rumen modifiers available for growing and finishing cattle in feedlots

Supplement	Typical examples	Modes of action and comments
Energy supplements	• Cereal grains (barley, wheat, oats, triticale) • Molasses • Grain legumes (peas, beans, lupins)	• Increased total energy intake • Substitution effect on pastures • Reduced fibre digestion • Potential acidosis • Legume grains operate as high energy and high protein supplements
Fibre supplements	• Oat husks • Rice husks • Wheat husks • Cotton hull waste • Grape marc	• Increase NDF to achieve optimal level in ration to maximise intake and stabilise rumen function • Increased feed intake if NDF too low • Reduced acidosis
Rumen degradable protein supplements	• Urea	• Unlikely response in growing/finishing rations as RDP already adequate
Undegraded Dietary Protein supplements	• Soybean meal • Cottonseed meal • Canola meal • Lupins • Field peas • Field beans	• Increased metabolisable protein supply possibly increased feed intake (total energy increase)
Rumen modifiers (ionophores)	• Monensin • Lasalocid • Narasin • Salinomycin • Tetronasin • Lysocellin • Laidlomycin	• Decreased acidosis • More uniform feed intake • Increased energy efficiency (increased propionate, reduced methane) • Feed intake reduced, daily gain maintained (increased feed conversion efficiency) • Reduced nitrogen excretion • Reduced phosphorus excretion
Rumen modifiers (other antibiotics)	• Virginiamycin • Tylosin • Bambermycin	• Increased daily gain and feed efficiency decreased lactic acidosis • Decreased laminitis • Improved mineral absorption
Rumen modifiers (pH buffers)	• $NaHCO_3$ • MgO • Na bentonite • $CaCO_3$	• Reduced acidosis • Increased feed intake
Rumen modifiers (miscellaneous)	• Garlic oil • Cinnamaldehyde • Eugenol • Capsaicin • Anise Oil • Tannins • Saponins	• Increased propionate/acetate ratio • reduced proteolysis/deamination • Reduced methane production

Source: Hynd (2014)

Table 7.12. The clinical signs for the two forms of hypomagnesaemia

Acute form	Chronic form
• Plasma Mg <1.5 mg/dL • Nervous apprehension • Staring eyes • Sudden bellowing and blind galloping • Stiff and stilted staggering • Muscle twitching especially in the face and ears • Extreme excitability and violent convulsions • Lateral recumbency and legs pedalling • Jaws grinding teeth • Death during convulsions or in coma	• Stiff gait and gradual loss of condition over several weeks • No reduction in milk yield or appetite • Often recovers but may progress to acute form • Hypersensitive to touch and sound • Frequent urination

Source: Underwood and Suttle (1999)

forms of the disease recognised: an acute form and a chronic form. Table 7.12 lists the clinical signs described by Underwood and Suttle (1999) for the two forms of hypomagnesaemia.

Aetiology of hypomagnesaemic tetany in beef cattle

Magnesium homeostasis, unlike calcium, is not tightly regulated by hormonal mechanisms. Magnesium is absorbed through the rumen epithelium and, other than the ability to excrete excess Mg through the kidneys, there is little capacity to regulate plasma Mg levels. Resorption from bone, where 60–70% of the total body Mg is stored, is very limited. So, any significant decrease in Mg absorption from the rumen results in hypomagnesaemia. Figure 7.2 outlines the absorption and excretion pathways for magnesium in ruminants.

Other than dietary Mg levels, the intake of K, soluble proteins and dietary fatty acids also influence Mg absorption. High levels of polyunsaturated fatty acids in some pasture forages are hydrogenated in the rumen to long-chain, saturated fatty acids. These combine with Mg and Ca ions to form insoluble soaps, which are not absorbed. Soluble plant proteins are rapidly degraded to ammonia in the rumen. Ammonia has direct inhibitory effects on the Mg transport channel (Martin-Tereso and Martens 2014), but also increases ruminal pH, which interferes with the ruminal transport channel. High levels of K in the forage also directly interferes with Mg absorption across the ruminal epithelium. Soils high in K or fertilised with potash and nitrogen are high-risk for hypomagnesaemic tetany.

Sudden reductions in Mg absorption then can occur if there is a sudden decline in feed intake due to inclement weather or if cattle are placed on short, grass-dominant pastures (<1000 kg DM/ha). Sudden stresses such as transport or cold snaps are also triggers. Grasses are particularly low in Mg and when they are growing rapidly, they can contain <0.2% Mg.

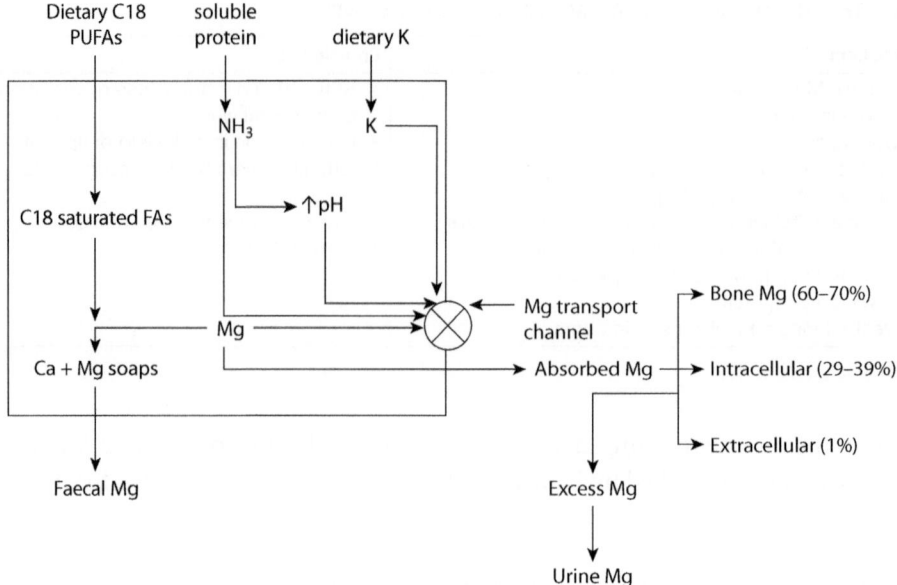

Fig. 7.2. Absorption of dietary Mg from the rumen and excretion through the kidneys.

Treatment of cattle with hypomagnesaemic tetany

Animals showing clinical signs should be treated immediately with combined solutions of calcium and magnesium because the two deficiencies usually occur together. Solutions can be given intravenous (slowly and with auscultation of the heart). For hypomagnesaemia alone, MgSO$_4$ can be administered subcutaneously, cattle moved off the risk pasture and given Mg-treated hay (60 g MgO/day). Prevention in high-risk animals requires daily supplements of MgO mixed with palatable ingredients such as molasses and concentrates and hay, or treatment with slow-release intraruminal bullets.

Acute ruminal lactic acidosis and subacute ruminal acidosis

Acidosis is a significant nutritional and metabolic problem in beef cattle and is generally considered to be of two major forms: Subacute Ruminal Acidosis (SARA) and Acute Lactic Acidosis (ALA). Some texts refer to ALA as clinical acidosis and SARA as subclinical acidosis, but SARA does present clinical signs at times. Cattle most likely to suffer these pathologies are those suddenly exposed to high grain intakes, cattle poorly adapted to high-concentrate rations in feedlots, cattle poorly adapted to lush, green pastures or brassica crops (such as kale, swedes, turnips), or cattle offered unusual feedstuffs with low fibre content (lollies, potatoes, fruits, breads). Cattle fed wheat grains are most susceptible to acidosis, followed by triticale, barley and oats. Lupins, which contain low levels of starch (3%), are the least prone to causing acidosis and can be safely fed to cattle provided the grains are cracked and not crushed.

Acidosis risk of grains

wheat > triticale > barley > oats > lupins

SARA and ALA have been extensively studied in dairy cattle, beef cattle and sheep (see Penner *et al.* 2009, 2010; Lettat *et al.* 2010, 2012; Aschenbach *et al.* 2019). In feedlots, SARA may be the more frequent and economically important problem but it is difficult to obtain data as by its nature it is difficult to quantify without careful observation, which is difficult in large feedlot pens.

Definitions of SARA and ALA

Attempts have been made to ascribe definitions to SARA and ALA on the basis of pH ranges and times below threshold values. Typically a pH of <5.5 is considered a critical threshold for defining SARA and <5.0 for ALA (Garrett *et al.* 1999). However, fluid pH is not uniform throughout the rumen, is influenced by the mode of sampling (rumenocentesis, stomach tube, fistula), and varies greatly with time after feeding (Aschenbach *et al.* 2019). There is also a poor correlation between rumen pH and clinical signs such as feeding depression (Uhart and Carroll 1967), suggesting that pH is a risk factor rather than a unique defining criterion. If pH is to be used as a defining criterion, definitions should include sampling method/site, time after feeding and a population description. For example, when '50% of the flock has ruminal pH <6.2 in fluid taken by rumenocentesis 2–4 h after feeding, a diagnosis of potential SARA can be made'. If more than 30% have a pH <5.5 it is likely that ALA is present. Alternatively multiple criteria might be used including pH, ammonia concentration, lactate concentration and VFA concentration (Bramley *et al.* 2008). Acidosis should be considered a continuum of pathologies associated with rapidly declining ruminal fluid pH as the major risk factor.

Aetiology of SARA and ALA

Figure 7.3 summarises the sequence of events leading to SARA or ALA in ruminants.

A sudden input of highly fermentable carbohydrates into the rumen of animals that have not been adapted to such substrates, results in rapid bacterial growth. The resulting high levels of VFA drop the fluid pH to less than 6.0, and those bacteria with pH optima between 6 and 7 are depressed in growth and activity. Given that many of the cellulolytic bacteria have pH optima in this range, cellulolysis is depressed and the rate of fibre digestion and voluntary feed intake fall. Provided

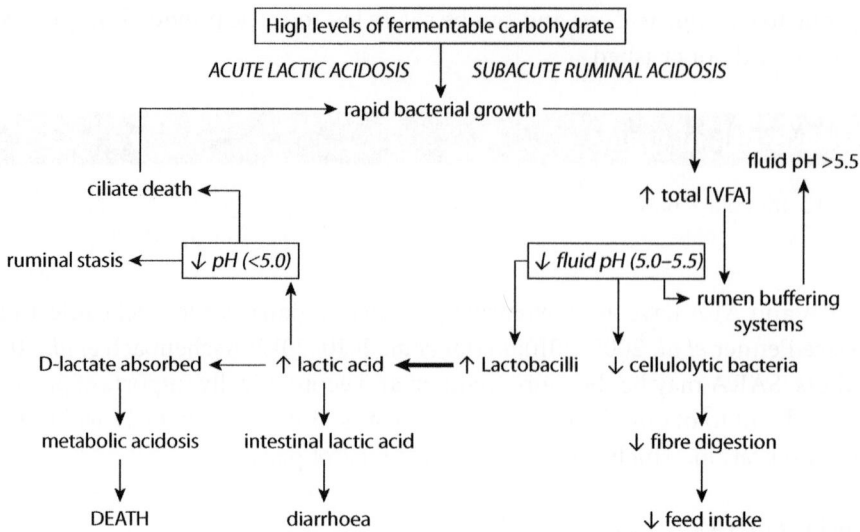

Fig. 7.3. Aetiology of subacute ruminal acidosis (SARA) and acute lactic acidosis (ALA). The right-hand side of this diagram traces the development of SARA in which total VFA concentrations are high but lactate does not accumulate. Proliferation of lactic acid-producing bacteria then precipitates acute lactic acidosis (left-hand side of the diagram). Loss of ciliates stimulates further bacterial growth and the sequence repeats.

the VFA levels are not too high, the rumen buffering systems will return the pH to pre-feeding levels (5.5–6.5) and the animal may enter this cycle of pH fluctuations on a daily basis. These buffering systems include:

- undissociated VFA are absorbed across the luminal epithelium, thereby removing the H$^+$ from the rumen (see Fig. 3.21)
- dissociated VFA are absorbed across the epithelium leaving the H$^+$ in the rumen. However bicarbonate is simultaneously released into the rumen thereby buffering the H$^+$ left behind and forming CO$_2$ and H$_2$O (Fig 3.21)
- uptake of protons by ammonia released from hydrolysed proteins to produce ammonium
- buffering of the protons by salivary buffers bicarbonates and phosphates
- dissociated VFA themselves take up protons as the pH declines towards their pK$_a$ of 4.8 (Fig. 3.20).

However, if VFA levels are too high these buffering systems begin to fail and the pH may fall below 5.5, triggering the following catastrophic cascade of events (Fig. 7.3). The low ruminal pH stimulates lactate-producing bacteria such as *Streptococcus bovis* and depresses lactate-utilising bacteria such as *Megasphaera elsdenii* and *Selenomonas ruminantium* (Russell and Hino 1985). Lactic acid begins to accumulate in the rumen fluid. The pK$_a$ of lactic acid is 3.86 so at pH 5.0–6.0 is largely in the dissociated form, leaving more protons in solution. This drops the pH markedly. The sudden pH decline to levels <5.0 has catastrophic effects on the rumen ecology as more gram-negative bacteria die. *Streptococcus bovis* also starts to die but is replaced rapidly by the other major lactate-producing bacteria, *Lactobacillus* spp. (a large gram + rod). High ruminal levels of lactic acid drop the pH to below 5.0 at which point the rumen movements decline and eventually stop. Both D- and L-lactate are absorbed into the portal system and transported to the liver. D-lactate is metabolised more slowly than the L-form and is largely responsible for the subsequent metabolic acidosis. Ruminal pH values below 5.0 also depress the ciliate populations. Ciliates consume large numbers of bacteria (Coleman and Sandford 1979) and starch, thereby preventing explosive increases in bacterial populations. Death of the ciliates takes the brake off bacterial proliferation, creating a snowballing fermentation.

The highly acidic rumen fluid damages the rumen epithelium and it becomes inflamed (chemically induced rumenitis), eroded and ulcerated. The damaged epithelium is invaded by fungi and bacteria (*Fusobacterium necrophorum* and *Arcanobacterium pyogenes*), producing mycotic and bacterial rumenitis and necrotic lesions. These bacteria enter the portal blood system and invade the liver causing hepatic abscesses. Bacterial endotoxins also exit the rumen and enter the systemic circulation. In prolonged cases of acidosis, the rumen epithelium becomes parakeratotic (thickened and heavily keratinised). The high levels of lactic acid in the rumen fluid draws water into the rumen from the blood stream, causing dehydration and haemoconcentration (high packed cell volume). High fluid flow into the gut produces profound watery diarrhoea and further dehydration. Polioencephalomalacia can result from a deficiency of thiamine caused by proliferation of bacteria that produce thaminases that break down the B-group vitamin, or the death of bacteria that produce the vitamin. Animals with acidosis are often lame as a result of the high levels of histamine produced in the rumen, bacterial endotoxins entering the systemic circulation and systemic acidosis. These cause changes in blood supply to the hoof and induce pressure changes that are extremely painful to the animal.

Clinical signs of ALA in cattle
Cattle with acute acidosis show signs within ~12 to 36 h after grain ingestion. Several hours after engorgement, some animals will be seen kicking at the abdomen, an indicator of abdominal pain. They stop eating, look depressed and have a severe diarrhoea that smells characteristically sweet/sour. There is often undigested grain in the faeces. Some animals will stand quietly while others will be lying down or staggering around. Body temperature is below normal (<38.5–39.6°C) and respiration is shallow. Heart rate is elevated. Severely affected animals die within 24 h but others die several days after apparently recovering, probably due to peritonitis associated with bacteria and fungi from the damaged rumen epithelium.

Prevention of acidosis in cattle

The following precautions should be taken to prevent acidosis occurring in cattle:

1. Ensure cattle do not have access to spilt grains, grain stores, bagged grains.
2. Ensure cattle have been adapted to grains before being placed on recently harvested cereal stubbles.
3. For feedlot rations use total mixed rations (TMR) containing roughage particles >2.5 cm to provide 'scratch factor' and maximal rumination.
4. Ensure cattle going onto a high-grain or concentrate ration (for feedlotting, drought feeding) are adapted to the ration as indicated in Table 7.9.
5. Include buffers such as sodium bicarbonate or commercial buffers such as AcidBuf® (Feedworks) in pelleted rations or total mixed rations containing high levels of grains or concentrates.
6. Include gram + antibiotics, such as virginamycin (Eskalin®), in high-concentrate rations.
7. Include ionophores in high-concentrate rations.
8. Use cereal grains that are less rapidly fermented, such as oats and barley, and not wheat. Use non-cereal grains such as the grain legumes and particularly lupins.
9. Ensure that trickle feeding (lick feeders) feeder bins are operating properly and that the flow of concentrate feed is not too high early in the adaptation period.
10. Expose calves to grains at a young age (while still consuming milk) to ensure their first experience does not produce variable intake and engorgement
11. Monitor the herd frequently during grain adaptation. Inspect behaviour, rumination, faecal consistency.

References

AFRC (1993) *Energy and Protein Requirements of Ruminants: An Advisory Manual Prepared by the AFRC Technical Committee on Responses to Nutrients.* Commonwealth Agricultural Bureaux, Slough and CABI, Wallingford, UK.

Aschenbach JR, Zebeli Q, Patra AK, Greco G, Amasheh S, Penner GB (2019) Symposium review: the importance of the ruminal epithelial barrier for a healthy and productive cow. *Journal of Dairy Science* **102**, 1866–1882.

Bramley E, Lean IJ, Fulkerson WJ, Stevenson MA, Rabiee AR, *et al.* (2008) The definition of acidosis in dairy herds predominantly fed on pasture and concentrates. *Journal of Dairy Science* **91**, 308–321.

Coleman GS, Sandford DC (1979) The engulfment and digestion of mixed rumen bacteria and individual bacterial species by single and mixed species of rumen ciliate protozoa grown *in vivo*. *Journal Agricultural Science, Cambridge* **92**, 729–742. doi:10.1017/S0021859600053971

CSIRO (2007) *Nutrient Requirements of Domesticated Ruminants.* CSIRO Publishing, Melbourne.

Fox DG, Tedeschi O, Tylutki TP, Russell JB, VanAmburgh M, Chase AN, *et al.* (2004) The Cornell Net Carbohydrate and Protein System model for evaluating herd nutrition and nutrient excretion. *Animal Feed Science and Technology* **112**, 29–78.

Freer M, Moore AD, Donnelly R (1997) GRAZPLAN: decision support systems for Australian grazing enterprises—II. The animal biology model for feed intake, production and reproduction and the GrazFeed DSS. *Agricultural Systems* **54**, 77–126.

Garrett EF, Pereira MN, Nordlund KV, Armentano LE, Goodger WJ, Oetzel GR (1999) Diagnostic methods for the detection of subacute ruminal acidosis in dairy cows. *Journal of Dairy Science* **82**, 1170–1178.

Grant SA, Suckling DE, Smith HK, Torvell L, Forbes TDA, Hodgson J (1985) Comparative studies of diet selection by sheep and cattle: the hill grasslands. *Journal of Ecology* **73**, 987–1004. doi:10.2307/2260163

Huber JT (1988) Vitamins in ruminant nutrition. In *The Ruminant Animal: Digestive Physiology and Nutrition*. (Ed. DC Church) pp. 326–341. Prentice Hall, Upper Saddle River, NJ, USA.

Hynd PI (2014) Growing and finishing beef cattle at pasture and in feedlots. In *Beef Cattle Production and Trade* (Eds D Cottle and L Kahn) pp. 381–399. CSIRO Publishing, Melbourne.

Lettat A, Noziere P, Silberberg M (2010) Experimental feed induction of ruminal lactic, propionic or butyric acidosis in sheep. *Journal of Animal Science* 88, 3014–3046.

Lettat A, Noziere P, Silberberg M, Morgavi DP, Berger C, Martin C (2012) Rumen microbial and fermentation characteristics are affected differently by bacterial probiotic supplementation during induced lactic and subacute acidosis in sheep. *BMC Microbiology* 12, 142.

Maekawa M, Baeuchemin KA, Christensen DA (2002) Effect of concentrate level and feeding management on chewing activities, saliva production, and ruminal pH of lactating dairy cows. *Journal of Dairy Science* **85**, 1165–1175. doi:10.3168/jds.S0022-0302(02)74179-9

Martin-Tereso J, Martens H (2014) Calcium and magnesium physiology and nutrition in relation to the prevention of milk fever and tetany (dietary management of macrominerals in preventing disease). *Veterinary Clinics: Food Animal Practice* **30**, 643–670.

NRC (2000) *Nutrient Requirements of Beef Cattle. Update 2000.* 7th revised edn. National Research Council, The National Academies Press, Washington, DC, USA.

NRC (2016) *Nutrient Requirements of Beef Cattle.* 8th revised edn. National Research Council, The National Academies Press, Washington, DC, USA.

Penner GB, Aschenbach JR, Gotthold G, Rackwitz R, Oba M (2009) Epithelial capacity for apical uptake of short chain fatty acids is a key determinant for intraruminal pH and the susceptibility to subacute ruminal acidosis in sheep. *The Journal of Nutrition* **139**, 1714–1720. doi:10.3945/jn.109.108506

Penner GB, Oba M, Gabel G, Aschenbach R (2010) A single mild episode of subacute ruminal acidosis does not affect ruminal barrier function in the short term. *Journal of Dairy Science* **93**, 4838–4845.

Russell JB, Hino T (1985) Regulation of lactate production in *Streptococcus bovis*: a spiraling effect that contributes to rumen acidosis. *Journal of Dairy Science* **68**, 1712–1721.

Santich BJ (2014) Beef consumption: historical overview, recent trends and contemporary attitudes In *Beef Cattle Production and Trade*. (Eds D Cottle and L Kahn) pp. 1–6. CSIRO Publishing, Melbourne.

Stallings CC (1998) Manure scoring as a management tool. In *Proceedings of the Western Canadian Dairy Seminar*, Red Deer, Alberta. Department of Agricultural, Food and Nutritional Science, University of Alberta, Edmonton, Canada, <https://wcds.ualberta.ca/wcds/wp-content/uploads/sites/57/wcds_archive/Archive/1998/ch25.htm>.

Tedeschi LO, Galvean ML, Beauchemin KA, Caton JS, Cole NA, Eismann JH, *et al.* (2016) The eighth revised edition of the Nutrient Requirements of Beef Cattle: development and evaluation of the mathematical model. *Journal of Animal Science* 94 (Supplement 5), 492–493.

Uhart BA, Carroll FD (1967) Acidosis in Beef Steers. *Journal of Animal Science* **26**, 1195–1198.

Underwood EJ, Suttle NF (1999) *The Mineral Nutrition of Livestock*. 3rd edn. Food and Agricultural CABI Publishing, Wallingford, UK.

8

Dairy cattle nutrition

Key points

- Good nutrition is a key driver of the productivity and profitability of dairy enterprises globally.
- Dairy cow milk is 85–88% water, 3.5–5.0% fat, 3.0–4.0% protein, 4.6–4.9% lactose and 0.7% ash (minerals). Composition varies with breed, stage of lactation, nutrition and disease status.
- Milk proteins are 80% caseins and 20% non-caseins (albumins and globulins). The globulins are particularly important in colostrum for calves.
- Milk fats are mainly triglycerides made up of straight chain, saturated fatty acids containing 4–18 carbons (65%) with 30% monounsaturated fatty acids (16:1 and 18:1) and 5% polyunsaturated fatty acids (18:2 and 18:3).
- Lactose is the main carbohydrate in milk.
- Milk production is driven by the rate of lactose synthesis because lactose synthesis draws water osmotically into the lumen of the mammary gland acinar. Osmotic equilibrium means the lactose content of milk is relatively constant unless the mammary epithelium is damaged.
- Milk components are made in the epithelial cells lining the mammary gland acinar. Milk fat is produced from circulating fatty acids (50%), and from fatty acids synthesised in the mammary epithelial cells from circulating acetate and ketone bodies (50%). Milk protein is synthesised by normal gene transcription and translation processes with amino acids derived from circulating amino acids. Lactose is synthesised from glucose and galactose synthesised from circulating propionate and amino acids, and circulating galactose.
- An 'ideal' dairy calendar involves a 305-day lactation, 85 days from calving to re-mating, and a 60-day 'dry' period to allow the mammary gland to recover.

Meeting this schedule requires excellent reproductive and nutritional management. Meeting body condition score targets at key stages of the cycle is the key to maintaining health and production.

- The nutritional requirements of dairy cows change throughout the lactation cycle, with changes in milk yield, milk composition and live weight status.
- Feeding practices should maximise feed intake in the lactating dairy cow.
- Replacement calves should be fed high-quality colostrum within the first 6 h of life.
- Replacement calves should be fed to establish early rumen development and solids feeding as early as possible. Ruminal papilla development requires high concentrations of volatile fatty acids (grain feeding) while development of the ruminal musculature requires roughage feeding, so mixtures of grain and hay should be offered early in life for optimal development of the rumen.
- Replacement heifers should be fed to meet live weight targets at weaning (100 kg), 12 months (290 kg), 15 months (345 kg) and 24 months (535 kg) of age.
- The majority (90%) of health problems in the dairy cow occur in a small window around calving. These include hypocalcaemia (milk 'fever'), ketosis, acidosis, subacute ruminal acidosis, hepatic lipidosis and left displaced abomasum.
- Milk fat depression in dairy cows is due to the formation of trans-10, cis-12 conjugated linoleic acid in the rumen. This compound depresses genes involved in milk fat synthesis.
- Hypocalcaemia is a consequence of failure of the homeostatic mechanisms that maintain plasma calcium levels during the period of high demand for calcium for milk production. Feeding anionic salts to create metabolic acidosis allows the bone-resorbing mechanism to operate.
- Ketosis and fatty liver syndrome result from rapid catabolism of adipose tissue during early lactation, combined with a depression in glucose availability, resulting in high levels of ketone bodies in the blood and deposition of fats in the liver.

Introduction

Global trends in dairy production

Dairy production in most countries began as family-owned, small farms producing milk and other milk-derived products for local markets. The dairy cows were often local breeds, or a mixture of breeds such as Jersey, Guernsey and Holstein Friesian, and breed dictated milk composition. Milk fat production was a dominant feature of milk pricing because many products such as butter, cheese and ice cream are fat-derived. Increasingly there was a shift towards whole milk production for liquid milk, and to derive products dependent on milk protein and volume. Globally, this meant a shift towards breeds capable of high-volume production, and the Holstein Friesian dominated the global dairy industries because this breed was capable of producing a large volume of high-protein milk. As the ability to shift dairy products around the world has grown, so has global competitive trade in milk and milk products. The European Union and countries such as New Zealand, Australia, and the United States of America compete in a global market for milk and, increasingly, free trade partnerships are opening up new markets for dairy products. On-farm there are clear trends across most producing countries towards a decline in the number of dairy operations, increasing corporatisation of dairy farms, a dramatic increase in the number of cows/farm and an increase in milk production/cow. More than 6 billion people globally consume milk products and the majority are from developing countries. The global increase in milk consumption is expected to be ~1.8% per annum. Milk and milk products contribute substantially to total protein intake in Asia and Africa (7%), and Europe (19%), and to dietary fat intake of 8% in Asia and African and 14% in Europe, Oceania and the Americas (FAO 2017).

There are more than 260 million dairy cows worldwide producing more than 600 million tonnes of milk each year (FAO 2017). The average milk production/cow globally is 2200 L/lactation, with some countries averaging over 10 000 L/cow (FAO 2017). A world record milk yield of over 35 000 L in one lactation has been recorded. Maintaining these high levels of production requires sound knowledge of the nutritional requirements of the lactating dairy cow throughout the dairy production cycle. The principles and practice of dairy nutrition are outlined in this chapter.

Drivers of profitability and trends in dairy production systems

The major determinant of the type of dairy production system (intensive grain-based versus pasture-based) is the price of milk relative to the price of grain-based compound feeds. In the USA, where the milk/compound feed price is high (2.5:1), intensive systems dominate and in countries where milk/grain price ratios are low (<1.5:1), pasture-based systems dominate (FAO 2017). There are no global statistics on the proportion of cows raised in intensive, housed, fully fed systems and those in forage-based systems, but in the EU, on average, 38% of cows are not grazed at all.

The most profitable pasture-based dairy enterprises are those that maximise forage quantity and quality (high-producing forages, high fertiliser input, regular forage renovation, optimal grazing management, efficient irrigation), maximise forage consumed per hectare (optimal stocking rate, grazing management to maximise forage production), optimal use of supplements and more milk/cow (best genetics, 300-day lactations) (e.g. Moran *et al.* 2000).

Overall then, the dairy industry is based on genetically high-producing Holstein Friesians in large herds in management systems, varying from pasture-based to intensive, grain-based enterprises. The nutritional management of these cows is the main determinant of the productive efficiency and profitability of the enterprises and is also the main source of the major production problems such as poor reproductive performance, and metabolic diseases such as acidosis, subacute ruminal acidosis, lameness, left displaced abomasum, hypocalcaemia, hypomagnesaemia and ketosis. This chapter explores the digestive anatomy and physiology of dairy cows, the biochemistry of milk synthesis, the nutrient requirements of dairy cows, diet formulation for dairy cows, nutrition of the replacement calf and heifer, pasture utilisation, and common nutritional and metabolic diseases of dairy cows.

Digestive anatomy and physiology of dairy cattle

The digestive system of the dairy cow is the same as that of other ruminants, as described in Chapters 3 and 6. The next sections refer to specific data for dairy cattle.

Anatomy and physiology of the mouth

Saliva in dairy cattle plays the particularly important role of buffering the rumen fermentation acids in addition to acting as a lubricant for chewing and swallowing, and for recycling urea as a source of nitrogen for ruminal microbes. Dairy cows produce between 100 and 300 L of saliva per day. This saliva contains ~400–1100 g of disodium phosphate and 1100–3200 g of sodium bicarbonate (Erdman 1988). These are the major buffers of the rumen and are critical to stable rumen function. The large quantity of sodium in saliva would deplete the cow's sodium pool if it were not recycled on passage through the digestive tract and would rapidly lead to dehydration, as is evident when calves suffer bacterial scours (diarrhoea).

The saliva of cattle contains little or no α-amylase but there may be some amylase activity in the watery secretions of the nasolabial glands, which the cow regularly licks. The saliva does contain salivary lipase secreted by the tongue epithelium (Sissons 1981). This salivary lipase releases short-chain fatty acids from water-insoluble triglycerides of dietary fat. The dentition of cattle is the same as that of beef cattle (Fig. 7.1). Cattle spend approximately one-third of their day grazing, one-third ruminating and one-third idling, but this varies greatly with their productivity and the amount and quality of forage available. On typical forage feeds, a cow will chew feed ~20 000–45 000 times/day, the extent of chewing being related to the mechanical toughness of the feed and to the total volume of saliva produced. Stimulating rumination and hence saliva production by ensuring roughage particle size is greater than 2.0–2.5 cm ('scratch factor') is critical to maintaining a stable ruminal pH and stable ruminal function.

Anatomy of the reticulo-rumen, omasum and abomasum

The capacity of the reticulo-rumen in cattle or any ruminant is difficult to quantify with precision because the compartments are never 'full' *per se*, because there is a large capacity to stretch further and there is always a gas space occupying approximately one-third of the rumen chamber. It is better to talk of the rumen 'volume' rather than capacity. The volume of digesta in the reticulo-rumen varies with the feed type, hydration state of the animal, and rate of production and intake. Typically, the reticulo-rumen holds ~10–30% of the bodyweight of the animal. For adult cows of 650 kg this equates to between 65 L and 200 L ruminal contents.

The epithelia of the stomach compartments in cattle are similar to those of other bovid ruminants, as are the microbes, motility patterns, fluid dynamics and particle dynamics (Chapter 3). The other compartments (reticulum, omasum and abomasum) have the same functions in cattle as in sheep and other bovids.

Milk composition

Milk is largely water (85–88%) with solids (fat 3.5–5.0%, protein 3.0–4.0%, lactose 4.6–4.9%, and vitamins and minerals 0.7%) suspended or solubilised in the liquid. The pH of fresh milk is ~6.5–6.7 and it is highly buffered against pH changes due to the amphoteric nature of proteins (i.e. they can act as acids or bases thereby buffering against changes in pH in either direction).

The composition of cow's milk depends on breed, diet and the stage of lactation (DePeters *et al.* 1995). Table 8.1 shows the yield of milk and the content of protein and fat (in %) of three breeds of cows fed and managed identically in a commercial herd (DePeters *et al.* 1995).

In general, as milk yield increases, the proportions of fats and proteins tend to decrease. This inverse relationship between milk yield and solids content holds throughout the lactation, so at peak lactation the solids content is lower (Fig. 8.1).

Table 8.1. Typical milk statistics for three dairy breeds

Milk characteristic	Breed		
	Holstein Friesian	Brown Swiss	Jersey
Yield (kg/day)	30.3	23.7	17.6
Fat (%)	3.74	4.22	4.81
Protein (%)	3.16	3.64	3.95
Lactose (%)[a]	4.9	5.0	4.9

[a]The lactose content was not reported in DePeters *et al.* (1995) but is a typical value for these species. Note there is little variation in percentage lactose because lactose is the osmotic 'driver' of milk yield (more lactose = more water = same concentration).

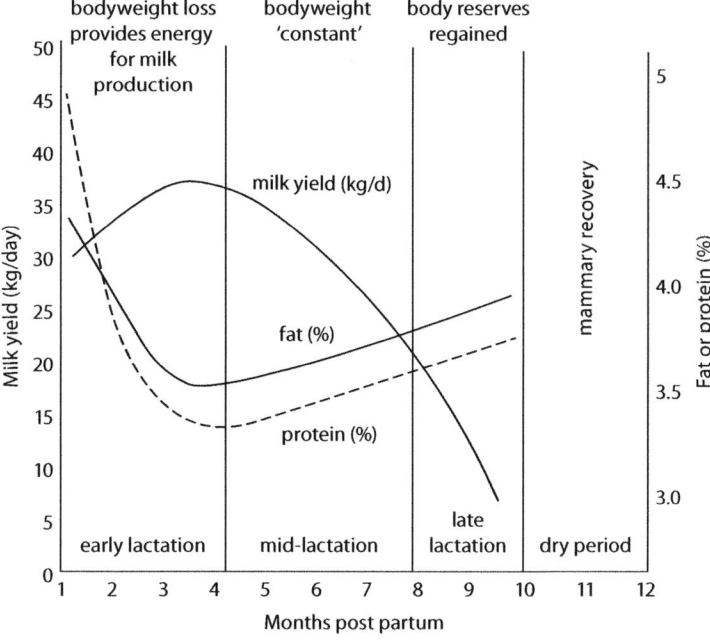

Fig. 8.1. Changes in milk yield (kg/day), milk fat (%) and milk protein (%) throughout a typical lactation for a high-yielding dairy cow.

Milk proteins

There are two major categories of milk protein: the casein family, which makes up ~80% of the total milk proteins; and the non-casein family (20%). The caseins contain phosphorus and are precipitated at pH 4.6, forming the curd during cheese production (Fig. 8.2). The non-caseins, or serum proteins, remain in solution at this pH and are called the whey proteins. The casein family comprises several types (α-s1, α-s2, β, κ and δ caseins). There are variations in the amino acid composition of each of these caseins and there are genetic variants in their composition. The β-casein variants are the source of a major marketing campaign based on health claims associated with the genetic variant A2 of the β-casein. The non-casein or whey proteins comprise lactalbumin (4%), lactoglobulin (11%), serum albumin and immunoglobulins (5%). The latter are high in the first milk post partum (colostrum).

Milk carbohydrates

The major milk carbohydrate is the sugar lactose, which is produced from glucose combined with galactose. Lactose is dissolved in the serum or whey protein phase of milk. The production of lactose in the epithelial cells of the mammary gland is the major driver of milk synthesis as it is osmotically active and draws water into the mammary gland (see Fig. 8.4).

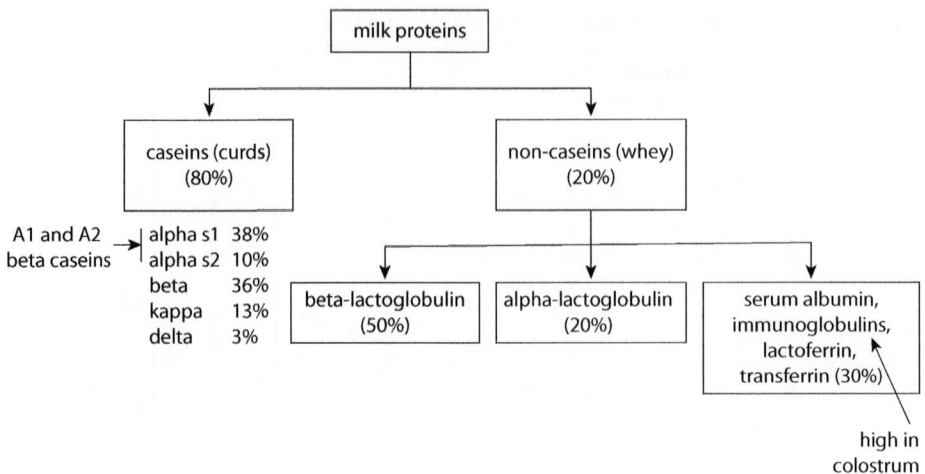

Fig. 8.2. Composition of the milk proteins.

Milk fats

Fats are made from fatty acids attached to glycerol, a 3-carbon backbone, to form a triglyceride (Fig. 8.3).

The wide range of fats in milk are a consequence of the combinations of fatty acids attached to the glycerol backbone. There are more than 400 individual fatty acids identified in milk fat, but only 15–22 fatty acids make up 90% of the milk fat. Most (65%) are straight-chain, saturated and contain 4–18 carbons (65%), with 30% monounsaturated fatty acids (16:1 and 18:1) and 5% polyunsaturated fatty acids (18:2 and 18:3). The fatty acid composition of milk varies throughout the lactation.

Vitamins and minerals in milk

Milk contains all the B-group vitamins and is a good source of thiamine, riboflavin and B12. Milk is a good source of calcium, phosphorus, potassium, selenium and zinc, but is low in copper, iron, manganese and sodium.

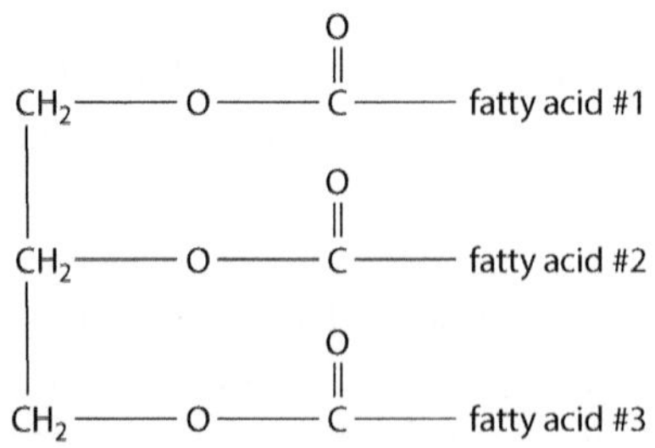

$$
\begin{array}{l}
\text{CH}_2 \text{---O---} \overset{\displaystyle \overset{O}{\|}}{C} \text{--- fatty acid \#1} \\[2ex]
\text{CH}_2 \text{---O---} \overset{\displaystyle \overset{O}{\|}}{C} \text{--- fatty acid \#2} \\[2ex]
\text{CH}_2 \text{---O---} \overset{\displaystyle \overset{O}{\|}}{C} \text{--- fatty acid \#3}
\end{array}
$$

Fig. 8.3. Triglyceride is the most common type of lipid in milk. Triglycerides comprise three fatty acids attached to a glycerol backbone.

Biochemistry of milk synthesis

The source of building blocks for synthesis of the components of milk are the nutrients available in the blood stream, which come from the gastrointestinal tract and liver, and from catabolism of body muscle and adipose tissue (Fig. 8.4).

Synthesis of milk fat

The triglycerides that make up milk fat are synthesised in the mammary epithelial cells (Mansson 2008). Approximately half of the fatty acids used to make the milk triglycerides are synthesised from circulating blood acetate, and β-OH butyrate from ruminal fermentation. The remaining 50% are non-esterified fatty acids produced from catabolism of body fat during weight loss. The fatty acids from these two sources then join with glycerol synthesised from propionate from rumen fermentation and amino acids from the intestines to form the milk triglycerides (Fig. 8.4). The fat droplets produced in the cytoplasm of the epithelial cells are then secreted from the apical membrane and enter the lumen of the acinus to form milk fat.

Synthesis of milk carbohydrate (lactose)

Glucose in ruminants is produced from circulating propionate from ruminal fermentation and amino acids from intestinal absorption (see Fig. 1.11). The

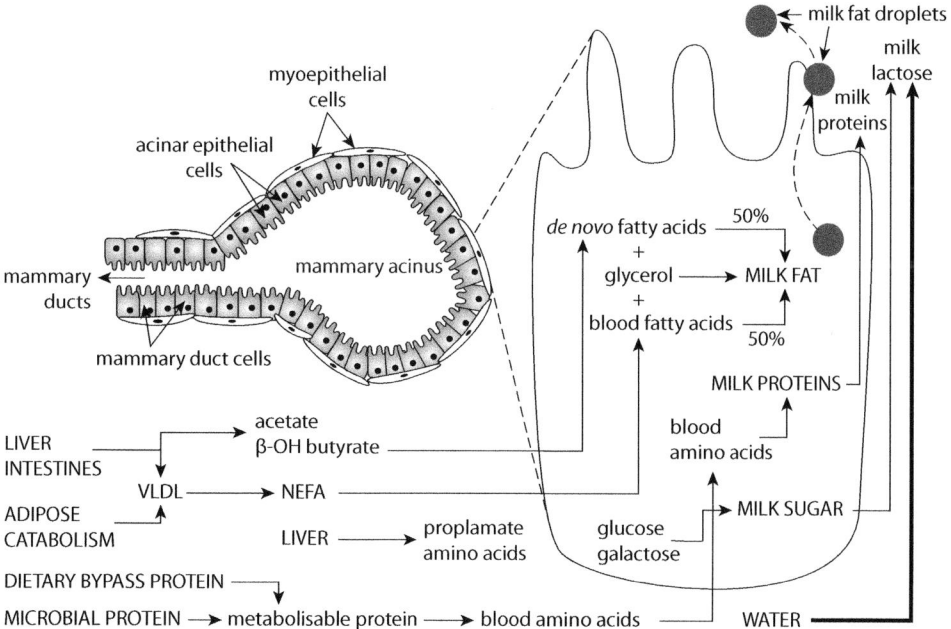

Fig. 8.4. Biosynthesis of milk fat, milk sugar (lactose) and milk proteins in the acinar epithelial cells lining the mammary ducts and glands. VLDL = very low density lipoproteins; NEFA = non-esterified fatty acids.

glucose taken up by the mammary epithelial cells combines with galactose from glucose synthesis in the body or within the mammary cell itself to form lactose under the activity of the enzyme lactose synthase. Note that as the lactose moves into the acinar lumen it draws water with it osmotically to maintain tonicity. This is important because it is the secretion of lactose that dictates the volume of milk produced, given milk is ~87% water. This relationship between lactose secretion and milk volume means that lactose concentrations in milk remain fairly constant unless there is a problem (e.g. infection or inflammation) with the integrity of the epithelial membrane (e.g. loss of tight junctions), which would allow water to enter the mammary gland ducts (e.g. during a mastitis infection). The rate of lactose production is limited by the supply of glucose precursors, which in turn depend on the supply of amino acids and propionate to the liver. The rate of propionate supply relates directly to the volatile fatty acid profile arising from ruminal fermentation, which depends on diet. High-grain diets generate high propionate, hence high glucose, high lactose and high milk volume, while high-roughage, acetate-type fermentations produce less milk, which is higher in fats.

Synthesis of milk proteins

The milk proteins are synthesised by the normal DNA/RNA/protein synthetic machinery operating in all cells of the body. The amino acids are derived from circulating amino acids and placed into the proteins on instructions from the genes encoding the caseins and the non-casein whey proteins. The protein composition of the milk, then depends on the breed of cow (hence the A2 casein variant for instance).

The energy, protein, vitamin and mineral requirements of dairy cattle

Ideal calendar of events on a dairy using a 365-day cycle

Before we estimate the nutrient requirements of dairy cows and their replacement heifers, it is essential that we understand a typical dairy calendar of events and the changes that take place in the animal's nutrient demands throughout the calendar year. Ideally, cows follow a 365 day calendar of events (Fig. 8.5), which allows 305 days of lactation and 60 days of 'rest'.

Ideally, replacement heifers are mated at 15 months of age to calve at 24 months of age. During this first pregnancy, the heifers are still growing quite rapidly as well as dealing with the energy demands of pregnancy. Her calf is removed 24–48 h after calving and the heifer enters the milking herd. After 85 days of milking she is mated at oestrous and hopefully is now pregnant and lactating (Note: during the first few years she is also still growing to her mature bodyweight). After 305 days of

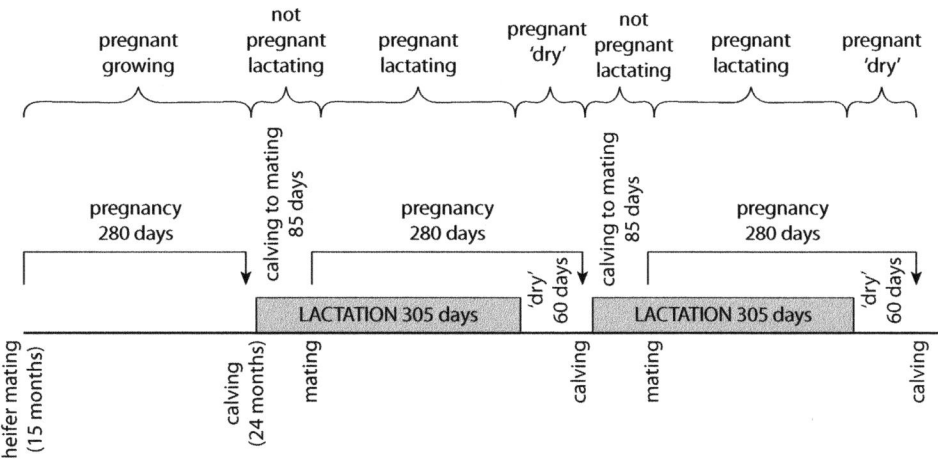

Fig. 8.5. 'Ideal' calendar for a 365-day cycle and 305-day lactation. The first mating is of a replacement heifer entering her first lactation at 24 months of age after mating at 15 months of age.

milking, she is 'dried-off' and returned to the dry herd until her next calf. This cycle is repeated three to six times, depending on the longevity of the cow and her reproductive and production performance. The precision and timing of this cycle is heavily dependent on reproductive success (i.e. getting her pregnant first time, every time, so there are no 'returns-to-service'). This depends on good nutritional management to keep the cow in ideal body condition for mating, and also on accurate detection of oestrous.

> The lactation 'cycle' changes the demands for energy, protein, vitamins and minerals, and requires careful management to ensure optimum nutrition at each stage.

Changes in nutrient demand throughout the lactation period

After calving, the milk yield/day increases to a maximum between weeks 4 and 10 after calving, depending on nutrition, breed and genetic potential (Fig. 8.6). High-producers peak later than low-producers. The higher the peak yield the higher the total lactation yield (every 1 kg peak yield equates to between 100 and 200 kg total lactation yield). In the case below, the peak yield is ~40 kg milk/day with a total lactation yield of ~8600 kg. The rate of decline after the peak is 4–5%/month, but this is also dependent on genetics. The rate of decline is called 'persistence' and is of great interest because there is variation between genotypes in the rate of decline in milk yield post-peak (Auldist *et al.* 2007), which can be

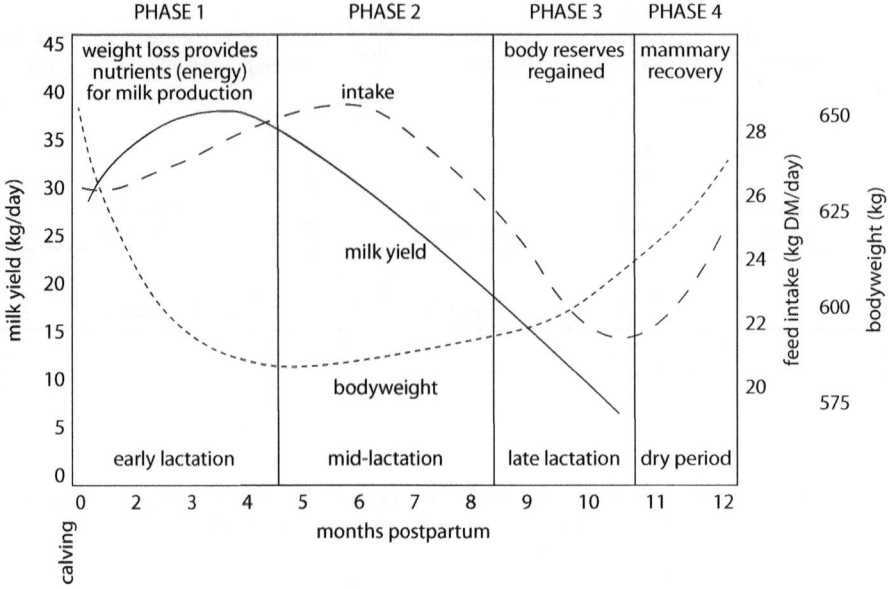

Fig. 8.6. Milk production, dry matter intake and bodyweight throughout a typical lactation in a moderate-producing dairy cow.

exploited in selection programs (see 'Common nutritional and metabolic diseases of dairy cows' later).

In early lactation (calving to peak lactation), the rate of dry matter and energy intake is insufficient to meet the high demands of high milk production. The animal loses weight and she is effectively 'milking off her back': that is breaking down fat tissues (and muscle proteins), to provide the non-esterified fatty acids (NEFA) and amino acids for milk fat and milk protein synthesis, and for the ATP required to synthesise milk components and secrete them into the mammary gland.

> Every kilogram of bodyweight loss during early lactation is equivalent to 28 MJ of feed Metabolisable Energy (this is the equivalent of the energy in ~2.3 kg grain).

High-producing dairy cows are capable of losing up to 2 kg of bodyweight daily, which provides the equivalent of 56 MJ of ME/day, or 4.5 kg of a high-concentrate ration. It is this ability to partition energy from body tissues to the mammary gland that is one of the marvels of the modern dairy cow and successful genetic selection for milk production.

As milk production declines in mid-lactation, intake has now responded, belatedly, to the high energy demands, and Net Energy balance is now positive, so bodyweight stabilises. Further declines in milk yield in late lactation reduce energy demands and, despite a decline in intake, the Net Energy available is positive and the animal gains weight again.

The final period of the lactation cycle is the 'drying off' period during which the animal is heavily pregnant, she is not lactating and her bodyweight has increased substantially. This is a critical period because it allows the epithelium of the mammary gland to rest and recover. This process is called involution and is characterised by a rapid decline in lactose synthesis and water ingress into the mammary gland lumen (Watson 2006).

Energy requirements of dairy cattle

The total ME requirement of a lactating dairy cow is calculated by adding up her requirements for maintenance, milk production, activity, pregnancy and then making adjustments for actual or desired changes in bodyweight and condition score. Bodyweight loss, particularly in early lactation, contributes positively to energy availability so this energy is subtracted from the total dietary requirement. Of all these components, the ME requirement for lactation is the greatest demand in high-producing cows (e.g. ME for maintenance = say 60 MJ/day; while ME for milk production (40 L/day) = 200 MJ/day)!

1. Metabolisable Energy requirement for maintenance

The ME requirement for maintenance of dairy cows differing in bodyweight is described by the equation (MAFF 1984):

$$\text{Dairy cow maintenance ME requirement (MJ/day)} = 0.52W^{0.75}$$

But this can be simplified to:

$$\text{Dairy cow maintenance ME requirement (MJ/day)} = 0.09W + 9.5$$

with very little loss of accuracy.

The ME requirement for maintenance of dairy cows differing in bodyweight using the simple equation are shown in Table 8.2.

2. Metabolisable Energy requirements for pregnancy

The additional energy required to carry the developing calf only become significant from months 6 to 9 (Table 8.3).

Add 8, 10, 15 and 20 MJ additional energy to the maintenance requirement over the last 4 months of gestation.

Table 8.2. ME requirement for maintenance of dairy cattle differing in bodyweight

Bodyweight (kg)	ME requirement (MJ/day)
100	19
150	23
200	28
250	32
300	37
350	42
400	46
450	51
500	55
550	60
600	65
650	68

Table 8.3. Additional ME required for pregnancy in months 6 to 9 in dairy cows

Month of pregnancy	Additional ME required (MJ/day)
6	8
7	10
8	15
9	20

Source: MAFF (1984)

3. Metabolisable Energy requirement for milk production

The ME required to produce 1 L of milk varies with the composition of the milk, and in particular its fat content (Table 8.4).

In general, allow 5 MJ/L of milk produced for Holstein Friesians and 6 MJ/L of milk for Jersey cows.

4. Metabolisable Energy requirements for activity of dairy cows

The minor activities of grazing and foraging are included in the maintenance energy requirement, but an additional activity estimate is for 1 MJ/km walking to and from the dairy. If the country is hilly, allow 5 MJ/km walked throughout the day.

Add 1 MJ for every kilometre walked to the milking parlour, or 5 MJ/km for walking in hilly country.

Table 8.4. Metabolisable Energy requirement to produce a litre of milk of varying protein and fat composition

Fat (%)	Protein (%)									
	2.6	2.8	3.0	3.2	3.4	3.6	3.8	4.0	4.2	4.4
3.0	4.5	4.5	4.6	4.7	4.8	4.8	4.9	5.0	5.0	5.1
3.2	4.6	4.7	4.7	4.8	4.9	5.0	5.0	5.1	5.2	5.2
3.4	4.7	4.8	4.9	4.9	5.0	5.1	5.2	5.2	5.3	5.4
3.6	4.9	4.9	5.0	5.1	5.1	5.2	5.3	5.4	5.4	5.5
3.8	5.0	5.1	5.1	5.2	5.3	5.3	5.4	5.5	5.6	5.6
4.0	5.1	5.2	5.3	5.3	5.4	5.5	5.5	5.6	5.7	5.8
4.2	5.3	5.3	5.4	5.5	5.5	5.6	5.7	5.7	5.8	5.9
4.4	5.4	5.5	5.5	5.6	5.7	5.7	5.8	5.9	6.0	6.0
4.6	5.5	5.6	5.7	5.7	5.8	5.9	5.9	6.0	6.1	6.2
4.8	5.6	5.7	5.8	5.9	5.9	6.0	6.1	6.1	6.2	6.3
5.0	5.8	5.8	5.9	6.0	6.1	6.1	6.2	6.3	6.3	6.4
5.2	5.9	6.0	6.0	6.1	6.2	6.3	6.3	6.4	6.5	6.5
5.4	6.0	6.1	6.2	6.3	6.3	6.4	6.5	6.5	6.6	6.7
5.6	6.2	6.2	6.3	6.4	6.5	6.5	6.6	6.7	6.7	6.8
5.8	6.3	6.4	6.4	6.5	6.6	6.7	6.7	6.8	6.9	6.9
6.0	6.4	6.5	6.6	6.6	6.7	6.8	6.9	6.9	7.0	7.1

Source: MAFF (1984)

5. Metabolisable Energy requirement to gain one body condition score

Body condition scoring of dairy cows is an essential part of managing the dairy herd to ensure good reproductive performance, milk production, feed conversion efficiency, cow health and welfare. A system of scoring between 1 and 8 is often used for dairy cows, but this varies with country. The USA and Ireland use a 5-point scale and New Zealand uses a 10-point scale. The same landmarks are used in all systems so it is a matter of adopting a system that suits you and then become proficient at using this system. Smartphone apps are now available that allow body condition score of cows to be determined from phone pictures (e.g. Bayer Cowdition; <http://animalhealth.bayer.com/en/treatment-care/farm-animals/bcs-cowdition-app/>), which allows different systems from different countries to be used).

Body condition scoring provides valuable information on previous nutrition, likely future productivity and future feed requirements. Cows that are too thin at calving are less fertile because they take longer to recommence cycling and have lower submission and conception rates. They also produce less milk because they partition more feed energy to body gain and away from milk production. Cows that are too fat at calving are prone to metabolic disorders (ketosis, milk fever) and eat less feed, and therefore produce less milk. Cows that lose too much condition during early lactation have reduced fertility.

Table 8.5. Calculating the ME requirement of dairy cows

1. **Estimate the ME required for maintenance**	$ME_m = 0.09W + 9.5$
2. **Estimate the energy required for milk production**	ME for milk (MJ) = milk yield (kg) × 5 (or from Table 8.4)
3. **Estimate the energy required for activity**	ME activity (MJ) = km walked × 1 MJ for flat or 5 (for hills)
4. **Estimate the energy available from weight loss**	ME from weight loss = −28 MJ × weight loss (kg)
5. **Estimate the energy required for weight gain**	ME for weight gain = weight gain (kg) × 44 (late lactation) or 55 (dry period)
6. **Estimate the energy required for pregnancy**	ME pregnancy = 8, 10, 15 or 20 MJ for months 6, 7, 8, 9

The aim should be to have the majority of cows, at calving, in condition score 4.5 to 5.5 in an 8-point system, and not to lose this condition by more than ~0.5 units by mating so the cows are still in condition score 5, or close to it, at mating. The following condition score (CS) targets for dairy cows at different stages of the cycle are based on a scores 1–8 system:

Calving
- No more than 15% below CS 4.5
- No more than 15% above CS 5.5

At mating
- Average decrease in CS since calving no more than 0.6 units
- No more than 15% lose more than 1 CS since calving
- Cows maintain or gain CS from commencement of mating

At drying off
- Cows maintain or gain CS during drying off
- Cows at CS 5 at calving.

To gain 1 condition score unit requires a weight gain of ~44 kg in a typical 550 kg cow, and this requires an additional 44 MJ of ME. Conversely, losing 1 kg of live weight provides ~28 MJ ME, which is useful during early lactation as a source of energy for high milk production

Estimating the total ME requirement of a dairy cow

The method for calculating the ME requirement of dairy cows is shown in the Table 8.5.

Example: A 550 kg cow, housed (no activity), in 7 month of pregnancy, producing 10 L milk/day of 3.6% fat and 3.2% protein and gaining 0.5 kg/day.

1. ME maintenance = $(550 \times 0.09) + 9.5 = 59$ MJ/day
2. ME milk = 5.1×10 L = 51 MJ/day
3. ME for activity = 0 MJ
4. Energy from weight loss = 0 MJ
5. Energy for weight gain = $0.5 \times 44 = 22$ MJ

TOTAL = $55 + 51 + 0 + 0 + 22 = 132$ MJ/day

Protein requirements of dairy cattle

The protein requirements of dairy cattle vary throughout the dairy cycle (Fig. 8.1) and during lactation the requirement is particularly high to provide the amino acids required for milk protein synthesis. The protein requirement of dairy cows is expressed in two ways. Dietary crude protein is, by definition, a crude but useful means of expressing requirements. The crude protein requirements at different stages of lactation are given in Table 8.6.

The second means of expressing protein requirements in dairy cows is to use the Metabolisable Protein system based on estimating the Rumen Degradable Protein (RDP) in the feed and the Undegraded Dietary Protein (UDP) or Undegraded Intake Protein (UIP). Feeds containing only RDP can only produce up to 12 L of milk/day, so high-producing cows producing perhaps 40 L/day in peak lactation require a high proportion of the dietary protein to be protected from ruminal degradation. Some feed proteins are particularly high in UIP and are therefore usefully included in dairy feeds.

> Microbial protein from ruminal fermentation is sufficient for only 12 L of milk/day in dairy cows. Additional milk production requires protein that escapes ruminal fermentation.

Total nutrient requirements for dairy cows at different stages of lactation

The nutrient requirements of cows at different stages of lactation are shown in Table 8.7 for a ration containing 11.0 MJ/kg dry matter.

Table 8.6. Crude protein requirement of dairy cows at different stages of lactation

Lactation stage	Dietary crude protein (%)
Early lactation	16–18
Mid-lactation	14–16
Late lactation	12–14
'Dry'	10–12

Source: Target10 (2002)

Table 8.7. Nutrient requirements of a 650 kg Friesian dairy cow at different stages of lactation

Nutrient requirements	Stage of lactation		
	Early	Mid	Late
Average milk yield (kg/day)	40	30	20
Dry matter intake (kg/day)	25	21–23	11–12
Water requirement (L/day)[a]	270	210	160
Metabolisable Energy (MJ/day)	245	209	159
Metabolisable Energy density (MJ/kg DM)	11–12	11	10
Crude Protein (% DM)	16–18	14–16	12–14
Metabolisable Protein (% DM)[b]	11.7	11.2	10.8
Rumen Degradable Protein (% CP)	55–60	65–70	75
Undegraded Dietary Protein (% CP)	35–40	30–35	25
NDF (%)	30–34	30–38	33–43
Effective fibre (% NDF)	25	25	25
Fat (% DM)	5–6	4–6	3–5
Calcium (% DM)	0.60	0.61	0.62
Magnesium (% DM)	0.21	0.19	0.18
Phosphorus (% DM)	0.38	0.35	0.32
Potassium (% DM)	1.07	1.04	1.00
Sodium (% DM)	0.23	0.22	0.23
Chlorine (% DM)	0.29	0.26	0.24
Sulphur (% DM)	0.20	0.20	0.20
Cobalt (mg/kg DM)	0.2–0.3		
Copper (mg/kg DM)	15–30		
Manganese (mg/kg DM)	55		
Zinc (mg/kg DM)	75		
Iodine (mg/kg DM)	0.8–1.4		
Iron (mg/kg DM)	50–100		
Selenium (mg/kg DM)	0.3		
Vitamin A (1000 IU/day)	100–200		
Vitamin D (1000 IU/day)	20–30		
Vitamin E (IU/day)	700	500	500

[a] Water requirements increase dramatically with ambient temperature
[b] See Fig. 3.22 for Metabolisable Protein calculation
Source: NRC (2001)

Designing and maintaining a feeding program for dairy cattle

Before we can design a feeding program for dairy cows at each stage of the lactation cycle, it is first necessary to determine the energy and protein requirements for maintenance, growth and lactation. We then need to estimate the likely intake of pasture, then balance the ration to meet the nutrient requirements and ensure herd health. These steps were described earlier.

Formulating balanced rations for dairy cows depends heavily on the production system operating. Feedlot dairies provide the opportunity to control the rations fed, while pasture-based dairies require formulations that are integrated with the nutrients consumed from the pasture. However, in both systems, cows are at different stages of lactation and body condition status, and therefore have different nutrient requirements. Even in feedlot dairies, unless animals are individually fed, it is difficult to provide the optimum nutrition to each individual cow. Ideally, cows should be fed according to their needs at their stage of the cycle. This is called 'phase feeding'. The four phases are shown in Fig. 8.6. The differing nutritional characteristics of each phase are described in Table 8.8.

Formulating rations for feedlot-fed dairy cattle

Now we know the nutrient requirements of the dairy cow throughout the dairy cycle, we can formulate rations to meet her needs at each stage of the cycle, provided we also know her production levels, bodyweight, stage of gestation and body condition. Rations can be formulated by hand or using computer programs. An example of a hand calculation is as follows:

Example: 650 kg Friesian cow producing 30 L of milk per day containing 3.2% protein and 3.5% fat. She is not pregnant and calved 20 days ago. She is in condition score 5/8 and is losing 0.5 kg/day. In a feedlot dairy she has little activity during the day.

Requirements for energy:
ME (maintenance) = 0.09W + 9.5 = +68 MJ/day (Table 8.2)
ME (milk) = 5.0 × milk yield (30) = +150 MJ/day
ME available from weight loss = 28 × 0.5 = –14 MJ/day
So her total ME requirement is 68 + 150 – 14 (because this is energy she is getting from weight loss and not needed from feed) = 204 MJ/day.

Requirements for crude protein:
Early lactation = 16–18% CP (35–40% UIP)
Tables of the composition of available feeds or results from feed test of your feeds can then be used to prepare a diet that will meet these requirements.
Having formulated a ration for this cow, it is essential that it is checked against the ability of the cow to consume that much of the ration. For example, if the ration contains, say, 11.5 MJ/kg dry matter and the cow's requirement is 204 MJ/day the cow has to eat 17.7 kg dry matter/day. Maximum dry matter intake can be estimated from the NDF content of the

Table 8.8. The nutritional characteristics of the four phases of the dairy cycle

Phase	Characteristics	Nutrient requirements	Potential nutritional issues
1 **Peak lactation** **0–100 days**	• Rapid increase in milk production • Delayed intake response • Bodyweight loss and mobilisation of fats and proteins from body tissues	• Ideal BCS at calving = 4.5–5.5 • Limit BCS losses to <0.5 • Maximise dry matter intake (high-quality forage, balanced ration, constant access to feed, low stress): Week 1: 2.5% bodyweight (16 kg) Week 2: 2.9% bodyweight (19 kg) Week 3: 3.4% bodyweight (21 kg) Week 4: 3.6% bodyweight (22 kg) Week 5: 4.0% bodyweight (24 kg) • Maintain good rumination and rumen pH >5.5 (NDF >28% with high effective fibre; 40% of ration as forage; particle length 50% >2.6 cm; avoid 'slug' feeding concentrates; feed forages before concentrates; feed always available) • Diet CP% = 16–18% (35–40% UDP) • Forage ME = 9.5 MJ/kg DM • Concentrate ME = 11–12 MJ/kg DM • Maximum of 7% total fat	• Acidosis • Subacute ruminal acidosis (SARA) • Ketosis • Hypocalcaemia/ hypomagnesaemia • Left displaced abomasum
2 **Mid-lactation** **100–200 days**	• Intake maximal (4% of bodyweight) • Attempt to maintain peak lactation as long as possible	• Maximise dry matter intake (high-quality forage at 40–45% of ration with high effective fibre; balanced ration; frequent feeding, minimise stresses) • Diet CP% = 14–16%	• Rapid decline in milk production • Low fat test • Ketosis • Silent heat (she is mated in this period)
3 **Late lactation** **200–300 days**	• Milk production is declining, the cow is pregnant and intake will exceed demands so cows put weight back on after early lactation • Young cows are still growing and require additional nutrients	• Sufficient energy intake to allow weight gain	• Few issues in this phase
4 **Dry period (includes the transition period (close up cow)** **300–365 days**	• Body maintenance/ growth to reach BCS 4.5–5.5 • Intake at 1% of bodyweight • Sufficient energy for fetal growth • Adaptation to the lactation diet and preparation for the high calcium requirement of early lactation	• Limit high-energy feeds to avoid over fattening avoid high Ca and P intakes • Introduce transition feed 2 weeks before calving • Introduce grain slowly to adapt rumen microbes (2–4 kg grain) • Provide anionic salts (DCAD management) • Maintain rumination (long effective fibre) • CP% = 14–15%	• Poor feeding in this phase can reduce lactational performance and produce metabolic problems • Poor adaptation produces acidosis, SARA, ketosis, hypocalcaemia in Phase 1

diet (= 120/NDF%; see Fig. 4.6). A typical dairy ration will contain 35% NDF so maximum intake of this cow would be 120/35 = 3.4% bodyweight = 22.1 kg which is greater than the 17.7 kg needed.

Formulating rations for dairy cows grazing forages

The reality now is that few dairy rations are formulated by hand and most are generated by computer programs such as the NRC (2001) model or the Cornell Net Carbohydrate and Protein System (CNCPS model). The later estimates pasture intake, and hence nutrient supply, from pasture and then balances these nutrients with supplements that can be fed in the milking parlour. More than 100 inputs related to pasture composition, environment, animal production and characteristics are required to run the model. However, the output then relates only to those inputs and, as the pasture, environment and animal all change with time, the inputs must also be changed. Forage changes with time must be monitored by regular sampling and feed testing.

One of the deficiencies in computer-based programs for grazing cattle is estimation of forage intake when the animals are supplemented. Supplements have a substitution effect, which is difficult to predict. Some supplements stimulate forage intake but most decrease forage intake to a varying extent (see Fig. 5.10).

Feed additives used in dairy production systems

Several feed additives are proposed with claims of economic benefits for dairy cattle (Table 8.9).

Grazing management of forages to maximise dairy productivity and profitability

Feed represents the highest variable cost in dairy enterprises and home-grown pasture forages are the cheapest source of nutrients on the dairy farm. Maximising the quantity and quality of forages is therefore a major priority in dairy nutrition and management. Ideally, a grazed pasture offered to cows would be 10–15 cm high and with total feed-on-offer of 1800–2200 kg of dry matter/hectare (see Chapter 5). The majority of the feed available would be leaf and when the cows have finished grazing the area it would be at 3–4 cm high and total feed on offer would be ~1300 kg dry matter/hectare. Maintaining an optimum height and density of pasture is essential to maximise the capture of light by photosynthesis and to minimise the losses associated with trampling, pugging (trampling into the mud) and other losses. Planned grazing (see page 144) allows the capture and harvesting of the highest amount of high-quality plant material as possible and dairy farmers are expert at managing this capture despite the vagaries of climate and season. To allocate the correct area for grazing you need to know the pasture growth rate, pasture available and the desired minimum biomass you want to achieve.

Table 8.9. Feed additives, claims about their efficacy, and relevant comments

Feed additive	Use and claim	Comments	References
Rumensin	• Increased feed efficiency • Reduced ketosis • Reduced LDA • Reduced methane output • Increased propionate production	• Feed to dry cows (metabolic disorders) and lactating cows	Duffield *et al.* (1998)
Rumen buffers (NaHCO$_3$)	• Reduced acidosis • More stable rumen • Increased fibre digestion • Increased milk fat percentage	• Feed in diets high in grain and soluble carbohydrates	Erdman (1988)
Yeasts	• Stimulate fibre digestion • Stabilise rumen environment • Utilise lactic acid	• 2 weeks pre-calving to 10 weeks post-calving • Calf starter feeds • Stress conditions	Guedes *et al.* (2008)
Niacin	• Improved lipid metabolism • Improved rumen fermentation • Increased milk efficiency • Increased milk fat and protein	• High-producing cows in negative energy balance • Ketotic-prone cows • 2 weeks pre-calving to 10 weeks post-calving	Schwab *et al.* (2005)
Anionic salts	• Reduce DCAD to −15 meq/100 g DM • Stimulate bone mobilisation mechanisms to supply calcium in early lactation	• Dry cows 2 weeks pre-calving to calving	Goff *et al.* (2004)

Maximising the feed intake of dairy cows

Feed intake regulation in ruminants is covered in Chapter 4 (see page 106). For dairy cattle, maximising intake is critical to meeting the high levels of production now demanded of the modern dairy cow. There are several strategies we can use to maximise energy intake of the dairy cow but still maintain health (Allen 2000):

1. **Optimise the NDF content and type in the feed:**
   ➤ optimise NDF levels (30–35%) (Fig. 4.6)
   ➤ increase the NDF digestibility
   ➤ highest producing cows and cows in peak lactation are most limited by gut fill so allocate the highest digestibility forages to these cows.
2. **Reduce propionate absorption:** although propionate is essential for milk synthesis, propionate is also a strong satiety signal. Reducing the fermentable starch in the ration or using a less rumen-fermentable starch source can greatly increase intake.
3. **Optimise protein intake:** inadequate protein (particularly Digestible Undegraded Protein) can limit energy intake. Provision of the first-limiting amino acids, methionine and lysine can also increase intake and production.

4. **Group feed cows at a similar stage of lactation:** grouping cows with similar energy requirements has been shown to increase intake and allows stage-specific diets to be fed.

Feeding replacement dairy heifers

The nutrition of dairy heifers is often the most overlooked aspect of dairy systems and yet it is vital to ensure heifers grow to target weights and condition scores as they approach the first pregnancy and lactation. The key periods of development include the calf-rearing phase in which mortalities from disease can be high, and the rumen-development phase in which the transition from milk feeding to solids feeding occurs, setting the animals up for continued steady growth to first mating.

Nutrition of the newborn calf

Well-fed calves are less likely to succumb to disease and particularly bacterial infections. They are more robust and more likely to grow well, so ensuring a good start to life is critical. Newly born calves require access to at least 4 L of high-quality colostrum-milk within the first 6 h of life. Colostrum 'quality' relates to the concentration of immunoglobulin antibodies. The concentration of antibodies in cow's milk increases with the age of the cow because she has been exposed to more infectious diseases in her lifetime. For this reason, replacement heifers born to first-calving cows may require additional colostrum stored or harvested from older cows. Colostrum can be stored fresh for 7–10 days or frozen for 18 months. Most calves receive their required colostrum from their mothers for the critical first 48 h of life, but any calf not suckled in the first 3–6 h should be hand fed colostrum from another cow or from stored colostrum. Sick or weak calves should have the colostrum administered by stomach tube. The critical period of colostrum feeding is the first 24 h, because the capacity of the intestines to absorb the long polypeptide immunoglobulin chains diminishes rapidly after this time. As many as 60% of calves left with their mothers for the first 24 h fail to receive sufficient colostrum, so intervention is often required. Care is required to ensure that cross contamination of calves does not occur in this critical window of immune transfer so scrupulous hygiene is important.

The colostrum-feeding phase is followed by the milk-feeding phase. Milk sucked from teats will bypass the rumen through the oesophageal groove: a muscular channel running from the base of the oesophagus to the omasum (Fig. 3.6). It is important to ensure that the milk does not enter the rumen because this can cause scours (diarrhoea). Calves should be fed milk at ~15% of bodyweight per day (~4–5 L) in two or more daily feeds until 4 weeks of age and then once/day.

The objective of the milk-feeding phase is to encourage the development of the reticulo-rumen as quickly as possible so that the calf can transfer to solids feeding by ~6 weeks. This is best achieved by offering the calves high-quality calf-rearing pellets or grain and good quality hay from week 1. The protein content of the pellets should be high (20–22%). The development of the ruminal epithelium is stimulated by the volatile fatty acid products of fermentation, particularly butyric acid (see Table 3.2). Pasture forages are inadequate for calves because they contain too much water and have a low density of energy/kg fresh material. They also contains inadequate fibre for rumen development. Provision of good-quality straw allows the calves to nibble the straw, which encourages rumen development rather than provide nutrients that are best obtained via the pellets or grain. Clean drinking water is critical at all times.

There are numerous calf-rearing systems but those that involve cafeteria-style teats from well-cleaned units in clean, dry, well-ventilated open sheds or hutches, with clean straw, with batches of same-age calves with access to clean water, and high-quality pellets or concentrates, work best. Getting the calf to transfer from milk to solids is economically important (see Table 8.10).

Clearly, the longer the milk or milk replacer is fed to the calf to weaning, the greater the costs.

> The objective of calf-rearing is to minimise disease, encourage early rumen development and transition from milk to solids as early as possible. Calf-rearing systems should produce calves weighing 100 kg at weaning at 12 weeks of age (growth rates of 0.7 kg/day average).

Table 8.10. Costs of energy and protein in milk versus concentrates

Feed	Cents/MJ Metabolisable Energy	Cents/kg crude protein
Whole milk	6.9	670
Milk replacer	12.0	1000
Pellets	2.9	475
Farm TMR mix	2.2	161

Source: Adapted from Moran (2005)

Feeding the weaned replacement heifers

Having successfully raised healthy, well-grown weaned heifers, it is common for producers to neglect these animals once they are turned out to pasture. Although these animals need less attention than calves, it is important that target weights are achieved for mating and first calving (Table 8.11).

Table 8.11. Target weights for US and Australian Friesian heifers at weaning, 12 months, mating and calving

Age (months)	Live weight (kg)	Wither height (cm)
1–3 (weaning)	90–110	90
12 months	270–300	124
15 months (mating)	330–360	130
24 months (calving)	520–550	142

Undersized heifers are permanently restricted in their milk-producing ability, have more calving difficulties and take longer to get back into calf during their first lactation. They are also still growing and use energy for growth instead of milk production. Puberty is related largely to live weight, not age, so meeting the targets in Table 8.11 dictates the time of first mating and pregnancy. Restricted live weights translate to very significant losses of total milk yield across the first three lactations. Heifers differing by 100 kg live weight at calving produce 1000 L less milk in their first lactation.

It is important, however, that growth rates are not too high, because rapid growth during udder development increases fatty tissue deposition in the udder, which permanently reduces milk production. High energy relative to protein intakes predispose heifers to fatty udder syndrome.

Energy and protein requirements of growing heifer replacements

Table 8.12 shows the recommended daily ME intakes and CP content of diets fed to growing heifers.

Pasture is the cheapest source of nutrients for dairy cows and will make up most of the heifer diet, with conserved fodders and concentrates making up the pasture shortfalls. To achieve the ME and protein intakes required for growth and udder development, pastures and supplements should contain 10–11 MJ/kg DM and 15–17% CP. The young heifers have lower total intake capacity so should be allocated the best quality pastures (11 MJ/kg DM).

Table 8.12. Metabolisable Energy and crude protein requirements of replacement heifers growing at the recommended rate of 0.6 kg/day

Live weight (kg)	ME requirement (MJ/day)	CP (%)
100	33	17
200	48	16
300	64	15
400	84	13
500	99	13

Source: NRC (2001)

Common nutritional and metabolic diseases of dairy cows

Many of the nutritional and metabolic diseases of dairy cows are related to the high yields of milk now achievable through genetic selection and the feeding of high-energy rations. Most of these diseases and syndromes occur during the transitional feeding periods (introduction to high-concentrate rations) and during periods of rapid weight loss and high milk yield (Table 8.8). Common problems during these phases are acute ruminal lactic acidosis, subacute ruminal acidosis, ketosis, hypocalcaemia, left displaced abomasa and milk fat depression (see review by Ingvartsen 2006).

Milk fat depression in dairy cows

Milk fat is the major energy component of milk and is a major influence on the processing properties of milk and milk products. Milk fat is also a component of many pricing systems. So, changes in milk fat content can influence the ME requirement of the cow, the quality of the milk and milk products, and the price/litre of milk. In ruminants, ~50% of milk fats are synthesised in the mammary gland *de novo* from the precursors acetate and betahydroxybutyrate derived from ruminal fermentation and 50% from circulating fatty acids derived directly from intestinal absorption of dietary and microbial fatty acids or from mobilisation of fatty acids from adipose tissue (Fig. 8.4). The depression in milk fat (MFD) can be dramatic, with concentrations falling below 2% in some cases (Bauman and Griinari 2003). MFD is typically seen in dairy cows on low-fibre, high-grain diets or when rations contain high levels of polyunsaturated oils such as full-fat seeds, oilseed meals or marine oils (Davis and Brown 1970). Oils from plant-based sources can only produce MFD when low-fibre diets are fed, but marine oils will induce MFD even when cows are on high-fibre rations (Chilliard *et al.* 1999).

There are three theories as to the causes of MFD in dairy cows:

1. There is inadequate rumen-derived acetate and butyrate for milk fat synthesis.
2. High levels of rumen-derived propionate and subsequently high rates of glucose synthesis (gluconeogenesis) induces high insulin levels, which reduce the availability of precursors for fat synthesis.
3. Biohydrogenation of products in the rumen produce unusual products that directly reduce fatty acid synthesis. This is known as the rumen biohydrogenation hypothesis.

The rumen biohydrogenation hypothesis is now widely accepted as the cause of milk fat depression. When dietary polyunsaturated fatty acids enter the rumen, they are hydrogenated by rumen microbes to saturated fatty acids such as stearic acid (C18:0). In this process, intermediate fatty acids such as conjugated

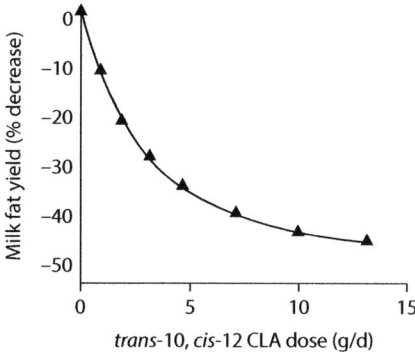

Fig. 8.7. Relationship between trans-10, cis-12 CLA and milk yield depression. CLA was infused per abomasum. Source: Adapted from Bauman and Griinari (2003).

linoleic acid (CLA) are formed. One of the intermediates in this process is trans-10, cis-12 CLA, and this compound has been shown to be a potent inhibitor of the genes involved in milk fat synthesis (Peterson *et al.* 2003). These compounds arise when the diet of the dairy cow contains highly fermentable carbohydrates and high levels of polyunsaturated fatty acids, such as plant oils and marine oils. In most pasture forage species the predominant polyunsaturated fatty acid is linoleic acid, with levels as high as 5% of the dry matter. The abundance of mRNAs encoding fatty acid synthases and the activity of fatty acid synthase enzymes is decreased by ~40–60% by CLA isomers and this corresponds to a 43% drop in milk fat percentage (Piperova *et al.* 2000). The relationship between the levels of trans-10, cis-12 conjugated linoleic acid and milk fat depression is shown in Fig. 8.7.

It is now widely accepted that milk fat depression in dairy cows on low-fibre, high-grain diets is due to the formation of unique fatty acid intermediates of hydrogenation of linoleic acid, particularly trans-10, cis-12 conjugated linoleic acid, which depresses genes involved in milk fat synthesis.

Hypocalcaemia in dairy cows (periparturient paresis, milk fever)

Calcium levels in blood are maintained by homeostatic mechanisms to ensure that the vital functions of calcium are maintained. These include nerve and muscle function, blood clotting, maintaining bone density, mitosis in cells and production of hormones. The 4-week period pre-calving and 4-week period post-calving is a period of transition from pregnant, non-lactating to non-pregnant, lactating (the periparturient period) and in the high-producing dairy cow is one of extreme physiological and metabolic change. Superimposed on this profound reproductive physiological change is a change in diet, energy demand, mineral status and live weight status. Homeostatic mechanisms operating to maintain 'constancy' of the body systems are working overtime. Failure of these mechanisms is reflected in disorders such as downer cow syndrome, displaced abomasum, acidosis, ketosis, udder oedema, metritis, hypomagnesaemia and hypocalcaemia. These disorders are often inter-related.

Hypocalcaemia is one such disorder, which is often referred to as milk fever even though no elevated core temperature is apparent. The more precise term for this disorder is periparturient hypocalcaemia and the predominant physical sign is periparturient paresis (literally weakness around the time of birth), referring to the inability of the cow to stand (recumbency) due to muscle weakness. However, there are many post-calving causes of recumbency and only ~40% are due to hypocalcaemia (Lean *et al.* 2006). Clinical hypocalcaemia is characterised by plasma Ca levels <1.4 mM and subclinical hypocalcaemia by plasma Ca between 1.4 and 2.0 mM. Both clinical and subclinical hypocalcaemia are risk factors for mastitis, displaced abomasum, ketosis, retained placenta, uterine prolapse and reduced milk yield. The incidence of hypocalcaemia is highly variable between herds, indicating the extent to which the condition can be manipulated by management and nutritional intervention. Reports in many countries reveal an incidence of hypocalcaemia of between 0 and 10%. A meta-analysis by Lean *et al.* (2006) revealed a mean incidence of 21%, but with a range from 0% to 83%.

Cows at highest risk of milk fever are:

- older cows (risk increases 9%/lactation)
- channel island breeds (Jerseys, Guernseys) but the heritability is low (DeGaris and Lean 2009)
- high-producing cows
- over-conditioned cows
- cows with hypomagnesaemia
- on diets with a high, positive dietary cation/anion difference
- on diets high in potassium
- on pre-calving diets high in calcium
- on pre-calving diets high in phosphorus.

Pathophysiology of hypocalcaemia in dairy cows

Plasma calcium in cows is maintained within tight limits (2.0–2.5 mM = 8.5–10 mg/100 mL) by homeostatic endocrine mechanisms (Fig. 8.8). When blood calcium levels fall below ~2.0 mM the parathyroid gland is stimulated to

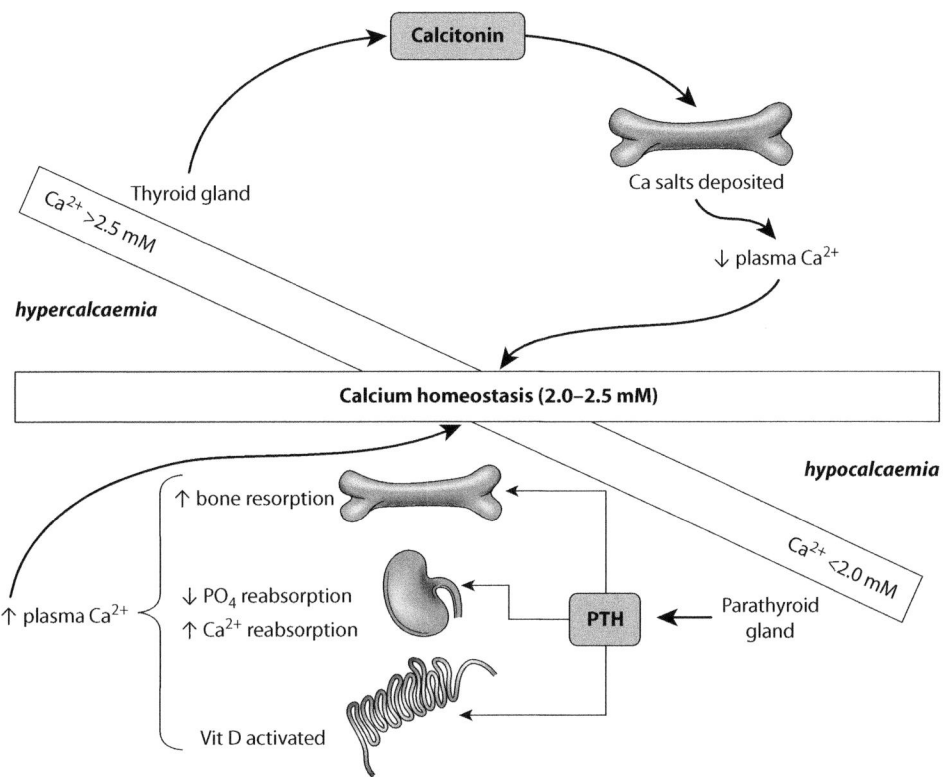

Fig. 8.8. Calcium homeostasis in dairy cows.

release the hormone, parathyroid hormone (PTH), which enters the circulation and interacts with its receptor on the surface of kidney cells and bone cells. This PTH/receptor complex triggers G-proteins within the cell and the adenylate cyclase complex is activated to produce cyclic AMP, a second messenger in the cytosol, which initiates bone calcium resorption and renal activation of vitamin D (1,25-dihydroxycholecalciferol). Vitamin D, in turn, stimulates the intestinal cells to increase calcium absorption and the kidneys to reabsorb calcium from the tubules. The kidneys also reduce PO_4 reabsorption, so blood phosphate levels decline, freeing up calcium previously bound as $CaPO_4$. Hypocalcaemia occurs if these calcium-scavenging mechanisms fail to extract sufficient calcium from bones and the diet to replace the calcium lost in milk. Before calving, the calcium losses in a dairy cow are ~30 g/day but calcium lost in milk each day can exceed 50 g, putting huge demands on calcium supply. Not surprisingly, ~25% of heifers and 50% of older cows have blood Ca levels <2.0 mM (Goff 2008).

Milk fever is not a nutritional deficiency of calcium. It is a failure of the homeostatic regulation of calcium when lactation demand for Ca is suddenly high.

Prevention of hypocalcaemia and periparturient paresis by dietary cation anion difference manipulation

Blood pH is affected by the concentration of dissolved CO_2, the concentration of weak buffers such as plasma proteins, and the relative concentrations of anions (particularly Cl^- and S^{2-}) and cations (particularly Na^+ and K^+) in the blood (Goff 2008; DeGaris and Lean 2009). Diets high in Na^+ and K^+ and low in Cl^- and S^{2-} (i.e. with a positive cation/anion difference) produce metabolic alkalosis, while those low in cations and high in anions produce mild metabolic acidosis. The PTH receptors in bone and kidney cells (Fig. 8.8) are more sensitive to PTH signalling when the animal is in a state of metabolic acidosis (Goff 2008).

The most common means of calculating the acidifying strength of a diet is to calculate the Dietary Cation Anion Difference (DCAD) as follows:

Dietary Cation Anion Difference (DCAD) determines the acidity of blood and the cow's sensitivity to the calcium homeostatic mechanisms.

$$DCAD \text{ (meq/100 g DM)} = [Na^+ + K^+] - [Cl^- + S^{2-}]$$

The ideal DCAD is −10 to −15 meq/kg DM (i.e. more anions than cations).

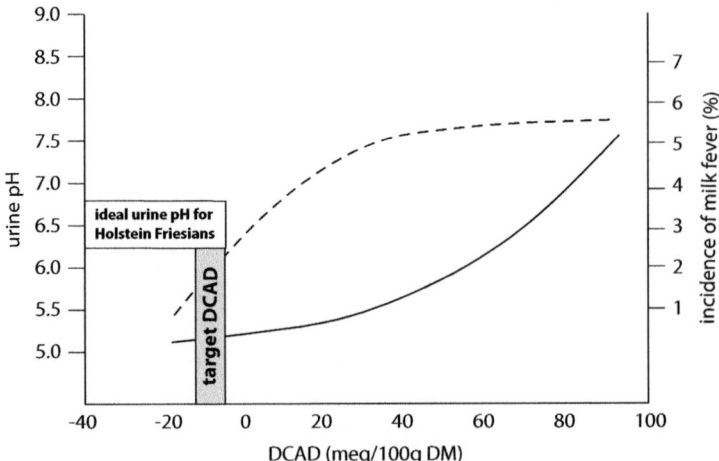

Fig. 8.9. Relationships between dietary DCAD (meq/100 g DM), urinary pH (dotted line) and incidence of milk fever (%, solid line). Source: Adapted from Degaris and Lean (2009).

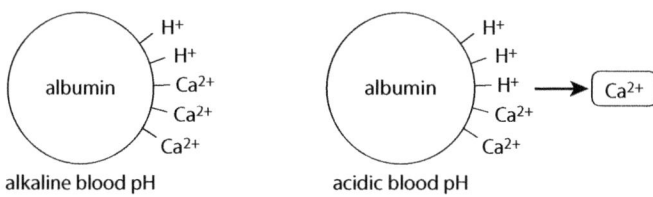

Fig. 8.10. Effect of blood pH on calcium release from weak buffers such as albumin. As blood pH declines and H^+ concentration increases, the H^+ ion displaces Ca^{2+} from the weak buffer, making it available for metabolism.

As the DCAD value decreases (i.e. relatively more anions than cations in the diet), the blood and urinary pH decrease and the incidence of milk fever decreases (Fig. 8.9), because the cow is able to respond to PTH more effectively. The bones can release Ca^{2+} more readily to meet the high demand for milk production.

Lower blood pH also allows H^+ ions to displace Ca^{2+} ions from plasma albumin (Fig. 8.10).

The level of magnesium in the blood also influences the development of milk fever because Mg is required for the interaction of PTH with its receptor. Hypomagnesaemia combined with metabolic alkalosis greatly reduces the effectiveness of PTH on bone resorption and hence Ca release into the blood. High dietary K interferes with Mg absorption from the rumen and can induce hypomagnesaemia (see Fig. 7.2) in addition to its effects on the DCAD.

Hypocalcaemia in dairy cows is a result of a combination of high calcium demand for milk production, and failure of the hormonal homeostatic mechanisms due to metabolic alkalosis induced by high levels of cations in the diet relative to anions. Ensuring diets have a dietary cation/anion difference of –10 to –15 meq/100 g DM reduces the incidence of hypocalcaemia.

Ketosis (acetonaemia) and fatty liver (hepatic lipidosis) in dairy cows

Ketosis and fatty liver are common metabolic disorders in early lactation. Ketosis and fatty liver arise as a result of a combination of reduced energy availability, rapid catabolism of adipose tissue, increased non-esterified fatty acids (NEFA) in the blood and reduced glucose availability (Bobe *et al.* 2004). Figure 8.11 shows the theoretical changes in plasma levels of the three key compounds: NEFA from adipose tissue catabolism, β-OH butyrate (a key ketone body) and glucose. As NEFA increase immediately post partum, the ketone bodies increase concomitantly with the depletion in glucose availability produced by inadequate gluconeogenic precursors and utilisation of glucose for lactose and milk synthesis.

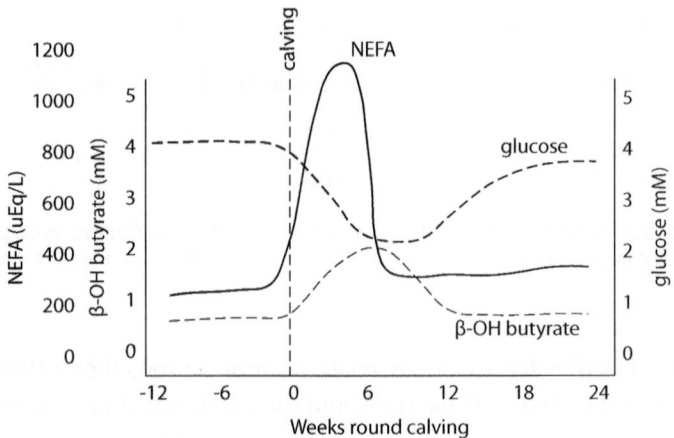

Fig. 8.11. Changes in plasma levels of NEFA (µEq/L), β-OH butyrate (mM) and glucose (mM) in dairy cows around calving. Source: Based on descriptions by Ingvarsten (2006).

A combination of rapidly increasing energy demands for fetal growth in the last few weeks of gestation, high glucose demand for lactose synthesis and a rapidly declining voluntary feed intake at this stage (typically dairy cows consume only 50–70% of their daily intake earlier in pregnancy), means energy requirements will not be met unless a high-energy ration has been introduced gradually over the transition period to lactation. About 7–14% of cows in early lactation show signs of ketosis. This energy imbalance is made worse if the cow has not been carefully transitioned to the new high-energy ration and acidosis or subclinical acidosis are present. Older cows and over-conditioned cows are particularly susceptible to this problem, because their intake decline is greater than heifers and cows in ideal body condition (BCS 4.5–5.5 in the 8-point scale) at calving. Weight loss and fat catabolism is desirable in allowing the cow to 'milk off her back', but too rapid a catabolism of fat unbalances the circulating levels of carbohydrate and fatty acids (NEFA) in the blood. Given the requirement for glucose from gluconeogenesis for functioning of the TCA cycle and utilisation of fatty acids (see Fig. 1.11) insufficient glucose to catabolise the NEFA results in excessive levels of fats in the circulation and in the liver where they are deposited as fat stores. Ketone bodies such as acetoacetate, β-hydroxybutyrate and acetone accumulate, because they cannot be metabolised to provide energy through the TCA cycle. Cows with ketosis, unlike most other animals, do not have concurrent metabolic acidosis.

Ketosis usually occurs just after parturition with cows showing the following clinical signs:

- low voluntary intake
- abnormal appetite (pica), abnormal licking
- bellowing, depression

- abnormal gait
- hard, dry faeces
- neurological signs
- decreased milk production
- acetone breath
- increased susceptibility to metritis, retained placenta, mastitis, abomasal displacement.

Cows with metritis, abomasal displacement and retained placenta are also at greater risk of developing ketosis. Ketosis can be clinical or subclinical, with the latter being the more common. Cows with subclinical and clinical ketosis show an increase in milk fat percentage and a decrease in milk protein percentage.

Prevention and treatment of ketosis and fatty liver in dairy cows

Ketosis is best prevented by managing the transition period in the 'close-up' period (21–29 days pre-calving) as follows:

1. Avoid ketogenic feedstuffs such as silages (high butyrate).
2. Aim for BCS of 4.5–5.5 out of 8 at calving.
3. Adapt cows to the concentrate feed carefully to avoid acidosis.
4. Monitor DCAD levels in the feed, as high anion levels can reduce feed intake.
5. Feed niacin supplements at levels of 6–12 g/day before calving.
6. Feed propylene glycol daily at 240–300 mL/day orally.
7. Incorporate ionophores in the transition ration.

Bolus IV injection of 500 mL of 50% dextrose solution is a common therapy to restore normal glucose levels in the blood, but recurrence is common. Long-acting glucocorticoids, such as dexamethasone, are effective in elevating glucose levels.

References

Allen M (2000) Effects of short-term regulation of feed intake by lactating dairy cattle. *Journal of Dairy Science* **83**, 1598–1624. doi:10.3168/jds.S0022-0302(00)75030-2

Auldist MJ, O'Brien G, Cole D, MacMillan KL, Grainger C (2007) Effects of varying lactation length on milk production capacity of cows in pasture-based dairying systems. *Journal of Dairy Science* **90**, 3234–3241. doi:10.3168/jds.2006-683

Bauman DE, Griinari JM (2003) Nutritional regulation of milk fat synthesis. *Annual Review of Nutrition* **23**, 203–227. doi:10.1146/annurev.nutr.23.011702.073408

Bobe G, Young JW, Beitz DC (2004) Invited review: pathology, etiology, prevention, and treatment of fatty liver in dairy cows. *Journal of Dairy Science* **87**, 3105–3124. doi:10.3168/jds.S0022-0302(04)73446-3

Chilliard Y, Chardigny JM, Charbot J, Ollier A, Sebedio JL, Doreau M (1999) Effects of ruminal or postruminal fish oil supply on conjugated linoleic acid (CLA) content of milk fat. *Proceedings of the Nutrition Society* **58**, 70A [abstract].

Davis CL, Brown RE (1970) Low milk-fat milk syndrome. In *Physiology of Digestion and Metabolism in the Ruminant* (Ed. AT Phillipson) pp. 545–565. Oriel, Newcastle-upon-Tyne, UK.

Degaris PJ, Lean IJ (2009) Milk fever in dairy cows: a review of pathophysiology and control principles. *Veterinary Journal* **176**, 58–69.

DePeters EJ, Medrano JF, Reed BA (1995) Fatty acid composition of milk fat from three breeds of dairy cattle. *Canadian Journal of Animal Science* **75**, 267–269. doi:10.4141/cjas95-040

Duffield TF, Sandals D, Leslie KE, Lissemore K, McBride BW, Lumsden JH, et al. (1998) Efficacy of monensin for the prevention of subclinical ketosis in lactating dairy cows. *Journal of Dairy Science* **81**(11), 2866–2873.

Erdman RA (1988) Dietary buffering requirements of the lactating dairy cow: a review. *Journal of Dairy Science* **71**, 3246–3266. doi:10.3168/jds.S0022-0302(88)79930-0

FAO (2017) *Food and Agricultural Organisation Dairy Outlook*. FAO, Rome, Italy, <http://www.fao.org/fileadmin/templates/est/COMM_MARKETS_MONITORING/Dairy/Documents/Food_Outlook_June_2017__Dairy_.pdf>.

Goff JP (2008) The monitoring, prevention, and treatment of milk fever and subclinical hypocalcaemia in dairy cows. *Veterinary Journal (London, England)* **176**, 50–57. doi:10.1016/j.tvjl.2007.12.020

Goff JP, Ruiz R, Horst RL (2004) Relative acidifying activity of anionic salts commonly used to prevent milk fever. *Journal of Dairy Science* **87**(5), 1245–1255. doi:10.3168/jds.S0022-0302(04)73275-0

Guedes CM, Goncalves D, Rodrigues A, Dias-da-Silva A (2008) Effects of a *Saccharomyces cerevisiae* yeast on ruminal fermentation and fibre degradation of maize silages in cows. *Animal Feed Science and Technology* **145**, 27–40.

Ingvartsen KL (2006) Feeding- and management-related issues in the transition cow: physiological adaptations around calving and strategies to reduce feeding-related diseases. *Animal Feed Science and Technology* **126**, 175–213.

Lean IJ, DeGaris PJ, McNeill DM, Block E (2006) Hypocalcaemia in dairy cows: meta-analysis and dietary cation anion difference theory revisited. *Journal of Dairy Science* **89**, 669–684. doi:10.3168/jds.S0022-0302(06)72130-0

MAFF (1984) *Energy Allowances and Feeding Systems for Ruminants*. 2nd edn. Ministry of Agriculture, Fisheries and Food. Her Majesty's Stationery Office, London, UK.

Mansson HL (2008) Fatty acids in bovine milk fat. *Food & Nutrition Research* **52**, [online]. doi:10.3402/fnr.v52i0.1821

Moran JB (2005) *Tropical Dairy Farming: Feeding Management for Small Holder Dairy Farmers in the Humid Tropics*. Landlinks Press, Melbourne.

Moran JB, Drysdale GR, Shambrook DA, Markham NK (2000) A study of key profit drivers in the Victorian Dairy Industry. *Asian-Australasia Journal of Animal Science* **13** (Supplement A), 54–57.

NRC (2001) *Nutrient Requirements of Dairy Cattle*. 7th edn. National Research Council. National Academy Press, Washington, DC, USA.

Peterson DG, Matitashvili EA, Bauman DE (2003) Diet-induced milk fat depression in dairy cows results in increased trans-10, cis-12 CLA in milk fat and coordinate suppression of mRNA abundance for mammary enzymes involved in milk fat synthesis. *The Journal of Nutrition* **133**, 3098–3102. doi:10.1093/jn/133.10.3098

Piperova LS, Teter BB, Bruckental I, Sampugna J, Mills SE, Martinl P, *et al.* (2000) Mammary lipogenic enzyme activity, trans fatty acids and conjugated linoleic acids are altered in lactating dairy cows fed a milk fat-depressing diet. *The Journal of Nutrition* **130**, 2568–2574. doi:10.1093/jn/130.10.2568

Schwab EC, Caraviello DZ, Shaver PAS (2005) Review: a meta-analysis of lactation responses to supplemental dietary niacin in dairy cows. *The Professional Animal Scientist* **21**, 239–247.

Sissons JW (1981) Digestive enzymes of cattle. *Journal of the Science of Food and Agriculture* **32**, 105–114. doi:10.1002/jsfa.2740320202

Target10 (2002) *Feeding Dairy Cows: A Manual for Use in the Target10 Nutrition Program.* 3rd edn. (Eds J Jacobs and A Hargreaves). Victoria Department of Natural Resources and Environment, Melbourne.

Watson C (2006) Involution: apoptosis and tissue remodelling that convert the mammary gland from milk factory to a quiescent organ. *Breast Cancer Research* **8**, 203. doi:10.1186/bcr1401

9

Camelid nutrition

Key points

- Camelids are soft-footed, even-toed ungulates with only three stomach compartments, C1, C2 and C3, although C3 has two distinct regions (the first 80% is a fermentation region, and the final 20% an acid region).
- The fermentation process is similar to that of ruminants but the structure and function of the fermenting chambers differ.
- Camelids are better adapted to low-quality, high-fibre, low-protein diets than ruminants.
- Camelids retain particles in the forestomachs for longer than ruminants, which increases digestibility of poor-quality forages, but decreases voluntary feed intake.
- Despite the slower turnover of solids in the reticulo-rumen, camelids have a higher rate of liquid turnover in the forestomachs and greater nitrogen recycling to the forestomachs than ruminants, and hence a higher efficiency of microbial synthesis.
- Camelids use amino acids extensively in metabolism to produce glucose. The urea produced by this is largely recycled to the forestomachs to maintain ammonia levels for microbes.
- Camelids require ~25% less energy and less protein for maintenance than ruminants and need less protein relative to energy.
- Camelids have consistently high blood glucose levels produced by continuous gluconeogenesis from amino acids and poor insulin sensitivity, so they are effectively constantly diabetic. Stress exacerbates this hyperglycaemia and makes them susceptible to hepatic lipidosis.
- The mineral requirements of camelids are similar to those of true ruminants.
- Alpacas are susceptible to hepatic lipidosis and vitamin D-induced hypophosphataemic rickets.

Introduction

Camelids are even-toed (Order Artiodactyla) ungulates (hooved animals) belonging to the suborder Tylopoda (pad-footed) in which there is only one family, the Camelidae. The camelids include camels (the single-humped *Camelus dromedarius* and the two-humped *Camelus bactrianus*), alpacas (*Vicugna pacos*), vicunas (*Vicugna vicugna*) and llamas (*Llama glama* and *Llama guanaco*). Camels weigh 300–1000 kg, *Llama glama* 130–200 kg, *Llama guanaco* 90 kg, alpacas 50–90 kg and vicunas 35–65 kg. The camelids play an important role in many harsh climatic regions of the world and are a major source of protein, meat, milk, fibre and load-carrying for millions of people in the Andean highlands and deserts of Africa and Asia. The camels play important roles as 'beasts of burden' (or 'ships of the desert') for nomadic pastoralists covering vast areas with little water and are increasingly being managed for milk production. The New World (South American) camelids live in the Andean zone some 4500–6000 m above sea level. They evolved on very poor-quality fibrous and woody plants. Although they do ruminate (chew cud), the structure and function of the compartmentalised forestomachs is very different to that of the ruminants.

These animals are of nutritional interest because they are foregut fermenters, like ruminants, but do not possess four-stomach compartments. Instead they have three distinct compartments, which has led to the convention of calling this group 'pseudoruminants'. Given the significantly different functionality of the camelid forestomach and their digestive superiority over ruminants when fed high-fibre diets, this term should be discontinued, because it implies a lesser status. Camelids evolved in the Altiplano of the Andes Mountains and survived on low-quality sedges and grasses. They are considered browsers and grazers. Their digestive efficiency on low-quality roughages is greater than that of sheep and they seem to be more resistant than ruminants to pH fluctuations. In this chapter, emphasis is placed on alpacas, given their economic importance in fibre production, but the same principles apply to all camelids.

Structure and function of the digestive system in camelids
Mouth and dentition of camelids

Camelids have a cleft upper lip (labial cleft or philtrum), which allows both sides of the lip to operate independently. This allows the animal to be very selective in its feeding behaviour and to select a diet significantly higher in nutritive value than the average of the feed on offer (Cebra 2009). Camelids do not use their tongue to grasp feed as cattle do.

The salivary glands in camelids comprise the large parotid glands (serous secretions), mandibular glands (mixed mucous/serous) and small glands (buccal, palatine, sublingual and labial glands producing mainly mucous secretions) (von Engelhardt and Holler 1987). The serous parotid secretion (containing bicarbonate-rich saliva) is mixed with the feed bolus between the molars, ensuring good buffering of the fermentation in the stomach. The mucus-rich secretions protect the oral cavity from abrasion by the coarse feedstuffs. The ionic composition of saliva of alpacas appears to be similar to that of sheep, but there is one report suggesting a greater rate of saliva production (San Martin and Bryant 1989). The pH of camelid saliva is 8.0–8.6 with a high concentration of bicarbonate (Eckerlin and Stevens 1973), which buffers the fermentation acids (see 'Volatile fatty acid production, VFA absorption and pH buffering in the camelid forestomachs' later).

Camelids have ~22 deciduous teeth but only ~16 of these erupt and are functional (Wheeler 1982). The permanent teeth erupt at different times depending on the teeth type (incisors, canines, premolars, molars) but most erupt from 2 to 5 years of age. Most camelids have 30 or 32 permanent teeth. The dental formulae for camelids is as follows:

Deciduous (temporary)

	Incisors	Canines	Premolars
Upper	1–2	1	2–3
Lower	3	1	1–2

Permanent

	Incisors	Canines	Premolars	Molars
Upper	dental pad	1–2	1–2	3
Lower	3	1	1–2	3

The upper jaw of camelids, like ruminants, has a dental pad, against which the incisors of the lower jaw press plant material. An odd-shaped incisor on the upper jaw is behind the lower incisors and joins with the canines to form the 'fighting teeth', which are larger in males than females.

Alpacas and vicunas have no enamel on the lingual (tongue) side but do have enamel on the labial side of the incisors. Alpaca teeth therefore continue to grow throughout life. Llamas and guanacos have enamel on both sides of the incisors, so their teeth do not continue to grow, but may still need trimming.

Camelids, like ruminants, chew their cud to break down feed particles in a swinging, rotational pattern of chewing, which grinds the feed between the molars on either side of the mouth (Koford 1957).

Gastrointestinal tract of camelids

The generalised anatomy of the camelid digestive tract is shown in Fig. 9.1.

The GI tract of camelids is similar to that of ruminants, except that there are three stomach compartments, not four. The third compartment ends at the pylorus

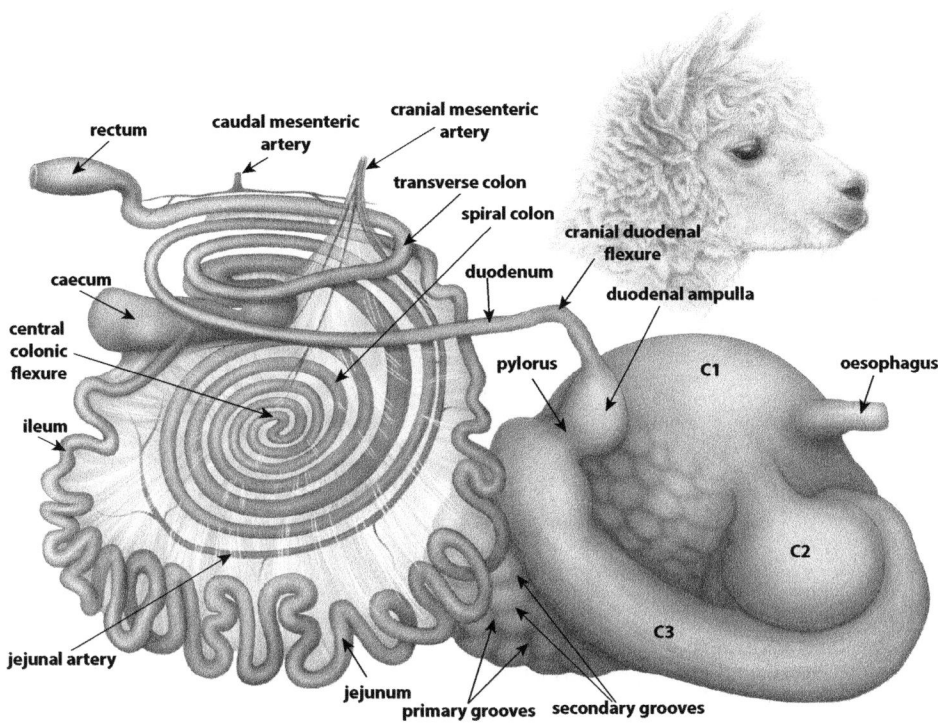

Fig. 9.1. The gastrointestinal tract of camelids.

where the duodenal ampulla is at the start of the small intestines. The duodenum, jejunum and ileum lead to the ileo-caeco-colic junction. Relative to ruminants, the camelids have a smaller caecum. The colon follows a spiral arrangement like that of ruminants, with a central colic flexure reversing the direction of the centrifugal and centripetal spirals (Fig. 9.1). The intestines are served by the cranial mesenteric artery, the jejunal artery and the caudal mesenteric artery.

The three compartments of the alpaca stomach are called C1, C2 and C3 (Figs 9.2 and 9.3). Fermentation takes place in compartments C1, C2 and in the proximal four-fifths of C3. In the adult, C1 holds ~83% of the total stomach contents, with C2 holding 6% and C3 ~11% (San Martin and Bryant 1989). C1 in a 44 kg alpaca holds 3–4 kg digesta, C2 70 g digesta and C3 ~600 g digesta (Perez *et al.* 2016) but these values probably vary greatly with diet and time after eating.

C1 is divided into a cranial blind sac and a caudal blind sac by a large transverse pillar (Vallenas *et al.* 1971). The ventral caudal sac is completely sacculated. The saccules are contained in groups of between one and three (usually two), lying between large parallel ridges running caudo-cranially from the transverse pillar. The ridges and the saccules between them end abruptly

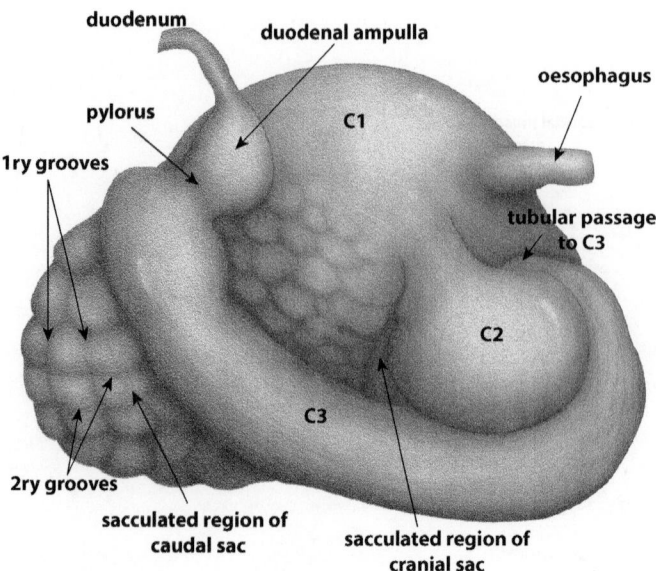

Fig. 9.2. External gross anatomy of the three camelid stomach compartments. The three major compartments are C1, C2 and C3. C1 is divided into cranial and caudal sacs; the ventral regions of these two sacs are sacculated by primary and secondary grooves. C2 is not sacculated. The final one-fifth of C3 has true gastric and pyloric glands.

about midway up the caudal sac (Fig. 9.2) so the dorsal sac contains no saccules. The cranial sac contains smaller saccules lying between small ridges running left to right. The ridges contain smooth muscle. Smaller, secondary ridges run at 90° to the primary ridges and also contain smooth muscle. There are no saccules in the cranial/ventral and cranial/dorsal sac right up to the transverse pillar.

The transverse pillar on the right side bifurcates into a pillar running dorsoventrally and a pillar running from left to right. These two pillars form C2. The whole of C2 contains small saccules in a random pattern bounded by small ridges (Fig. 9.2). A ventricular groove of smooth muscle runs from the cardia of the oesophagus caudally to the tubular entrance to C3. It is thought the ventricular groove operates in the same way as the oesophageal groove (reticular groove) in ruminants, to allow milk to bypass the fermentation chambers. The canal between C2 and C3 is very narrow and may play a role in regulating the size of feed particles flowing to the lower tract. C3 is a tubular compartment that terminates at the pylorus (torus pyloricus).

The recessed saccules in C1 and C2 are lined by glandular mucosa and the exposed non-recessed epithelium is stratified squamous. In camels, the stratified squamous epithelium is keratinised like ruminal epithelium (Hansen and Schmidt-

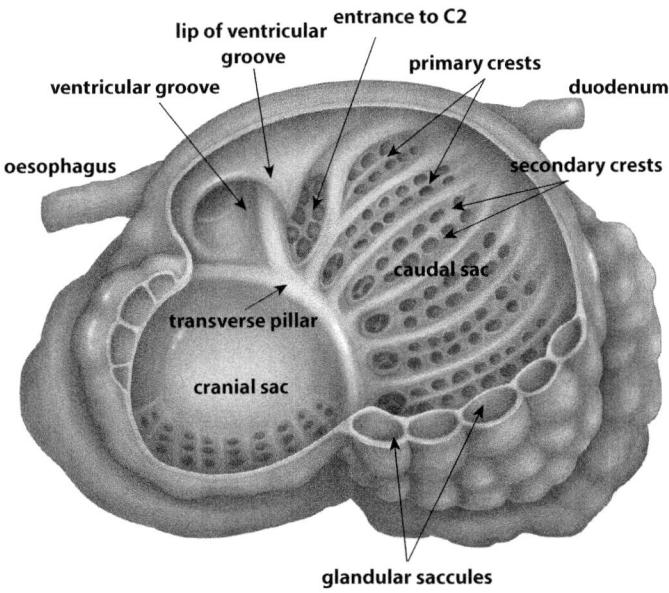

Fig. 9.3. Internal gross anatomy of the left side of the camelid stomach compartment C1.

Nielsen 1957), but the squamous epithelium in new-world camelids (alpacas and llamas) is not keratinised. The stratified epithelium does not contain papillae. The glandular mucosa of the recessed saccules consists of simple columnar epithelial cells with microvilli and mucus cells, similar to small intestinal mucosa. The saccules appear to carry out a similar role to ruminal papillae in increasing surface area. The small saccules in C2, unlike those in C1, contain numerous small papillae (Vallenas *et al.* 1971).

The epithelium in C3 is entirely glandular. The first one-fifth is a continuation of the net-like structure of C2. The mucosa in the middle three-fifths is arranged in permanent longitudinal pleats. The terminal fifth is thickened and contains gastric glands similar to that of the abomasum in ruminants (Vallenas *et al.* 1971). This terminal end of C3 is considered distinct from the preceding four-fifths and is referred to as the hind-stomach (Vallenas *et al.* 1971). Camels differ from the alpacas and llamas in that the saccules of C1 are fewer and larger (Vallenas *et al.* 1971).

Camelids have a three-compartment stomach designed for efficient digestion of very low-quality forages. Fermentation takes place in C1, C2 and most of C3. The last part of C3 contains true gastric mucosa. Prominent saccules in C1 and C2 play a key role in regulating microbial activity and absorption of VFA and water.

Digestive physiology of camelids

Camelids, like ruminants, ferment masticated feed in a symbiotic relationship with microorganisms. There have been fewer studies of the microbial populations of camelids compared with ruminants. They show that many of the bacteria and ciliated protozoans are co-hosted in sheep and alpacas (Ghosal *et al.* 1981; Del Valle *et al.* 2008) but that differences in diversity and abundance of bacteria exist between the two families (Pei *et al.* 2010). When sheep and alpacas were consuming a diet of *Medicago sativa* (lucerne or alfalfa), the percentages of methanogens and the cellulolytic bacterium *Ruminococcus flavefaciens* were lower in alpacas than sheep (Pei *et al.* 2010). Alpacas had a higher percentage of fungi and of the bacterium *Fibrobacter succinogenes*.

Camelids ruminate (i.e. chew a cud of regurgitated material), but the digestive processes, residence time and fractional outflow rate of fluid and particles differ from ruminants. These differences in the kinetics of digestion significantly impact on digestive efficiency, the efficiency of microbial protein synthesis, the rate of methane production, the extent of pH buffering and the extent to which nitrogen is recycled to the microbes, as discussed in the next sections.

Volatile fatty acid production, VFA absorption and pH buffering in the camelid forestomachs

Although the production of VFA by anaerobic microbes in the reticulo-rumen and omasum of ruminants and C1 and C2 of camelids is the same, the rate of absorption of the VFA differ. In ruminants, the papillae provide a large surface area for absorption and this area can be modified by changes to papillae shape and size in response to the VFA themselves. Camelids, on the other hand, have a high absorptive surface area created by the saccules. These fill with fluid and then evert, expelling the fluid back into the lumen (Vallenas *et al.* 1971). This constant exposure of fluid to the absorptive mucosal surface plays a critical role in VFA absorption and fluid dynamics. It is probable that camelids absorb VFA as undissociated acids and in exchange with bicarbonate ions, as is the case with ruminants (see Fig. 3.21). Secretion of bicarbonate and phosphate ions has been demonstrated in the saccules (Eckerlin and Stevens 1973). This system of secreting bicarbonate buffers into C1 and C2, and absorbing undissociated VFA from C1 and C2, is an effective buffering system. The pH of alpaca digesta in C1 is higher than that of ruminal digesta (Vallenas *et al.* 1973), suggesting greater pH buffering and more rapid acid absorption in alpacas. The rate of VFA and solute absorption in alpacas is two to three times faster than in ruminants (Rubsamen 1978). The higher pH of alpaca digesta would benefit cellulolysis, which is inhibited by low pH (Russell 1985). There is evidence that alpacas have a more stable pH than ruminants and that sudden introduction of grains does not elicit the marked pH decline and subsequent acidosis experienced by sheep and cattle (Robinson *et al.* 2013).

Forestomach motility in camelids

Forestomach motility in camelids, like in ruminants, is controlled by the vagus nerve and follows a cyclical sequence of events that can be divided into phases A and B. However, substantial differences exist between ruminants and camelids in the patterns and frequency of forestomach motility (Heller *et al.* 1984). Camelids have a more continuous and regular pattern of motility than ruminants. The sequence of events involved in forestomach motility in camelids follows a pattern of a single A-wave contraction followed by numerous B-wave contractions. The entire cycle lasts ~80 s (Heller *et al.* (1984). The A-wave contraction begins with a contraction of the canal joining C2 with C3. C2 then contracts vigorously, followed by a contraction of the caudal portion of C1. The B-wave involves contractions of C1 (caudal and cranial) repeated six or seven times, followed by a pause before the next A-wave. Eructation in camelids occurs at the peak of a caudal sac contraction, and can be as frequent as three to four times/cycle. This frequent eructation cycle may account for the low incidence of bloat in camelids compared with ruminants, and may contribute to the more thorough mixing of the forestomach contents in camelids compared with the heterogeneous nature of the rumen contents. There is no floating 'raft' of digesta in camelids. Rumination in camelids occurs during the contraction of the cranial sac of C1 (Vallenas and Stevens 1971). Onward passage of digesta from C2 to C3 occurs during the A-wave contractions (Heller *et al.* 1984). Motility in C3 is peristalsis-like, moving contents onwards towards the anus (Heller *et al.* 1984). These peristaltic contractions occur at high frequency (5–10/s). The role of contraction of the saccules, which contain smooth muscle, is not well described or understood. Individual saccules fill with fluid and then evert to eject the fluid, but the extent to which the movements are part of the normal contractions of C1 and C2 versus individual motility is not clear.

Rate of passage, fluid flows and particle kinetics in camelids

In ruminants, particles are retained in the reticulo-rumen until they are of a size and density that places them in an 'escapable' pool presented to the reticulo-omasal orifice (see Fig. 3.12). Several studies have shown that particles are retained for longer in the forestomachs of camelids than in ruminants but that the rate of outflow of fluid is greater in camelids (Florez 1973; Clemens and Stevens 1980; San Martin 1987). This makes camelids very efficient at digesting low-quality, low-digestibility forages, because the long retention time of particles allows more thorough digestion. At the same time, the rapid fractional outflow rate of fluid will increase microbial efficiency because faster outflow rates require microbes to grow and divide rapidly, otherwise they are flushed out of the stomachs. Faster growth means energy will be used for growth and relatively less for maintenance of the microbial population (Isaacson *et al.* 1975). Differences between ruminants and camelids in solid and fluid turnover rates might be expected to generate differences in methane production. Long retention times of solids would increase methanogenesis, but faster liquid turnover rates would generate lower methane outputs. The effect of diet, then, on methane output in camelids will depend on the net effect on the relative solid versus liquid flow rates from the stomach compartments. Pinares-Patiño *et al.* (2003) showed that alpacas produced more methane than sheep on ryegrass/clover and lotus pastures, but there was no difference in methane production on a diet of lucerne hay. There is evidence that camelids utilise non-protein nitrogen more efficiently than sheep. On a low-protein ration, llamas recycle body urea to the forestomachs and excrete less urea through the kidneys than sheep and goats (von Engelhardt and Schneider 1977).

Diet selection and forage digestibility in camelids versus ruminants

Camelids are better adapted to low-quality, low-protein forages than the ruminants. Of the camelids, the llamas are better adapted than alpacas to poor-quality feeds (San Martin and Bryant 1989). The longer retention time of particles in, and faster liquid flow from, the forestomachs of camelids allows greater digestibility of poor-quality, fibrous forages than in ruminants (Fig. 9.4). However, the slow particle flow also restricts voluntary feed intake.

Llamas are better adapted to, and tend to select for, poorer quality feeds than alpacas (Pfister *et al.* (1989). Alpacas and llamas tend to select grasses even when legumes such as clovers are available, in contrast to sheep, which select 2.5× more legumes than alpacas. The digestibility of C4 plants (high-fibre) in alpacas is higher (57%) than in goats (44%).

> Camelids have a high rate of fluid flow from the stomachs (hence high efficiency of microbial synthesis), but a low particle flow rate (hence high digestibility of high-fibre roughages). They are thus well adapted to high-fibre, low-protein roughages.

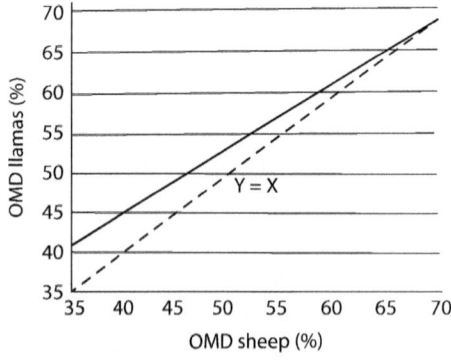

Fig. 9.4. Relationship between organic matter digestibility (OMD) of diets consumed by sheep and llamas. Llamas have a higher digestibility of roughages than sheep, particularly on very low-quality, low-digestibility roughages. Source: After Dulphy *et al.* (1998).

Metabolism in camelids

Camelids differ from ruminant animals in having an unusual metabolism of glucose and amino acids. Ruminant animals typically have a blood glucose concentration of 2.2–3.3 mM, compared with camelids at >4.7 mM. The blood urea nitrogen concentration in camelids is also higher than in ruminants. These differences arise because camelids use amino acids rather than propionate to produce glucose by gluconeogenesis (see Fig. 1.12). The carbon skeleton of the amino acids produces glucose, and the ammonia removed by deamination of the amino acids produces urea. The rate of clearance of urea in the kidneys is lower in camelids than in ruminants (Hinderer and von Engelhardt 1975). Much of this urea is recycled to the forestomachs to stimulate microbial protein synthesis. This use of amino acids to produce glucose is probably an adaptation to high-fibre diets, which produce less ruminal propionate.

The apparent 'hyperglycaemia' in camelids is increased during mild stress where blood glucose can exceed 11 mM (Cebra 2009). It appears camelids have a slow insulin response in addition to insulin resistance, rendering them effectively diabetic. This metabolic condition makes camelids susceptible to hepatic lipidosis (see 'Hepatic lipidosis and hyperlipidaemia in camelids' later).

> Camelids have high circulating levels of glucose produced from glucogenic amino acids. They are effectively constantly insulin-resistant and diabetic. The ammonia from deaminated amino acids is recycled as salivary urea to C1 and C2 for microbial protein synthesis.

Energy and protein requirements of camelids

The few studies of the energy requirements of camelids suggest that they have a lower maintenance energy requirement than ruminants (Van Saun 2006a; 2009). An average value of 0.305 $MJ/W^{0.75}$ has been adopted as the ME_m for camelids. This is ~25% lower than the ME_m for sheep of ~0.400 $MJ/W^{0.75}$ if we assume a moderate-quality feed is consumed. The maintenance energy requirement for camelids can be estimated simply using the following equation with little or no loss of accuracy:

Maintenance energy requirement of camelids:

$$ME_m \text{ (MJ/day)} = 0.08W + 1.7$$

where W = bodyweight (kg).

To account for different grazing conditions corrections to this ME_m are made as in Table 9.1.

Table 9.1. Energy and protein requirements for maintenance, growth and pregnancy of camelids grazing in different environments

Activity/grazing environment	Maintenance energy requirement (MJ/day)	Dietary crude protein (%DM)
Maintenance Zero grazing (confined feeding) Moderate to good quality pasture Rangeland pasture Sparse grassland, mountainous	$ME_m = 0.08W + 1.7$ $1.25ME_m$ $1.5ME_m$ $1.75ME_m$	9–10%
Growth	Maintenance + 3 MJ/100 g weight gain	14–16% CP
Pregnancy 9th–10th month 10th –11th month	 Maintenance + 3.1 MJ/day Maintenance + 5.4 MJ/day	11–16% CP
Lactation Early	 Maintenance + 6 MJ/day	12–16% CP

Source: Adapted from Van Saun (2006a, 2009)

Example: How much poor-quality meadow hay (6.8 MJ/kg DM) is required to maintain a 60 kg alpaca?

$$ME_m = 0.08W + 1.7 = 0.08 \times 60 + 1.7 = 6.5 \text{ MJ/day}$$

Feed DM required/day = 6.5/6.8 kg/day = 0.96 kg DM/day
= 1.06 kg 'as-fed'/day (assuming DM = 90%)

Assuming a maximum intake of ~1.5% of bodyweight = 0.9 kg/dayDM.

The alpacas may not be able to maintain live weight on this ration and may require supplementation. A small quantity of a higher quality hay would ensure maintenance was met: for example, adding 0.2 kg of high-quality hay (10.5 MJ/kg DM) adds 2.1 MJ/day plus poor-quality hay now at, say, 0.7 kg (4.8 MJ/day) = 6.9 MJ/day, which meets the requirement.

Mineral and vitamin requirements of camelids

Given the tendency for camelids to select higher fibre diets than ruminants, the potential for mineral deficiencies is high, but there appear to be no specific deficiencies in camelids not reported in ruminants (Van Saun 2009). Table 9.2 indicates the suggested mineral concentrations in feeds for camelids based on sheep and cattle requirements and corrected for the relatively lower intake/bodyweight of camelids (Van Saun 2009).

There are cases reported of urolithiasis (Kock and Fowler 1982), metabolic bone disease (Van Saun *et al.* 1996), zinc-responsive dermatosis (Rosychuk 1994), vitamin D seasonal deficiency and hyophosphataemic rickets, which are indicative of mineral deficiencies or disorders. Hypocalcaemia, hypokalaemia, hypophosphotaemia and hypomagnesaemia have all been reported in camelids and

Table 9.2. Suggested macro- and micro-mineral concentrations in diets for camelids

Macromineral	Requirement (%DM)	Micromineral	Requirement (mg/kg DM)
Calcium	0.2–0.75	Copper	9.0–12.0
Phosphorus	0.17–0.38	Cobalt	0.12–0.14
Magnesium	0.13–0.22	Iron	47.0–72.0
Potassium	0.6–0.96	Iodine	0.6–0.13
Sodium	0.07–0.14	Manganese	24.0–64.0
Chloride	0.15–0.25	Selenium	0.35–0.48
Sulphur	0.19–0.23	Zinc	35.0–54.0

Source: Van Saun (2009)

are responsible for recumbency (Belknap 1994). Selenium deficiency has been reported in camels but not in llamas or alpacas to date. Camelids are prone to copper toxicity, but do not appear to be as sensitive as sheep to clinical copper toxicity (Van Saun 2006b). There are few reports of vitamin deficiency in camelids other than an apparent high sensitivity to vitamin D deficiency (see 'Vitamin D deficiency and hypophosphataemic rickets' below).

Common nutritional and metabolic diseases of camelids
Hepatic lipidosis and hyperlipidaemia in camelids

Hepatic lipidosis is common in camelids and occurs in all classes and ages of animals. Although hepatic lipidosis in pregnant and lactating camelids resembles that occurring in ruminants, with hyperlipaemia, increased blood NEFAs, increased blood ketones and hypoglycaemia, in non-pregnant, non-lactating animals hyperglycaemia is common, suggesting there must be other triggers for lipid mobilisation or impaired fat use (Cebra 2009). Anorexia, combined with continued use of amino acids for gluconeogenesis, may result in an amino acid deficiency associated with decreased mobilisation of hepatic fat (Cebra 2009). In contrast to ruminants, where pregnancy and lactation are major predisposing factors to ketosis and hepatic lipidosis, male camelids of all ages also display the condition. It appears that camelids are particularly prone to stress-induced anorexia and subsequent fat accumulation.

Vitamin D deficiency and hyophosphataemic rickets

Camelids are more susceptible to seasonal vitamin D deficiency than ruminants, and young llamas and alpacas 3–6 months of age develop hypophosphataemic rickets syndrome (Fowler 1990). Presumably evolution of the small camelids in high altitude environments with high UV input, has reduced the vitamin D-synthesising capacity of the skin of these animals, or vitamin D retention in the body. The incidence of this syndrome is greatest in high latitudes and particularly in crias born in autumn, as they do not obtain vitamin D from sunshine over the first 6–10 months of life. Animals with darkly pigmented and heavy fleeces have lower vitamin D levels in blood. Clinical signs of this syndrome include slow growth, humped-back stance, enlarged joints (especially the carpus), hypophosphataemia and irregular growth plates. Injections of vitamin D (1000–1500 IU/kg bodyweight) provide 3 months protection.

> Camelids are sensitive to vitamin D deficiencies because they evolved at high altitudes with high UV input. Dark-coated alpacas are the most susceptible to vitamin D deficiency.

Polioencephalomalacia

Pre-gastric fermenters usually obtain sufficient thiamine from microbial synthesis in the forestomachs, but reduced microbial synthesis, the presence of thiaminases in feeds, thiaminases produced by certain bacteria and the presence of high level of sulphur in the diet or water supply can also induce a deficiency. Polioencephalomalacia (PEM) is a neurological disorder produced by a deficiency of thiamine. Clinical signs include opisthotonus (stargazing), depression, seizures, hyperesthesia and blindness. PEM has been reported in young and adult llamas and alpacas (Van Saun 2006b). Abrupt changes to the diet towards concentrate feeding were major causes of the disorder.

> Alpacas should be introduced to high-grain diets gradually, and the levels of sulphur in the diet and water supply should be monitored to reduce the incidence of polioencephalomalacia.

Urolithiasis

Urolithiasis is common in males and is a consequence of their small diameter urethra and its sigmoid flexure. The urethra blocks with mineral crystals, making urination difficult or impossible. If a complete blockage is present, the bladder will burst producing 'waterbelly' (fluid in the abdomen) and death. Grain feeding can produce an imbalance in the Ca:P ratio. Crystals of struvite (magnesium-ammonia-phosphate) form when the Ca:P ratio falls below 2:1. Alternatively, the presence of oxalates and silicates in certain plants can induce formation of the uroliths. Ensuring a Ca:P ratio of between 2:1 and 4:1 and ensuring high water intake are the best preventive actions. Treatment includes acidification of the urine by supplementing with ammonium chloride.

References

Belknap EB (1994) Medical problems of llamas. *The Veterinary Clinics of North America. Food Animal Practice* **10**, 291–307. doi:10.1016/S0749-0720(15)30563-6

Cebra CK (2009) Disorders of carbohydrate or lipid metabolism in camelids. *The Veterinary Clinics of North America. Food Animal Practice* **25**, 339–352. doi:10.1016/j.cvfa.2009.02.005

Clemens ET, Stevens CE (1980) A comparison of gastrointestinal transit time in ten species of mammal. *Journal of Agricultural Science, Cambridge* **94**, 735–737. doi:10.1017/S0021859600028732

Del Valle I, de la Fuente G, Fondevila M (2008) Ciliate protozoa of the forestomach of llamas (*Llama glama*) and alpacas (*Vicugna pacos*) from the Bolivian Altiplano. *Zootaxa* **1703**, 62–68.

Dulphy J-P, Dardillat C, Jailler M, Jouany J-P (1998) Intake and digestibility of different forages in llamas compared to sheep. *Annales de Zootechnie* **47**, 75–81. doi:10.1051/animres:19980106

Eckerlin RH, Stevens CE (1973) Bicarbonate secretion by the glandular saccules of the llama stomach. *The Cornell Veterinarian* **63**, 436–445.

Florez JA (1973) Rate of passage and digestibility in alpacas and sheep. BSc thesis. Universidad Nacional Mayor de San Marcos, Lima, Peru.

Fowler ME (1990) Rickets in llamas and alpacas. *Llamas* **4**, 92–95.

Ghosal AK, Tanwar RK, Dwarakanath PK (1981) Note on rumen microorganisms and fermentation pattern in camel. *The Indian Journal of Animal Sciences* **51**, 1011–1012.

Hansen A, Schmidt-Nielsen K (1957) On the stomach of the camel with special reference to the structure of its mucous membrane. *Acta Anatomica* **31**, 353–375. doi:10.1159/000141291

Heller R, Gregory PC, von Engelhardt W (1984) Pattern of motility and flow of digesta in the forestomach of the Llama (*Lama guanaco* f. *glama*). *Journal of Comparative Physiology. B, Biochemical, Systemic, and Environmental Physiology* **154**, 529–533. doi:10.1007/BF02515158

Hinderer S, von Engelhardt W (1975) Urea metabolism in the llama. *Comparative Biochemistry and Physiology* **52**, 619–622. doi:10.1016/S0300-9629(75)80012-0

Isaacson HR, Hinds FC, Bryant MP, Owens FN (1975) Efficiency of energy utilization by mixed rumen bacteria in continuous culture. *Journal of Dairy Science* **58**, 1645–1659. doi:10.3168/jds.S0022-0302(75)84763-1

Kock MD, Fowler ME (1982) Urolithiasis in a 3-month old llama. *Journal of the American Veterinary Medical Association* **181**, 1411.

Koford CB (1957) The vicuna and the puna. *Ecological Monographs* **27**, 153–219.

Pei CX, Liu Q, Dong CS, Li HQ, Jiang JB, Gao WJ (2010) Diversity and abundance of the bacterial 16S rRNA gene sequences in forestomach of alpacas (*Lama pacos*) and sheep (*Ovis aries*). *Anaerobe* **16**, 426–432. doi:10.1016/j.anaerobe.2010.06.004

Perez W, Konig HE, Jerbi H, Clauss M (2016) Macroanatomical aspects of the gastrointestinal tract of the alpaca (*Vicugna pacos* Linnaeus, 1758) and dromedary (*Camelus dromedarius* Linnaeus, 1758). *Vertebrate Zoology* **66**, 419–425.

Pfister JA, San Martin F, Rosales L, Sisson DV, Flores E, Bryant FC (1989) Grazing behaviour of llamas, alpacas, and sheep in the Andes of Peru. *Applied Animal Behaviour Science* **23**, 237–246. doi:10.1016/0168-1591(89)90114-7

Pinares-Patiño CS, Ulyatt MJ, Waghorn GC, Lassey KR (2003) Methane emission by alpaca and sheep fed on lucerne hay or grazed on pastures of perennial ryegrass/white clover or birdsfoot trefoil. *The Journal of Agricultural Science* **140**, 215–226. doi:10.1017/S002185960300306X

Robinson TF, Harris BW, Johnston NP (2013) Initial compartment 1 pH response to grain supplementation in alpacas (*Vicugna pacos*) fed alfalfa and grass hay. *Journal of Animal Science Advances* **3**, 354–360.

Rosychuk RAW (1994) Llama dermatology. *The Veterinary Clinics of North America. Food Animal Practice* **10**, 228–239. doi:10.1016/S0749-0720(15)30557-0

Rubsamen K (1978) Bicarbonate secretion and solute absorption in the forestomach of the llama. *The American Journal of Physiology* **235**, E1–E6.

Russell JB (1985) Factors influencing competition and composition of the rumen bacterial flora. In *Herbivore Nutrition in the Subtropics and Tropics* (Eds EMC Gilchrist and RI Mackie) pp. 222–243. The Science Press, Pretoria, South Africa.

San Martin F, Bryant FC (1989) Nutrition of domesticated South American llamas and alpacas. *Small Ruminant Research* **2**, 191–216. doi:10.1016/0921-4488(89)90001-1

San Martin FA (1987) Comparative forage selectivity and nutrition of South American camelids and sheep. PhD dissertation. Texas Tech University, Lubbock, TX, USA.

Vallenas AP, Stevens CE (1971) Motility of the llama and guanaco stomach. *The American Journal of Physiology* **220**, 275–282.

Vallenas A, Cummings JF, Munnell JF (1971) A gross study of the compartmentalised stomach of two new-world camelids, the llama and guanaco. *Journal of Morphology* **134**, 399–423. doi:10.1002/jmor.1051340403

Vallenas A, Esquerre J, Valenzuela A, Candela E, Chauca D (1973) Volatile fatty acids and pH in the first two stomach compartments in the alpaca and sheep. *Rev. Invest. Pecu. (IVITA)* **2**, 115–130.

Van Saun RJ (2006a) Nutrient requirements of South American camelids: a factorial approach. *Small Ruminant Research* **61**, 165–186. doi:10.1016/j.smallrumres.2005.07.006

Van Saun RJ (2006b) Nutritional diseases of South American camelids. *Small Ruminant Research* **61**, 153–164. doi:10.1016/j.smallrumres.2005.07.007

Van Saun RJ (2009) Nutritional requirements and assessing nutritional status in camelids. *The Veterinary Clinics of North America. Food Animal Practice* **25**, 265–279. doi:10.1016/j.cvfa.2009.03.003

Van Saun RJ, Smith BB, Wastrous BJ (1996) Evaluation of vitamin D status of llamas and alpacas with hypophospotaemic rickets. *Journal of the American Veterinary Medical Association* **209**, 1128–1133.

von Engelhardt W, Holler H (1987) A survey of the salivary and gastric physiology of camelids. *Animal Research and Development* **26**, 84–94.

von Engelhardt W, Schneider W (1977) Energy and nitrogen metabolism in the llama. *Animal Research and Development* **5**, 68–72.

Wheeler J (1982) Aging llamas and alpacas by their teeth. *Llama World* **1**, 12–17.

10

Deer nutrition

Key points

- Deer belong to the group of ruminants referred to as 'intermediate feeders', with feeding habits between grazers and browsers. They select diets significantly higher in nutritive value, soluble carbohydrates, energy, protein and minerals than the average feed on offer.
- Deer are a highly variable group of animals with species ranging in adult live weight from 5 kg to 800 kg. There are native species on all continents except Antarctica and Australia. They are farmed on all continents for their low-fat meat (venison) and velvet (the blood-rich skin covering the newly formed antler before calcification).
- The reticulo-rumen of deer is relatively smaller than that of sheep and cattle and their reticulo–omasal orifice is relatively larger. This means the outflow rate of larger, low-digestibility particles is higher, and hence the extent of digestion of poor-quality feeds is lower. Highly digestible components, however, are more efficiently digested because they are flushed into the intestines.
- Voluntary feed intake in deer is highly seasonal, with a large increase in summer and decrease in winter when live weight, particularly in males, can decline by 30%.
- The Metabolisable Energy requirement of deer varies with age and season. Maintenance energy requirement is 0.45 MJ/kg$^{0.75}$ for calves 3–11 months of age and 0.57 MJ/kg$^{0.75}$ for adults.
- Pregnancy increases ME requirement by 30–40%, but lactating deer require double the maintenance requirement.

- Dietary crude protein requirements of adult deer at maintenance are low (6–10%) but there is a high requirement for growth, antler development and reproduction (13–16%) and 17% for young deer.
- The mineral and vitamin requirements of deer are similar to those of sheep and cattle. Vitamin A appears to be particularly important for antler development and there is evidence that selenium and copper are important in the first year of life.
- Deer are less susceptible to bloat and acidosis than sheep and cattle, presumably due to the high rumen outflow rate.

Introduction

Archaeological evidence indicates that venison has been part of human diets for between 5000 and 50 000 years, much longer than sheep meat and beef. A significant proportion of the venison consumed was from hunting wild deer but there is evidence that deer were sourced from managed populations. Unlike sheep and cattle, there was no selective pressure placed on breeding deer for increased fatness, which, in a health-conscious 21st century, has made lean venison a healthy alternative red meat. Farming deer presents several difficulties. They are still relatively undomesticated and have behavioural characteristics that make handling difficult, and at times dangerous. Physiologically, many species are highly seasonal in reproduction, energy metabolism, digestive function and pelage (coat hair) growth, which means nutritional data from sheep and cattle are not always applicable.

Deer are a highly diverse group of even-toed ungulates (Artiodactyla) belonging to the Suborder Ruminantia and Family Cervidae. The Cervidae is further divided into the Subfamilies of Old World deer (Subfamily Cervinae) and the New World deer (Caprolinae). There are 43 species. They vary widely in phenotype (bodyweight varies from 3–6 kg for the northern pudu to 800 kg for the moose), seasonality of reproduction, browsing and grazing habits. Native deer occupy all continents except Australia and Antarctica. Large deer are commonly found in temperate deciduous forests, coniferous forests in mountainous regions, savannas and tropical seasonal forests. Formulating diets or managing pastures to meet the nutrient requirements of farmed deer is difficult because little research has been conducted on their nutrient requirements. Little is known also of the impact of season, activity, browsing habits and antler growth on nutrient requirements.

The most common farmed species globally are the red deer from Western Europe (*Cervus elaphus elaphus*), red deer from Eastern Europe (*Cervus elaphus montanus*), fallow deer (*Dama dama*), rusa deer (*Cervus timorensis*), wapiti/elk (*Cervus elaphus canadensis*), chital (*Axis axis*) and sambar (*Cervus unicolor*). There are ~4 million farmed deer globally. Table 10.1 shows the distribution of farmed deer.

Table 10.1. Approximate numbers of farmed deer globally

Country/region	Deer numbers	Species
Europe	411 000	Red, fallow[a]
USA	250 000	Fallow (34%), red (24%), axis (13%), whitetail (15%), elk (4%), sika (3%)
Canada	98 650	Elk (35%), red (13%), whitetail (11%), fallow (29%), reindeer (10%)
New Zealand	1 600 000	Red, fallow
China	500 000	Red, wapiti, sika
Far East CIS (Russia)	400 000	Sika, wapiti
North and South Korea	150 000	Sika, wapiti, red
Other Asia	70 000	Red, fallow, sika, sambar, rusa

[a]Relative numbers of farmed red versus fallow deer not available. Varies greatly with European country
Source: After Tuckwell (2003)

Although deer are ruminants with many digestive similarities to the bovids (sheep and cattle), their dietary habits range from that of true grazers to browsers as defined by Hoffmann (1985). In contrast to domesticated ruminants such as sheep and cattle, deer exhibit marked seasonal changes in metabolic rate, voluntary food intake (VFI), endocrine profile and pelage growth (Loudon *et al.* 1989).

In this chapter the differences in digestive anatomy and physiology of deer compared to other ruminants are considered along with the latest information available on the nutrient requirements of deer and practical feeding systems for optimising venison production per hectare. Nutritional aspects of the production of antler velvet are also considered along with common nutritional and metabolic disorders of farmed deer.

The terminology used to describe different classes of deer varies with the species. For most species (red, wapiti, rusa, sambar and chital) the male is a stag or bull, a female is a cow or hind and the young are calves (except rusa). However, for fallow deer a male is a buck, the female is a doe and the young are fawns (as for rusa). The nutrient requirements of female deer are greatly affected by reproduction and lactation, which are highly seasonal in most species (exceptions are the sambar, chital and the Javan rusa, which breed all year). Gestation length in farmed deer is 230 to 255 days depending on species.

Structure and function of the digestive system in deer
Mouth and dentition of deer

Deer, like other ruminants, have no incisors on the upper jaw. They are replaced by a dental pad. The dental formula for most deer is as follows:

	Incisors	Canines	Premolars	Molars
Upper	0	0–1	3	3
Lower	3	1	3	3
Total = 32–34 teeth				

Unlike other ruminants such as sheep and cattle, deer replace their milk teeth incisors (deciduous incisors) at ~11 months of age rather than in a progression from the middle pair. This makes ageing deer more difficult and dependent on experience related to the degree of wear and colour of the premolars and molars (Tuckwell 2003). For deer that are intermediate or browse species, fine diet selection is achieved by having a long tongue and a narrow jaw.

Saliva production and composition in deer

Browsers and intermediate feeders, such as deer, often consume diets containing high levels of condensed tannins or polyphenolic compounds. An evolutionary adaptation to this habit has been the production of a large volume of saliva containing tannin-binding proteins. In deer, the tannin-binding protein is a small glycoprotein containing large amounts of proline, glycine and glutamate/glutamine (Austin et al. 1989). Grazing ruminants such as sheep and cattle do not produce this protein. A comparison of roe deer, fallow deer and sheep showed that roe deer secreted the highest volume of both parotid and mixed saliva per unit body mass and the largest amount of salivary proteins, which bind to tannins (Fickel et al. 1998). Presumably this adaptation of deer to produce tannin-binding proteins in their salivary secretions not only negates negative effects of tannins on digestion, but may also reduce the astringent taste of tannins. Once bound to the tannin-binding protein, subsequent binding to taste bud proteins to produce the astringent taste would be reduced in much the same way as adding milk to strong tea makes it less bitter because the caseins bind to the tannins. This would greatly increase the range of feeds acceptable to deer relative to sheep and cattle.

Digestive strategies: grazing versus browsing

Ruminants are classified broadly as browsers (those that select high-quality feed components from the feeds on offer; usually from shrubs and bushes), grazers (those that consume large quantities of high-fibre feeds) and intermediate feeders (those that carry out both types of behaviours) (Hoffmann 1985). Generally, browsers have large salivary glands, tannin-binding proteins in their saliva, a relatively small reticulo-rumen, large omasal orifice allowing larger particles to flow, and a well-developed caecum (Table 10.2). Grazers tend to have a broad muzzle, flat incisors, small salivary glands, hold a large quantity of material in the reticulo-rumen, and retain particles in the reticulo-rumen for long periods. This

Table 10.2. Two major types of feeding behaviour in ruminants

Browsers	Grazers
• Rely on browse shrubs (foliage in summer and twigs in winter) • Large salivary glands, a high volume of saliva containing tannin-binding glycoproteins high in proline and glutamine • Small reticulo-rumen • Large reticulo-omasal orifice allowing large forage particle to escape the rumen • Well-developed caecum	• Adapted to grasses/sedges • Broad muzzle and flat incisors to graze short pastures • Small salivary glands • Large reticulo-rumen capacity • Small reticulo-omasal orifice so long retention time for large particles

Source: Hoffman (1985)

means high-fibre materials remain in the reticulo-rumen until digestion is completed. Generally, deer used for domestic production are intermediate feeders with digestive system characteristics intermediate between browsers and grazers (see Table 10.2). This means they lack the specialisation of browsers and grazers, but benefit from greater nutritional flexibility.

Digestive physiology of deer

The digestive tract of deer is similar to that of other ruminants, with four stomach compartments (see Fig. 3.1). There are differences in the relative capacities of the compartments and the rates of outflow of fluids and solids, depending on the diet, the deer species and the season. Deer can consume and digest a similar range of diets as sheep and cattle, but there are several features of their digestive process that differ. Deer digest low-quality roughages less completely than sheep but digest concentrate rations more completely (Milne *et al.* 1978). Deer have a relatively smaller reticulo-rumen and a larger reticulo-omasal orifice than sheep. This results in a shorter retention time of material in the reticulo-rumen and faster rate of passage of larger, indigestible fibres (Nagy and Regelin 1975). Smaller species, such as fallow deer, are unable to fully digest poor-quality roughages such as hays for this reason. Other studies have shown that large deer, such as red deer, digest total fibre and lignin, better than sheep, especially in summer (Domingue *et al.* 1991). The digestive physiology of deer reflects their intermediate feeding behaviours and selection of high-quality diets from the feed on offer. High rates of fluid outflow from the rumen allows the high water-soluble carbohydrate content of their diets to escape rumen fermentation and subsequently be digested in the intestines. This may also explain why deer are less susceptible to bloat and acidosis than other ruminants.

Deer have a marked seasonal cycle of voluntary feed intake with a marked increase (~20%) in summer and a decrease in winter in temperate species of deer

(Milne *et al.* 1978; Fennessy *et al.* 1981; Kay 1985). This cycle is driven by hormonal changes. The reduction in intake in winter, particularly in males, results in a loss of bodyweight of up to 20% of summer weight, but this is countered by a large increase in appetite in summer and compensatory gain in weight. Usually, when ruminants consume more feed, the rate of passage of digesta increases and the dry matter digestibility decreases, negating some of the benefits of increased intake. However, several studies have shown that deer appear to maintain the digestibility of feed and retention time of feed in the gut by increasing the total ruminal pool sizes of dry matter and liquid despite the increased voluntary intake (Milne et al. 1978; Domingue *et al.* 1991; Fraudenberger *et al.* 1994). In summer the voluntary intake of feed by deer increases by ~35% and this is accompanied by an increase in the rumen pool size of ~50% (Domingue *et al.* 1991) with no reduction in the digestibility of the feed.

Deer differ from other ruminants in ruminal digestion:
- The capacity of the reticulo-rumen is lower.
- The rate of solids outflow is greater.
- The digestibility of low-quality roughages is lower.
- The rate of outflow of liquids is greater, which flushes water-soluble components to the intestines where they are digested more efficiently.
- When intake increases, deer increase the total ruminal pool size with no reduction in digestibility.

The microbial populations in the reticulo-rumen of deer differ widely between species, reflecting the different feeds consumed by predominantly browsers, grazers and intermediate feeders. Larger deer such as red deer have a higher density of ciliates and more genera of ciliates than smaller species such as fallow and roe deer (Prins and Geleen 1971). The red deer can digest cellulose more effectively than the smaller species.

Matching the highly seasonal voluntary feed intake of deer, their seasonal metabolic rates (see Figs 10.1, 10.2 and 10.3) and highly seasonal reproductive cycles (in males and females), with highly seasonal pasture quantity and quality in many climatic environments, presents a very significant challenge to nutritionists formulating rations for farmed deer. In addition, although the growth of pelage is normally included in the maintenance nutrient requirements of ruminants, for some species of deer this pelage growth can make a significant contribution to the maintenance energy, protein and mineral requirements (Dryden 2011). Antler growth can also be considered a part of the maintenance requirement and greatly influences mineral requirements.

Metabolism in deer

The metabolic rate of deer varies markedly with season, reflecting a combination of seasonal changes in diet type and availability, the voluntary intake differences noted earlier, ambient temperature, and pelage insulation changes, which alter the critical temperature.

Energy requirements of deer

The energy requirements of deer are expressed as Metabolisable Energy (ME) requirements using the usual feeding standard method of estimating the Net Energy (NE) requirement by adding up all the components of maintenance (bodyweight, age, species, reproduction, lactation, activity, sex, growth of pelage and, uniquely for deer, the effect of season and antler growth). To convert the NE requirement to ME the 'k factors' for efficiency of conversion of ME to NE are used (see Chapter 4). This factorial approach is reviewed in detail by Dryden (2011).

Energy requirements for maintenance

Given the large variation between species in adult bodyweight, it is essential to use metabolic scaling to estimate maintenance energy requirements of deer. Shin *et al.* (2000) suggested the following equations be used for deer:

$$ME_m \text{ (MJ/day)} = 0.45 \text{ MJ/kg}^{0.75} \text{ (for 3–11 months of age)}$$

$$ME_m \text{ (MJ/day)} = 0.57 \text{ MJ/kg}^{0.75} \text{ (for adults)}$$

Energy requirements for growth

The maintenance requirement of deer changes with season due to a change in the efficiency of utilisation of ME for growth (Milne *et al.* 1987). Table 10.3 shows the

Table 10.3. Daily ME requirements (MJ/day) of deer for maintenance plus growth at different rates of live weight gain. Note the impact of season on energy requirements for growth of calves.

Live weight (kg)	Season	Live weight gain (g/day)				
		0	50	100	150	200
40 (calf)	Autumn	7.2	10.0	12.7	15.5	18.2
50 (calf)	Winter	8.5	12.9	17.2	21.6	25.9
60 (calf)	Spring/Summer	9.7	12.1	14.6	17.0	19.4
80 (hind)		15.2				
100 (hind)		18.0				
150 (stag)		24.4				
250 (stag)		35.8				

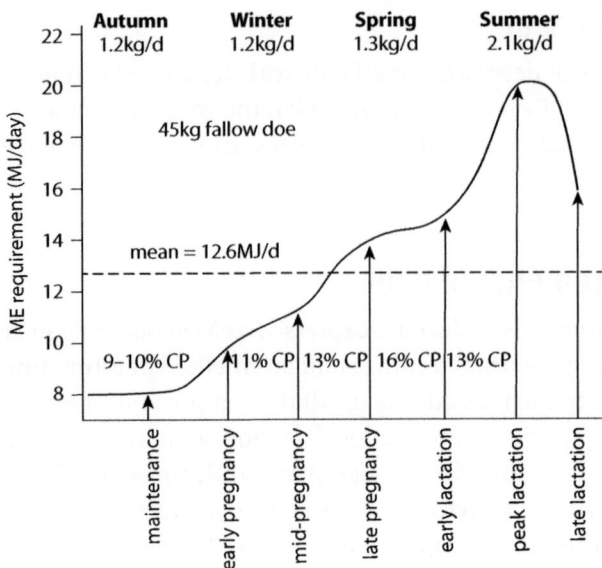

Fig. 10.1. Metabolisable Energy and crude protein requirement of a 45 kg fallow doe through a typical breeding cycle. Typical daily dry matter intakes are also indicated. Sources: Fennessy *et al.* (1981); Haigh and Hudson (1993); Shin *et al.* (2000); Mulley and Flesch (2001); Tuckwell (2003); NRC (2007); Dryden (2011).

ME requirements of red deer, assuming the maintenance energy requirements above for different ages, and the ME for growth changes with season as follows: autumn (56 MJ/kg gain); winter (88 MJ/kg gain); and spring/summer 48 MJ/kg gain).

Energy requirements for pregnancy and lactation

Pregnancy increases the ME requirement of deer by ~30–40% but lactation greatly increases ME requirement (by 200%). Failure to provide sufficient ME to meet this high demand for lactation results in a 30–60% reduction in milk yield and poor weight gain in the calf (Shin *et al.* 2000).

Protein requirements of deer

A supply of protein is essential for growing deer and inadequate protein at a young age can influence subsequent growth and antler production (Dryden 2011). Although a diet of 6–7% can maintain rumen function, typically ~13–16% crude protein is required for growth, antler development and reproduction (French *et al.* 1956). The requirements for crude protein by pregnant and lactating females are shown in Figs. 10.1, 10.2 and 10.3. For growing deer, young calves or fawns (3–5 months) require high protein (17%) in autumn and in spring/summer, and ~10% in winter (Adam 1994).

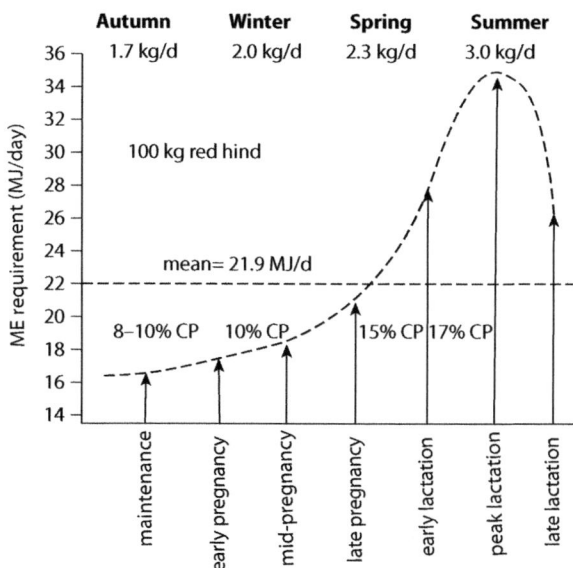

Fig. 10.2. Metabolisable Energy and crude protein requirement of a 100 kg red hind through a typical breeding cycle. Typical daily dry matter intakes are also indicated. Sources: Fennessy *et al.* (1981); Haigh and Hudson (1993); Shin *et al.* (2000); Mulley and Flesch (2001); Tuckwell (2003); NRC (2007); Dryden (2011).

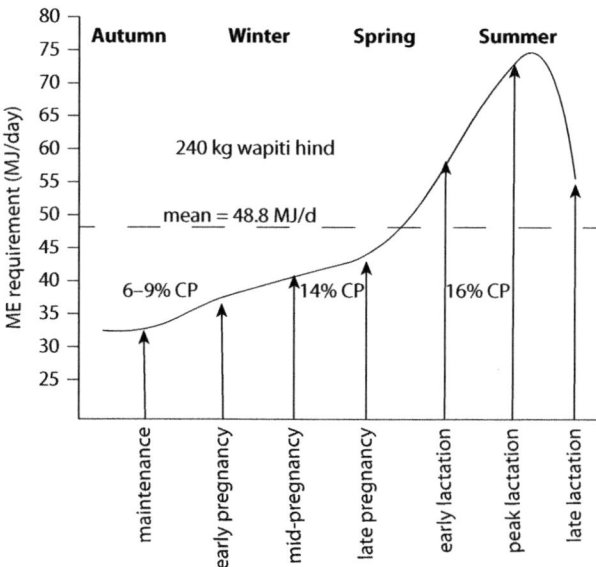

Fig. 10.3. Metabolisable Energy and crude protein requirement of a 100 kg wapiti/elk hind through a typical breeding cycle.

Mineral and vitamin requirements of deer

Although precise estimates of the mineral requirements of deer have not been made, the few estimates that are available suggest that deer differ little in their mineral requirements from other ruminants such as sheep. Nearly all of the estimates made by Haigh and Hudson (1993), Adam (1994) and Puntenney (1995) are within the ranges accepted for sheep. A summary of these estimates is provided in Table 10.4.

Little research has been conducted on the vitamin requirements of deer but it seems reasonable to assume they are similar to that of other ruminants. This means that the B-group will be provided by the ruminal microbes as long as the rumen is functioning properly (note young deer will need B-group supplementation until the rumen is developed). Vitamins A, D and E are required in the feed. Vitamin A can be synthesised from β-carotene and careful diet selection by deer ensures high levels of these vitamins are usually present in their diet. As for the other ruminants, long periods on dry, desiccated feeds may result in vitamin A deficiency and long periods on lush, green feeds with little irradiated feeds may result in a vitamin D deficiency. Vitamin A is particularly important for antler development.

One study showed that young deer are susceptible to vitamin E deficiency but are resistant to selenium deficiency even when diets are low in selenium (Brady *et al.* 1978). Neither vitamin E nor selenium deficiency in adult deer increased mortality in this study. However, another study has shown that Se supplementation

Table 10.4. Mineral requirements of deer for maintenance, growth and reproduction (diet dry matter basis)

	Maintenance	Velvet	Growth	Gestation	Lactation
Macrominerals					
Ca (%)	0.35	1.40	0.6	0.6	0.70
P (%)	0.25	0.70	0.3	0.4	0.40
K (%)	0.65	1.00			1.00
Mg (%)	0.20	0.40			0.25
NaCl (%)	0.15	0.20			0.20
Microminerals					
Cu (mg/kg)	15				15
Mn (mg/kg)	40	As for maintenance	As for maintenance	As for maintenance	40
Zn (mg/kg)	50				50
Fe (mg/kg)	50				50
I (mg/kg)	0.30				0.60
Co (mg/kg)	0.10				0.20
Se (mg/kg)	0.20				0.20

Sources: Puntenney (1995); Haigh and Hudson (1993); Adam (1994); Smits and Haigh (1990); Shin *et al.* (2000)

of black-tailed free-ranging deer increased fawn survival significantly (Flueck 1994), leading this author to conclude that assumptions that deer have evolved to be resistant to trace element deficiencies are wrong. Given that deer in their first year of life appear to be particularly susceptible to depletion of selenium and copper levels in plasma (Pareja-Carrera *et al.* (2018), it is best to assume that deer are equally susceptible to trace element and vitamin deficiencies as other ruminants and should be supplemented accordingly.

References

Adam CL (1994) Feeding. In *Management and Disease of Deer*. (Eds TL Alexander and D Buxton) pp. 44–54. The Veterinary Deer Society, London, UK.

Austin PJ, Suchar LA, Robbins CT, Hagerman AE (1989) Tannin-binding proteins in saliva of deer and their absence in saliva of sheep and cattle. *Journal of Chemical Ecology* 15, 1335–1347. doi:10.1007/BF01014834

Brady PS, Brady LJ, Whetter PA, Ullrey DE, Fay D (1978) The effect of dietary selenium and vitamin E on biochemical parameters and survival of young among white-tailed deer (*Odocoileus virginianus*). *The Journal of Nutrition* 108, 1439–1448. doi:10.1093/jn/108.9.1439

Domingue BMF, Dellow DW, Wilson PR, Barry TN (1991) Comparative digestion in deer, goats and sheep. *New Zealand Journal of Agricultural Research* 34, 45–53. doi:10.1080/00288233.1991.10417792

Dryden GMcL (2011) Quantitative nutrition of deer: energy, protein and water. *Animal Production Science* 51, 292–302. doi:10.1071/AN10176

Fennessy PF, Moore GH, Corson ID (1981) Energy requirements of red deer. *Proceedings of the New Zealand Society of Animal Production* 41, 167–173.

Fickel J, Göritz F, Joest BA, Hildebrandt T, Hofmann RR, Breves G (1998) Analysis of parotid and mixed saliva in roe deer (*Capreolus capreolus* L.). *Journal of Comparative Physiology. B, Biochemical, Systemic, and Environmental Physiology* 168, 257–264. doi:10.1007/s003600050144

Flueck WT (1994) Effect of trace elements on population dynamics: selenium deficiency in free-ranging black-tailed deer. *Ecology* 75, 807–812. doi:10.2307/1941736

Fraudenberger DO, Toyakawa K, Barry TN, Ball AJ, Suttie JM (1994) Seasonal changes in rumen physiology in red deer during summer. *British Journal of Nutrition* 71, 489–499.

French CE, McEwen LC, Magruder ND, Ingram RH, Swift RW (1956) Nutrient requirements for growth and antler development in white-tailed deer. *The Journal of Wildlife Management* 20, 221–232. doi:10.2307/3796954

Haigh JC, Hudson RJ (1993) *Farming Wapiti and Red Deer*. Mosby, St Louis, MI, USA.

Hoffmann RR (1985) Digestive physiology of deer – their morphophysiological specialisation and adaptation. In *Proceedings of the International Conference on the Biology of Deer Production*. 13–18 February, Dunedin. (Eds PF Fennessy and KR Drew) pp. 393–407. Royal Society of New Zealand, Wellington, New Zealand.

Kay RNB (1985) Body size, patterns of growth, and efficiency of production in red deer. In *Proceedings of the International Conference on the Biology of Deer Production*. 13–18 February, Dunedin. (Eds PF Fennessy and KR Drew) pp. 411–422. Bulletin 22, Royal Society of New Zealand, Wellington, New Zealand.

Loudon ASI, Milne JA, Curlewis JD, McNeilly AS (1989) A comparison of the seasonal hormone changes and patterns of growth, voluntary food intake and reproduction in juvenile and adult red deer (*Cervus elaphus*) and Père David's deer (*Elaphurus davidianus*) hinds. *The Journal of Endocrinology* **122**, 733–745.

Milne JA, Macrae JC, Spence AM, Wilson S (1978) A comparison of the voluntary intake and digestion of a range of forages at different times of the year by the sheep and the red deer (*Cervus elaphus*). *British Journal of Nutrition* **40**, 347–357.

Mulley RC, Flesch JS (2001) 'Nutritional requirements for pregnant and lactating red and fallow deer: a report for the Rural Industries Research and Development Corporation'. RIRDC, Canberra.

Nagy JG, Regelin WL (1975) Comparison of digestive organ size of three deer species. *The Journal of Wildlife Management* **39**, 621–624. doi:10.2307/3800407

NRC (2007) *Nutrient Requirements of Small Ruminants: Sheep, Goats, Cervids and New World Camelids*. National Research Council, The National Academies Press, Washington, DC, USA.

Pareja-Carrera J, Rodriguez-Estival J, Martinez-Haro M, Ortiz JA, Mateo R (2018) Age-dependent changes in essential elements and oxidative stress. *Science of the Total Environment* **626**, 340–348. doi:10.1016/j.scitotenv.2018.01.072

Prins RA, Geleen MJH (1971) Rumen characteristics of red deer, fallow deer and roe deer. *The Journal of Wildlife Management* **35**, 673–680. doi:10.2307/3799772

Puntenney S (1995) *Health and Feeding Needs for Wapiti*. Foster Mills, Delta, CO, USA.

Shin HT, Hudson RJ, Gao XH, Suttie JM (2000) Nutritonal requirements and management strategies for farmed deer – review. *Asian-Australasian Journal of Animal Sciences* **13**, 561–573.

Smits JEG, Haigh JC (1990) Specialized livestock – Game farming practice. Saskatchewan Agriculture and Food, Regina, SK, Canada.

Tuckwell CD (2003) Species in Australia. In *The Deer Farming Handbook* (Ed. CD Tuckwell) pp. 12–22. Rural Industries Research and Development Corporation, Canberra.

11
Dog and cat nutrition

Key points

- An evidence-based approach to dog and cat nutrition is essential given the plethora of sources of often-misleading and conflicting views.
- Dogs evolved from the grey wolf but during domestication acquired a suite of genes associated with starch digestion and metabolism. Dogs are not carnivores. They are best described as meat-eating omnivores capable of digesting and assimilating a wide variety of feedstuffs of both plant and animal origin.
- Dogs and cats are mono-gastric, post-gastric fermenters with simple and relatively short gastrointestinal tracts and small caecae.
- Cats are strict carnivores and have a high requirement for animal protein. They require pre-formed vitamin A, taurine, arachidonic acid, and high dietary levels of niacin, arginine and sulphur amino acids.
- Cats are poor at regulating protein metabolism and rapidly show adverse clinical signs on low-protein diets.
- More than 40% of dogs and cats in developed countries are overweight (30–35%) or obese (5–10%). These animals are at increased risk of developing several pathologies (diabetes mellitus, pancreatitis, osteoarthritis, hip dysplasia, pulmonary disease, cardiovascular disease, reduced tolerance to exercise and heat, possibly increased neoplasms). Over-fatness also makes veterinary examination, palpation, diagnostic imaging and surgery more difficult.
- Overweight dogs have a significantly reduced lifespan (30% decrease) and reduced quality of life.
- Risk factors for overweight dogs include breed, neutering, female, increased age, high feeding frequency, larger number of people in the household, feeding of

treats, obese owners, low socioeconomic status of owners, lack of exercise, hypothyroidism, hyperadrenocorticism and increased availability of low-priced food.

- For cats, provision of dry food *ad libitum* is the major risk factor for obesity.
- Feeding programs for dogs overestimate the energy requirements and overestimate the impact of modest exercise. Different dog breeds differ in their energy requirements for maintenance.
- A new feeding system for dogs is presented in this chapter based on dog breed and tendency to obesity. It uses estimation of target weights based on optimum fatness and the Metabolisable Energy required to maintain this target weight, taking into account life stage and reproductive status. Continual reassessment of the body condition score and health of the dog and readjustment of the program is essential.
- Teaching cat and dog owners to condition score their animals accurately is an essential feature of a holistic health pet care program.
- Homemade diets are almost always unbalanced, suggesting it is difficult for owners to formulate a nutritionally adequate diet from a few simple ingredients. There is no evidence of nutritional benefit of BARF® diets and some evidence (poor quality) of a nutritional risk. There is strong evidence that raw diets present a pathogen risk to both dogs and owners.
- It is possible to formulate a nutritionally adequate vegetarian diet for dogs but the challenges presented by the high levels of antinutritional factors in plant foods are such that it would be difficult for a non-professional to formulate a balanced and palatable vegetarian diet for dogs.
- Adverse food reactions in dogs are common and can be identified by well-designed elimination and replacement trials.

Introduction

Dog nutrition – background

The remarkable diversity in modern dogs in terms of morphology, behaviour, physiology and metabolism arose from a common ancestor ~10 million years ago (Vila *et al.* 1999). There is little doubt that domesticated dogs evolved from a grey wolf ancestor, and sequence divergence with wolves appeared ~100 000 years ago. However, it appears that continued admixture of wolf and dog genetic information provided a source of genetic variance on which artificial and natural selection was founded (Vila *et al.* 1997). It appears that after the separation of dogs from wolves, continued backcrossing events provided highly variable genetic material for the extraordinary phenotypic variance apparent in the domestic dog today. This high diversity presents challenges to nutritionists that are not present for many other domesticated species, where phenotypic variance is relatively small (e.g. sheep, cattle, horses). Diversity relevant to nutritional considerations in dogs includes:

- variance in bodyweight (1.5–100 kg)
- variance in appetite drive (e.g. labrador versus toy dogs)
- variance in body shape and composition (e.g. greyhound versus bassett hound)
- variance in propensity to obesity (e.g. labrador versus greyhound)
- variance in exercise-drive and ability (e.g. border collie versus lapdogs)
- variance in work potential (Alaskan husky versus pekingese).

These wide variances in phenotype are driven by underlying genetic differences in metabolic and physiological function, which present unique challenges for dog nutritionists.

Nearly everyone has strongly held opinions and beliefs about dog nutrition. This widespread interest and passion about dog nutrition stems from the close relationships we have with these pets, to the point that dogs are regarded as members of the family. As such, concerns about the health of the dog are foremost, and nutrition is a major factor the owner can control to ensure good health. However, the average dog owner is confronted with a plethora of often conflicting 'information' about dog nutrition, making it difficult to make informed decisions about diets. Conflicting advice comes from:

- businesses with a thoroughly reasonable interest in maximising profit (this includes multinational, but also many smaller, special-interest businesses) selling commercial products into a highly competitive market. The global pet food market was worth US$75 billion in 2016, with an annual growth of ~4% per annum. Dog and cat food dominate this market. There are 10 companies that control more than 50% of the market including Mars Petcare (US$17 billion), Nestle Purina Petcare (US$12 billion.), Big Heart Pet Brands (US$2.3 billion), Hills Pet Nutrition (US$2.2 billion), Diamond Pet Foods (US$1.2 billion), Blue Buffalo (US$1.0 billion), Spectrum Brands/United Pet Group (US$800 million.), Unicharm Corporation (US$722 million.), Deuerer (US721 million) and Heristo AG (US$700 million).
- special interest groups, such as animal rights groups, with an interest in promoting meat-free diets
- people espousing values-based ideals such as vegetarianism, based on ethical, environmental, animal rights and supposed 'health' reasons.

In developed countries, the proportion of dogs that are classified as overweight or obese has risen to 35–45% and appears to be increasing in direct synchrony with the obesity epidemic in humans, probably for the same reasons. Being overweight or obese is probably the largest single predisposing factor for a wide range of pathologies including cardiovascular disease, metabolic disorders (fatty liver, pancreatitis, diabetes), orthopaedic problems and a reduction in lifestyle and exercise. There is a direct link between body condition score at middle age and age at death, with dogs at condition score >3.5/5 having a reduction in lifespan of 30% (Laflamme 2006).

> Given the diversity of dog phenotypes, the plethora of conflicting views on nutrition, and the impact of nutrition on dog health and longevity, dog nutrition is an area in which application of **evidence-based nutritional science** is essential.

Cat nutrition – background

The domestication of cats differed significantly from that of dogs. The house cat is probably not a domesticated wildcat (Driscoll *et al.* 2009). Wild cats are solitary animals and very territorial and unlikely to have been attracted to human settlements. Cats do not perform directed tasks and would therefore have been of little use to humans. It is more likely that humans tolerated wild cats and over time these animals diverged from their wild ancestors. This occurred ~10 000 years BP, probably in the Fertile Crescent. The cats may have been attracted to human populations due to the presence of small rodents attracted to grains produced in agriculture. The cat was therefore a product of natural, not artificial, selection (with the exception of the last 200 years in which some controlled breeding has taken place). This parallel development of the domestic cat and wild cat (sympatry) has produced differences in digestive function and nutrition. Developing with humans, the domestic cat has developed shorter intestines than wild cats, reflecting a less strictly carnivorous diet due to feeding on meal scraps (Darwin 1890). The domestic cat is a small, frequent meal eater, unlike the larger wild cats, which are intermittent, large meal eaters. Being a solitary hunter and only socialising at mating times, the domestic cat has not had to compete for food with others and is therefore a slower, and perhaps more selective (fussy) feeder. Cats provided food *ad libitum* throughout the day eat up to 16 meals/day.

Given its evolutionary background, the domestic cat is best served by a diet similar in composition to that provided by a small rodent (see Table 11.1).

Cats (*Felis catus*), like dogs, belong to the Order Carnivora, but unlike dogs, cats are obligate meat eaters and this has implications for the development of feeding programs that are consistent with health and longevity. Cats also have several metabolic idiosyncrasies that are of importance in designing diets (Box 11.1).

Table 11.1. The nutrient composition of an adult, small field mouse (*Mus domestica*) – a guide to the 'ideal' nutrient profile for domestic cats

Nutrient	Content (DM basis)
Dry matter	33%
Crude protein	56%
Fat	24%
Ash	11.8%
Gross Energy	22 MJ/kg
Metabolisable Energy (estimated)	15 MJ/kg
Calcium	2.98%
Phosphorus	1.73%
Ca:P ratio	1.7:1

Source: Adapted from Corva and Medrano (2000) and Dierenfeld *et al.* (2002)

Box 11.1. Unique features of feline nutrition

- Cats utilise protein as a major energy source and have a very high protein turnover rate.
- Cats utilise protein more for maintenance than for growth and so have a very high protein requirement.
- Cats are poor at regulating their protein metabolism with changes in protein intake, so are prone to deficiencies and toxicities of amino acids.
- Cats cannot synthesise taurine, so taurine is an essential dietary component.
- Cats have a high dietary requirement for the essential amino acid arginine to drive the urea cycle to remove excess ammonia from meat protein metabolism.
- Cats cannot convert sufficient amino acid tryptophan to vitamin B3 (niacin) and so require high levels of dietary preformed niacin.
- Cats cannot synthesise arachidonic acid from linoleic acid and so require preformed dietary arachidonic acid.
- Cats are deficient in glucokinase, an enzyme in the glycolysis pathway, and so cats are poor at utilising simple sugars.
- Cats do not adjust their water intake to feed water content very well.

These metabolic specialisations mean that cats are not small dogs and should not be fed diets designed for dogs. The absolute requirements for preformed nutrients means that cats have much more-stringent nutritional requirements than dogs. The unique metabolic adaptations of cats also predispose them to several metabolic challenges and increase their susceptibility to certain pathologies such as diabetes mellitus (Type 2 diabetes).

The cat food market, as for dogs, is a multi-billion industry, with more than 90% of pet owners in the US choosing to feed a commercially prepared cat food due to its reliability, consistency and convenience. As for dogs, there is a small proportion of consumers with concerns about the wholesomeness of commercial pet foods, their quality, safety and for some ethical issues around the use of animals as food. Vegetarian diets for cats are produced and, although it is possible to produce cereal-based rations for cats by supplementing the essential preformed nutrients, it is clear that cats evolved to be meat eaters and that meat and animal tissues should form a major part of their diet.

> If cat owners have ethical objections to the use of animals as food for their carnivorous cat, then perhaps they should get a pet rabbit!

In this chapter the digestive systems and processes of dogs and cats are described, followed by a description of their energy, protein, vitamin and mineral requirements. The process of formulating, monitoring and adjusting a feeding and exercise program to maintain cats and dogs in good health is then described. Finally, major issues in cat and dog nutrition are discussed including obesity in dogs, diabetes in cats, adverse food reactions in dogs, vegetarian diets for cats and dogs, raw food diets for dogs and coprophagy in dogs.

Digestive anatomy and physiology of dogs and cats

The permanent teeth arrangement for cats and dogs is shown below.

Cats	Incisors	Canines	Premolars	Molars
Upper	3	1	4	2
Lower	3	1	4	3
Dogs				
Upper	3	1	3	1
Lower	3	1	2	1

Although both cats and dogs have the same number of incisors and canines on each jaw (six incisors/jaw and two canines/jaw), the number and size of premolars and molars are very different. Dogs have a total of 16 premolars and 10 molars while cats have only 12 premolars and four molars. These teeth are designed for chewing and crushing plant-derived materials and, along with the genomic evidence, now confirm the dog as a true omnivore. The blade-like carnassial teeth in dogs (the 4th premolars) are designed for shearing muscle and the large canines are designed for gripping prey and puncturing flesh.

Saliva in dogs and cats

Saliva in both species acts a lubricant to facilitate swallowing and to dissolve water-soluble components that can then interact with the taste buds. It is difficult to find verifiable data on taste bud numbers in different species. The number of taste buds in cats has been estimated as 2600 (Robinson and Winkles 1990) and 1150 (Hayes and Elliot 1942) and for dogs the widely quoted, but not validated number is 1700. Dogs and cats are sensitive to the tastes of amino acids, nucleotides and certain organic acids, all present in animal tissues, but neither has a strong preference for salty foods, unlike most herbivores and humans. Cats have lost the ability to detect sweetness and this has been traced to the absence of expression of the TasR2 receptor gene (Li *et al.* 2005). This appears to be feature of true carnivores and may reflect a focus on tastes associated only with animal tissues and not sugars, which are plant compounds.

Volatile substances are also released and impact on the olfactory system. In dogs, but not cats, saliva also acts as an evaporative cooling system. Salivary glands in dogs and cats are classified as major (parotid, mandibular (or submaxillary), sublingual and zygomatic glands) or minor (buccal, labial, lingual, tonsillar, palatine and molar). The positioning of the major salivary glands is indicated in Figs. 11.1 and 11.2.

The mandibular (submaxillary) and sublingual glands produce large amounts of mucus, while the parotid glands produce a serous fluid. The relative amounts of serous and mucus saliva depends on the type of food consumed and the rate of saliva secretion. A 20 kg dog produces ~0.5–1.0 L of saliva daily (Goff 2015). The electrolyte composition and osmolarity of saliva resemble that of plasma but with

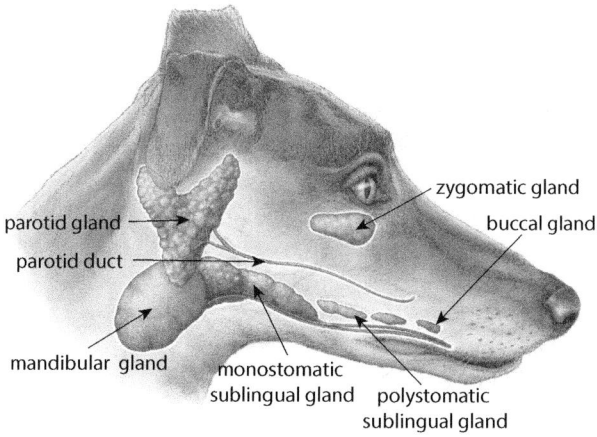

Fig. 11.1.   Major salivary glands of the dog (*Canis familiaris*).

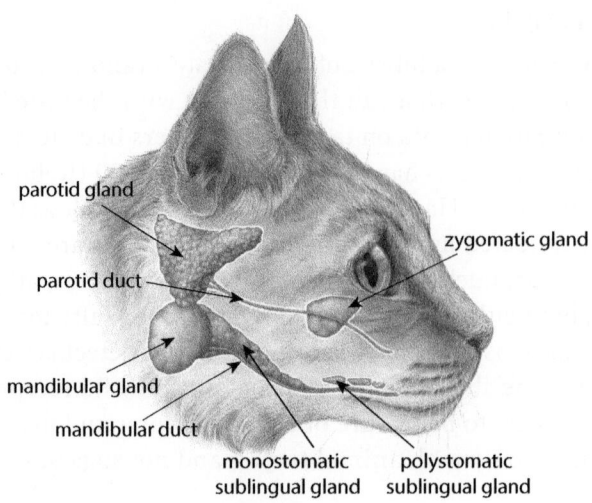

Fig. 11.2. Major salivary glands of the cat (*Felis catus*).

higher bicarbonate ions and higher pH. The pH of dog saliva is 7.3–7.8. Dog saliva also has high levels of Ca, Na and K and, together with the high pH, may account for the lower incidence of dental caries in dogs but also higher rates of gingivitis and periodontal disease than humans (Lavy *et al.* 2012). The presence of amylase in dog saliva has been debated for decades. Goff (2015) in *Duke's Physiology of Domestic Animals* refers to the parotid secretion as being 'laden with amylase'. However, almost all other reports are that the secretions of the mandibular gland contain trace amylase activity, while those of the parotids are devoid of any amylase activity (Chauncey *et al.* 1963). Similarly there is no detectable amylase activity in cat saliva. There are also no reports of lipase activity in dog saliva. Enzyme activity in the mouth of dogs and cats is of little importance in their digestive function.

Saliva contains histatins, nerve growth factor and lysozymes, which have antibacterial properties, but dog and cat saliva does contain significant pathogenic bacteria such as *Pasteurella* and *Salmonella* and the parasite *Giardia*, which have been shown to infect humans. Dog and cat saliva also contains ~12 allergens, which are transferred to fur during licking, and when the saliva dries the allergens become airborne and can cause allergy in humans.

Gross anatomy of the digestive tract of dogs and cats

The digestive tracts of dogs and cats are relatively simple, both comprising a simple gastric stomach, small intestines (duodenum, jejunum and ileum), large intestines (small caecum and colon) and rectum (Figs. 11.3 and 11.4).

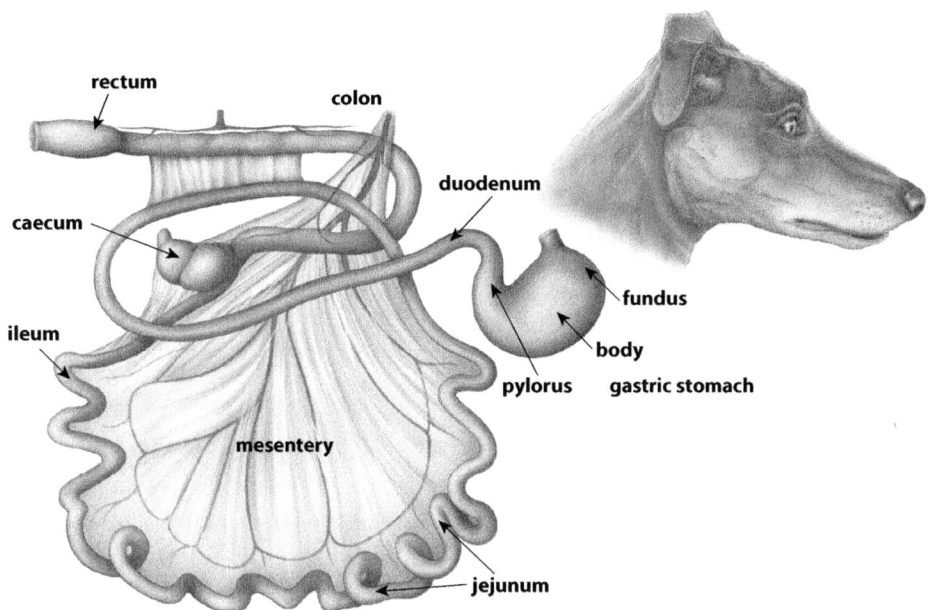

Fig. 11.3. Simplified diagram of the digestive tract of the dog (*Canis familiaris*).

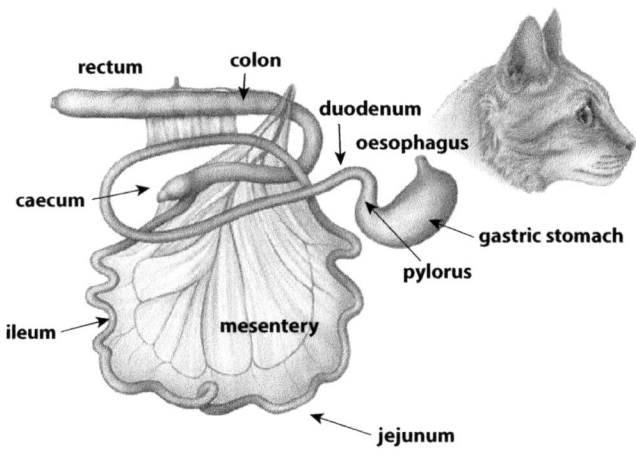

Fig. 11.4. Simplified diagram of the digestive tract of the cat (*Felis catus*).

Length and capacity of the digestive compartments in dogs and cats

The lengths and capacities of the various segments of the gastrointestinal (GI) tract of a beagle dog are shown in Table 11.2. The relative lengths normalised for bodyweight are similar for cats.

The average pH of the cat stomach contents is 2.5 ± 0.07 (Brosey *et al.* 2000).

Table 11.2. Length, volume, pH of various segments of the gastrointestinal tract of a beagle dog (bodyweight 10 kg)

Segment	Length (cm)	Volume (mL)	pH
Stomach	12	450 (47%)	1.1–2.5
Duodenum	25	250 (26%)	4.0–7.0
Jejunum	220		6.0–8.0
Ileum	15		6.0–8.0
Ascending colon	5	250 (26%)	
Transverse colon	7		6.0–7.0
Descending colon	12		

Source: Mahar *et al.* (2012)

Digestion in the stomach, small intestines and large intestine of dogs and cats

The stomach stores meals and begins the initial stages of digestion by secreting pepsinogen, lipase and HCl acid. During meal eating, the proximal part of the dog stomach expands, allowing a large intake of a food, which is important in a large intermittent feeder, unlike the cat, which consumes frequent small meals. Rhythmic contractions of the middle circular layer of smooth muscle in the wall of the stomach macerate the feed and mix it with the acid chyme. The pH of dog stomach contents is widely reported to be more acidic than humans and other animals and this is said to protect the dog from pathogens in meat. There is no evidence that dog stomach pH is lower than humans and in fact it appears to often be higher (Sagawa *et al.* 2009). The average pH of dog stomach digesta is between 1.5 and 2.0 (Mahar *et al.* 2012), compared with humans at 1.1. In contrast to humans, however, the pH of dog gastric digesta stays acidic after feeding. There are also large spikes in the pH of dog digesta to above pH 6.0, which may reflect salivation in anticipation of feeding (Mahar *et al.* 2012). There are few reports of detailed analysis of the gastric pH in cats, but it is widely repeated that cats and dogs have very low gastric pH and that this is important in digestion of meats and bones and in protection from pathogens. Little evidence is available to support any of these statements.

Detailed discussion of the digestive processes in mono-gastric animals is provided in Chapter 2.

During domestication, dogs acquired genes for metabolism of starches. Dogs are not carnivores and are well equipped to digest and metabolise plant foods, and should be classified as 'meat-eating' omnivores.

Cats are carnivores and require pre-formed essential nutrients found in animal tissues. They can digest carbohydrates but are predisposed to diabetes.

The energy, protein, vitamin and mineral requirements of dogs and cats

Energy requirements of dogs for maintenance, growth, pregnancy, lactation and exercise

Given the wide range of dog phenotypes (bodyweight, shape, fatness, athletic ability, temperament and metabolic rate), it is impossible to develop simple predictive equations for estimating the nutrient requirements of all dogs. The task becomes even more difficult if we add the effects of exercise activity, neutering, age, environment, reproduction, lactation and growth stage, on requirements for energy, protein, vitamins and minerals. It is not surprising that the published estimates of the ME requirements for dogs vary widely. For example, the maintenance energy requirements in the literature range from 62 $kcalW^{0.75}$/day for dogs that have lost weight after obesity (German *et al.* 2011) to 856 $kcalW^{0.75}$/ day for racing endurance sled dogs (Loftus *et al.* 2014). For dogs, the absolute energy intake is important because most dog feeds are formulated to meet minimum and maximum levels of nutrients based on a per kilogram of feed or per kilocalorie of ME basis (Dzanis 1994). Ensuring the dog receives the correct amount of energy/day also ensures the dog is receiving the correct levels of other nutrients.

The standard method for formulating a feeding program for a dog is to use the dog's current bodyweight and then estimate the quantity of a commercial feed of known ME density to determine the quantity of the feed to be provided to the dog each day. Factors are applied to account for life stage, reproductive status, activity level, lactation and age. There are several problems with this approach:

- Basal metabolic rate varies with bodyweight raised to an exponent such as 0.75 (metabolic bodyweight), not with bodyweight *per se* (see Chapter 1).
- Different breeds vary in basal metabolic rate at the same bodyweight (NRC 2006). The problem with using current bodyweight in the MER for maintenance equations is that the dog may already be above a healthy body fat percentage, which has two effects: (1) fat is metabolically less active than lean body mass so on a per unit weight basis overfat dogs use less energy than lean dogs of the same weight; and (2) feeding to the already existing overfat weight is overfeeding the dog. To overcome these problems, it is recommended that an 'ideal' bodyweight or 'target' bodyweight is estimated using either body condition scoring (BCS) or other methods of estimating body fat percentage. The target weights of overfat dogs are adjusted to an 'ideal' value of ~20% fat using BCS as a predictor of fat percentage.
- Corrections for activity levels are very inaccurate, because most people overestimate the total exercise activity of their dog.

The author has devised a simple system for estimating the ME requirement for maintenance for dogs differing in body fat percentage, breed, activity level and physiological state. This approach can be easily taught to pet owners or can be used to advise them of a good starting point for a feeding plan to manage the dog's health.

A new, simple approach to formulating dietary plans for dogs

The following is a simple method for designing a feeding program for pet dogs to ensure they reach and maintain a healthy level of body fatness. It takes into account differences between breeds in propensity to obesity, allows an 'ideal' target weight to be identified based on a simple estimation of body fat percentage, and accounts for overestimation of exercise activity. The following simple steps allows a dietary plan to be formulated for dogs varying in age, sex, breed, activity level, reproductive status and lactation.

1. Weigh the dog

It is best to take an average over several weighings in a 1–2 week period. This can be done at a veterinary practice or at home using bathroom scales. If the dog can be lifted, weigh yourself, pick the dog up, reweigh, then subtract your weight.

2. Estimate the body fat percentage of the dog using body condition scoring or body measurements that are related to actual body fat percentage

Body condition scoring is a very useful tool in dog nutrition and pet owners should be taught to frequently assess their dog's condition, particularly if the breed is obesity-prone and/or covered in hair, which makes visual assessment of condition difficult (Laflamme 2006). With some training, the repeatability and accuracy of body condition scoring by novices improves markedly, and commonly they begin to make further divisions of the 5-point scoring system. This effectively makes the 5-point system equivalent to the 9-point system. Given the ease of learning the 5-point system, this seems a good place to start (see Table 11.3).

However, body condition scoring is inaccurate when used in dogs with scores over ~3/5 (Witzel et al. 2014). This study showed that BCS scoring of dogs scoring 3, 4 or 5 on a 5-point scale was poorly related to actual DEXA-measured fat percentage. This is partly due to differences in fat distribution between breeds, and because dogs scoring 5/5 can vary in actual fat percentage from 40 to 70% of bodyweight. An alternative system for measuring body fat percentage of dogs is to measure various part of the dog's body (morphometrics) that are related to body fat percentage (Box 11.2).

Table 11.3. A 5-point scoring system for estimating body condition score in dogs[a]

Score (/5)	Visual	Description
1 **Very thin**		• Ribs and vertebrae easily seen at distance • Sharp bony protrusions felt over vertebrae and ribs sharp and shallow • Very obvious 'waist' from above • Starved appearance
2 **Underweight**		• Ribs, vertebrae easily felt • Obvious 'waist' from above • Skinny appearance
3 **Ideal**		• Ribs and vertebrae obvious • Individual ribs can be just discerned • 'Waist' just apparent from above • Healthy appearance
4 **Overweight**		• Ribs and vertebrae have a smooth feel • Hard to distinguish individual ribs • 'Waist' not apparent • Abdomen sagging from the side • Rounded, fat appearance
5 **Obese**		• Can't distinguish ribs or vertebrae • Fat covers all bones • No waist from above • Sagging abdomen from side • Overweight, fat appearance

[a]Half scores can be used as the skill level and confidence increases

Box 11.2. Estimation of body fat percentage using morphometric measurements

The following equation estimates body fat percentage (as measured in a DEXA machine) with an accuracy of 82% (Witzel et al. 2014) and is particularly useful in dogs with a body fat >40%:

Body fat percentage = 0.71 × TC – 0.1 × (PC/6)2 – 5.78 × HLL 0.8 + 26.56 × PC/HC) + 2.06

Where TC = thoracic circumference; PC = pelvic circumference; HLL = hindlimb length; HC = head circumference (all measurements in cm).

Figure 11.5 shows the sites of morphometric measurement used in this equation.

In practical situations, it is probably unlikely that dog owners will make the morphometric measurements, but teaching them body condition scoring is very valuable.

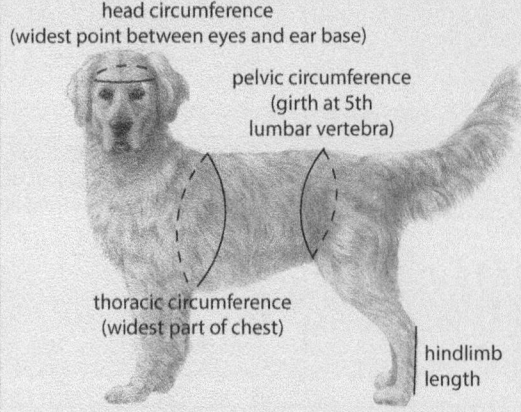

Fig. 11.5. Measurements used to estimate body fat percentage in dogs. (hind limb length = from the proximal aspect of the metatarsal pad to the dorsal tip of the calcaneal tuber; pelvic circumference = body girth at the level of the 5th lumbar vertebrae; thoracic circ. = body girth at the widest part of the chest immediately caudal to the elbow; head circ. = widest part of head between eyes and ears). Note: this equation was derived from a large cohort of dogs representing 20 breeds and bodyweights ranging from 5 to 74 kg. However, in breeds with angular limb deformities (chondrodystrophia) such as the dachshund, corgi, bulldog, pug and pekinese, or dogs with large heads such as Staffordshire terriers, this equation may not hold. Source: Adapted from Witzel et al. (2014).

3. Estimate the ideal target weight for the dog

Table 11.4 shows the relationships between body condition score, body fat percentage and ideal target weight for the dog.

Using the dog's current bodyweight and its body condition score or body fat percentage from body measurements, the ideal weight for the dog can be determined. For growing dogs, or dogs below their optimum fat percentage, target weights and growth rates need to be chosen from adult weight targets and growth

Table 11.4. Target bodyweight of dogs estimated from its current bodyweight and either body condition score (1–5 or 1–9 systems) or body fat percentage estimated from morphometric measurements[a]

BCS (/5)	1	1.5	2	2.5	3	3.5	4	4.5	5	5	5	5
BCS (/9)	1	2	3	4	5	6.0	7	8	9	9	9	9
Body fat %	<15	<15	15	15–19	20–24	25–29	30–34	35–39	40–45	50	60	70
Current weight (kg)						Target weight (kg)						
3	Dogs in this range should be fed to meet a target weight appropriate to the dog breed			3	3	2.7	2.5	2.3	2.1	1.9	1.5	1.1
4				4	4	3.6	3.3	3.1	2.8	2.5	2.0	1.5
5				5	5	4.5	4.2	3.9	3.6	3.2	2.5	1.9
6				6	6	5.5	5.0	4.6	4.3	3.8	3.0	2.3
7				7	7	6.4	5.8	5.4	5.0	4.4	3.5	2.7
8				8	8	7.3	6.7	6.2	5.7	5.0	4.0	3.0
9				9	9	8.2	7.5	6.9	6.4	5.7	4.5	3.4
10				10	10	9.1	8.3	7.7	7.1	6.3	5.0	3.8
12				12	12	10.9	10.0	9.2	8.5	7.6	6.0	4.6
14				14	14	12.7	11.7	10.8	9.9	8.8	7.0	5.3
16				16	16	14.5	13.3	12.3	11.4	10.1	8.0	6.1
18				18	18	16.4	15.0	13.9	12.8	11.3	9.0	6.8
20				20	20	18.2	16.7	15.4	14.2	12.6	10.0	7.6
22				22	22	20.0	18.3	16.9	15.6	13.9	11.0	8.4
24				24	24	21.8	20.0	18.5	17.0	15.1	12.0	9.1
26				26	26	23.6	21.7	20.0	18.5	16.4	13.0	9.9
28				28	28	25.5	23.3	21.6	19.9	17.6	14.0	10.6
30				30	30	27.3	25.0	23.1	21.3	18.9	15.0	11.4
32				32	32	29.1	26.7	24.6	22.7	20.2	16.0	12.2
34				34	34	30.9	28.3	26.2	24.1	21.4	17.0	12.9
36				36	36	32.7	30.0	27.7	25.6	22.7	18.0	13.7
38				38	38	34.5	31.7	29.3	27.0	23.9	19.0	14.4
40				40	40	36.4	33.3	30.8	28.4	25.2	20.0	15.2
44				44	44	40.0	36.7	33.9	31.2	27.7	22.0	16.7
48				48	48	43.6	40.0	37.0	34.1	30.2	24.0	18.2
52				52	52	47.3	43.3	40.0	36.9	32.8	26.0	19.8
56				56	56	50.9	46.7	43.1	39.8	35.3	28.0	21.3
60				60	60	54.5	50.0	46.2	42.6	37.8	30.0	22.8
64				64	64	58.2	53.3	49.3	45.4	40.3	32.0	24.3
68				68	68	61.8	56.7	52.4	48.3	42.8	34.0	25.8
72				72	72	65.5	60.0	55.4	51.1	45.4	36.0	27.4
76				76	76	69.1	63.3	58.5	54.0	47.9	38.0	28.9
80				80	80	72.7	66.7	61.6	56.8	50.4	40.0	30.4

[a]The use of the morphometric system to estimate body fat percentage has the advantage that body condition scoring fails to predict body fat percentage above 40–45%

Source: Adapted from German *et al.* (2009) and Witzel *et al.* (2014)

paths (see Table 11.6 for growing dogs). The target bodyweight is then used to estimate the ME requirement for maintenance of the dog at optimum body fat percentage, as described in the section 'Examples of how to use the simple system to estimate the ME requirement of dogs'

4. Estimate the Metabolisable Energy requirement for maintenance of the dog from its target bodyweight

There are significant differences between dog breeds in their metabolic rate and hence energy requirement for maintenance. Table 11.5 has been developed from

Table 11.5. Breed differences in maintenance Metabolisable Energy requirement (MER)

MER group/breed	MER (kcals/day)
High MER (obesity-resistant)[a] Bearded collie Dalmatian English foxhound Flat-coated retriever German boxer Great dane Jack Russell terrier Rhodesian ridgeback Sight hound Small munsterlander	$113W^{0.75}$
'Standard' pet dogs[b]	$95W^{0.75}$
Low MER (obesity-prone)[a] Airedale terrier American Staffordshire terrier Bassett hound Beagle Bichon Boxer Bulldog Cocker spaniel Collies Dachshund English mastiff Golden retriever Labrador retriever Newfoundland Rottweiler West Highland white terrier Yorkshire terrier	$82W^{0.75}$
Working dogs[c] Kelpie Border collie Blue heeler	$188W^{0.75}$
Greyhounds (in racing mode)[d]	$143W^{0.75}$

Sources: [a]Thes *et al.* (2016); [b]NRC (2006); [c]Bermingham *et al.* (2014); [d]Hill (1998)

Table 11.6. MER for gestation and lactation in bitches and growth in puppies

Stage	Energy requirement (kcals/day)	Notes
Gestation (late)	1.25–1.5 × MER	
Lactation	3 × MER	Depends on number in litter and week of lactation
Growth Weaning 40% adult weight (3–4 months) 80% adult weight (4.5–8 months)	 2 × MER 1.6 × MER 1.2 × MER	The aim is to optimise growth and minimise obesity and developmental orthopaedic disease (particularly in larger breeds such as great danes: for these, slower growth reduces skeletal problems)

literature values for different breeds, differing in metabolic rate. Simple formulas are shown for estimating the ME requirement for maintenance of obesity-resistant breeds, 'standard' pet dog breeds and obesity-prone breeds.

5. Estimate the effect of exercise activity on MER

Exercise has a large effect on the MER of dogs, but activity reporting by pet dog owners is seriously overestimated and may account for poor alignment of energy intake with energy utilisation (Butterwick and Hawthorne 1998). Even pet dogs receiving 1–3 h of exercise daily had a mean MER of only 110 kcals/$W^{0.75}$, which is ~20% lower than the NRC (2006) recommendation for active dogs (132 kcals/$W^{0.75}$). The MER for athletic dogs (racing, hunting, sled pulling and working) depends heavily on the duration and intensity of exercise, and on the nature of the exercise (sprinting versus endurance). MER values of 1000 kcals/$W^{0.75}$ have been recorded for sled dogs competing in a 3-day race at low temperatures (Hinchcliff *et al.* 1996).

For the standard pet dog it is probably best to assume zero effect of exercise activity unless the dog is **very** active. Use this system and then iteratively adjust the energy intake to reach the 'ideal' target weight.

6. Determine the effect of gestation, lactation and growth in puppies on the ME requirement of dogs

The factors used to calculate the MER of gestating, lactating or growing dogs are shown in Table 11.6.

Examples of how to use the simple system to estimate the ME requirement of dogs

We can now easily estimate the ME requirement of most dogs. Some examples follow.

Example 1: How much of an AAFCO-approved food is required to maintain a 35 kg golden retriever with a BCS of 4.5/5, at its ideal weight and fatness? The dog is being fed an AAFCO-approved food.

$$\text{Ideal weight (from Table 11.5)} = 29.3 \text{ kg}$$

$$\text{MER (maintenance)} = 82W^{0.75} \text{ (obese-prone breed)} = 0.398 \text{ kcals/day}$$

AAFCO foods contain 4000 kcals/kg
  So the dog needs 1032/4000 kg/day = 0.258 kg/day.

Example 2: How much of an AAFCO-approved feed is required to maintain a greyhound weighing 28 kg and with a BCS of 3/5, at its ideal bodyweight?

Ideal weight = 28 kg (i.e. the dog is already at ideal condition given its body condition score is 3/5)

$$\text{MER maintenance} = 28^{0.75} \times 113 = 1375 \text{ kcals/day}$$

AAFCO feed contains 4000 kcals/kg
  So the dog needs 0.343 kg/day.

Note that in these two examples the required actual ME and feed intake for the labrador is less than that of the greyhound despite the labrador's target weight being higher. Note also that if we simply used the labrador's current weight as recommended by dog food companies and the NRC (2006) ME requirement value, we would be feeding the labrador 0.341 kg of the feed/day, which is 32% more than the amount estimated using the system above.

Example 3: How much of an AAFCO-approved dog food is required to maintain a lactating bitch ('standard' pet) weighing 25 kg and with a BCS of 3/5, feeding four puppies in early lactation?

$$\text{Ideal weight} = 25 \text{ kg}$$

$$\text{MER maintenance} = 95W^{0.75} = 1062 \text{ kcals/day}$$

$$\text{ME for lactation} = 3 \times \text{MER} = 3 \times 1062 = 3186 \text{ kcals/day}$$

AAFCO feed contains 4000 kcals/kg
  So the dog needs 3186/4000 = 0.800 kg/day.

Checking the performance of your feeding program

Having calculated your dog's feed requirement, regularly check its body condition score, bodyweight and general health. Adjust the quantity fed to maintain a healthy bodyweight and general good health (Fig. 11.6).

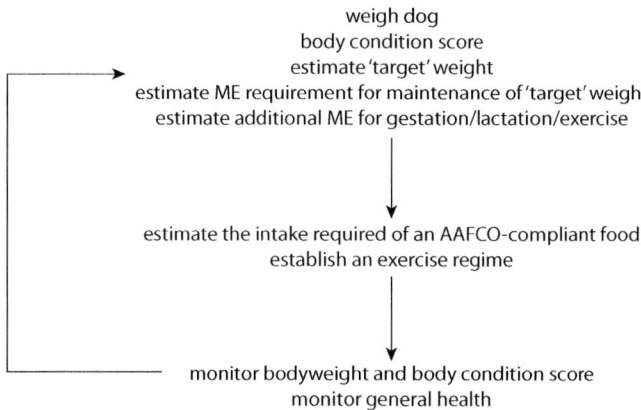

weigh dog
body condition score
estimate 'target' weight
estimate ME requirement for maintenance of 'target' weight
estimate additional ME for gestation/lactation/exercise

estimate the intake required of an AAFCO-compliant food
establish an exercise regime

monitor bodyweight and body condition score
monitor general health

Fig. 11.6.   Establishing and maintaining a feeding and exercise regime for dogs to maintain ideal body fatness and good health.

Protein requirements of dogs for maintenance, growth, gestation and lactation

The total amount of protein required by dogs depends on the quality of the protein in the feed (digestibility of the protein and its essential amino acid composition). The minimum protein requirements (NRC 2006) have been adjusted by AAFCO (see Table 11.7), assuming the digestibility of the protein in dog foods is 80%.

An alternative means of evaluating the daily protein requirement is:
$$\text{Protein required per day (g)} = 3.28 \times W^{0.75}$$
where W is the target weight in kg.

Source: NRC (2006).

Summary of implementation of a dietary regime for dogs

Figure 11.6 summarises the steps involved in establishing a dietary and exercise regime for dogs to maintain good health and an ideal body fatness.

Table 11.7.   Minimum levels of crude protein in the diets of dogs for maintenance, growth, gestation and lactation (Dzanis 1994)

Life stage	CP (% of dry matter)[a]	CP (g/1000 kcalME)
Maintenance	18.0	45.0
Puppies (post weaning)	22.5	36.3
Late gestation/early lactation	22.5	56.3

[a]Assumes the ME density of the diet is 4000 kcal/kg: if the energy density is >4000 kcals/kg, multiply these values by the actual energy density/4000 (e.g. 4300 kcals/kg: CP required for maintenance = 4300/4000 × 18.0 = 19.4% of dry matter). Note: Feeds vary greatly in dry matter percentage, so this needs to be taken into account.

Vitamin and mineral requirements of dogs

Vitamin and mineral requirements of dogs are expressed relative to the Metabolisable Energy content of the feed, assuming that the standard dog feed contains 4000 kcal ME/kg feed DM. The reasons for expressing requirements this way are discussed in 'Energy systems for cats and dogs' in Chapter 4.

Table 11.8 shows the NRC (2006) recommended daily allowance standards for dogs for growth or reproduction and for maintenance of an adult dog.

Table 11.8. Nutrient requirements for growth, late gestation and peak lactation of dogs

Vitamin or mineral	Units	Nutrient requirements (units/kg DM) assuming ME = 4000 kcals/kg	
		Growth, late gestation and peak lactation (minimum)	Adult maintenance (minimum)
Calcium	g	8.0	4.0
Phosphorus	g	5.0	3.0
Ca:P ratio		1:1	1:1
Potassium	g	3.6	4.0
Sodium	g	2.0	0.8
Chloride	g	3.0	1.2
Magnesium	g	0.6	0.6
Iron	mg/kg	70	30
Copper	mg/kg	12.4	6.0
Manganese	mg/kg	7.2	4.8
Zinc	mg/kg	96	60
Iodine	µg/kg	880	1000
Selenium	µg/kg	350	350
Vitamin A[a]	RE/kg	1515	1515
Vitamin D	µg/kg	13.8	13.8
Vitamin E	mg/kg	30	30
Vitamin K	mg/kg	1.6	1.63
Thiamine	mg/kg	2.25	2.25
Riboflavin	mg/kg	5.3	5.25
Pantothenic acid	mg/kg	15	15
Niacin	mg/kg	17	17.0
Pyridoxine	mg/kg	1.5	1.5
Folic acid	mg/kg	0.27	0.27
Vitamin B12	µg/kg	36	35
Choline	mg/kg	1700	1700

[a]RE = recommended retinol equivalents where 1 RE = 3 International Units of activity
Source: Adapted from NRC (2006)

Clearly a greater density of most nutrients is required in the feeds of growing puppies, pregnant or lactating bitches, although they will be fed at higher rates of energy intake to meet their higher energy needs and this also provides additional nutrients.

Energy requirements of cats for maintenance, growth, pregnancy, lactation and exercise

As for dogs it is important to know how to body condition score the cat (Table 11.9).

It was originally thought that the energy requirements of cats could be estimated directly from bodyweight, given the low range of bodyweights of cats. However, it became apparent that for larger cats this overestimates the energy requirement. The equations for estimating the Metabolisable Energy Requirements for maintenance (MER) of adult cats are shown in Table 11.10. Note the changing exponents of bodyweight in each equation.

Protein requirements for cats for maintenance, growth, gestation and lactation

Cats are unusual in their protein metabolism for the following reasons:

1. In most animals fed low-protein diets, the activity of enzymes involved in amino acid catabolism, gluconeogenesis and nitrogen disposal (urea cycle) are down-regulated to conserve the amino acids for essential functions. These adaptive mechanisms also act to protect the animal from toxicity of amino acids when the animal is consuming high-protein diets. Cats, in contrast, do not have these protective mechanisms, so on low-protein diets they continue to use amino acids for energy and on high-protein diets they are prone to amino acid toxicities (Rogers *et al.* 1977).
2. Amino acid catabolism in cats occurs at a very high rate regardless of protein intake, which means they have a high obligatory loss of nitrogen even when fed protein-free diets. This high obligatory loss of nitrogen in cats is particularly important when cats are ill or anorexic because long-term food deprivation results in very high losses of urinary nitrogen than other species. This unusual feature of amino acid metabolism means that cats use most of their protein for maintenance (60%) compared with growth (40%), in contrast to dogs, which only use 30% of their protein requirement for maintenance and 70% for growth (Russell *et al.* 2003).
3. Taurine (2-aminoethanesulphonic acid) is an unusual amino acid that is not incorporated into proteins but is found as a free amino acid or in small peptides in tissues. Taurine is required in all animals for conjugation of bile, osmoregulation, membrane stabilisation, antioxidation of free radicals and

Table 11.9. A body condition scoring system for domestic cats

Score (/5)	Visual	Description
1 Very thin		• Ribs and vertebrae easily seen at distance • Sharp bony protrusions felt over vertebrae and ribs sharp and shallow • Very obvious 'waist' from above • Starved appearance
2 Underweight		• Ribs, vertebrae easily felt • Obvious 'waist' from above • Skinny appearance
3 Ideal		• Ribs and vertebrae obvious • Individual ribs can be just discerned • 'Waist' just apparent from above • Healthy appearance
4 Overweight		• Ribs and vertebrae have a smooth feel • Hard to distinguish individual ribs • 'Waist' not apparent • Abdomen sagging from the side • Rounded, fat appearance
5 Obese		• Can't distinguish ribs or vertebrae • Fat covers all bones • No waist from above • Sagging abdomen from side • Overweight, fat appearance

Table 11.10. Daily Metabolisable Energy requirements of cats for maintenance, growth, gestation and lactation

Type of cat	MER (kcals/day)
1. Maintenance adults Domestic cats (lean) (BCS <3/5 or <5/9) Domestic cats (overweight) (BCS > 3/5 or 5/9)	$100W^{0.67}$ $130W^{0.40}$
2. Growth weaned kittens (mature weight 5 kg) 1.0 kg 2.0 kg 3.0 kg 4.0 kg 5.0 kg	200 284 326 330
3. Gestating queens (allows for some weight gain in pregnancy as queens lose weight in lactation regardless of diet)	$145W^{0.75}$
4. Lactating queens (3–4 kittens/litter) Week 1 and 2 Week 3 and 4 Week 5, 6, 7	$100W^{0.67} + 54W$ $100W^{0.67} + 72W$ $100W^{0.67} + 60W$

Source: Adapted from NRC (2006)

stabilisation of membranes. It is particularly important in heart function and functioning of the retina. Taurine is synthesised as follows (Fig. 11.7). The activity of the two synthetic enzymes, cysteine dioxygenase and cysteinesulphinate decarboxylase, is low in cats and insufficient to meet the requirements. Cats fed on taurine-deficient diets develop specific deficiency

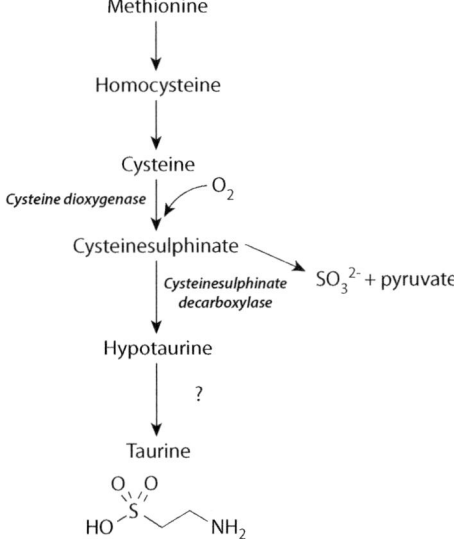

Fig. 11.7. Synthesis of taurine from methionine and cysteine. The dioxygenase and decarboxylase enzymes have low activity in cats.

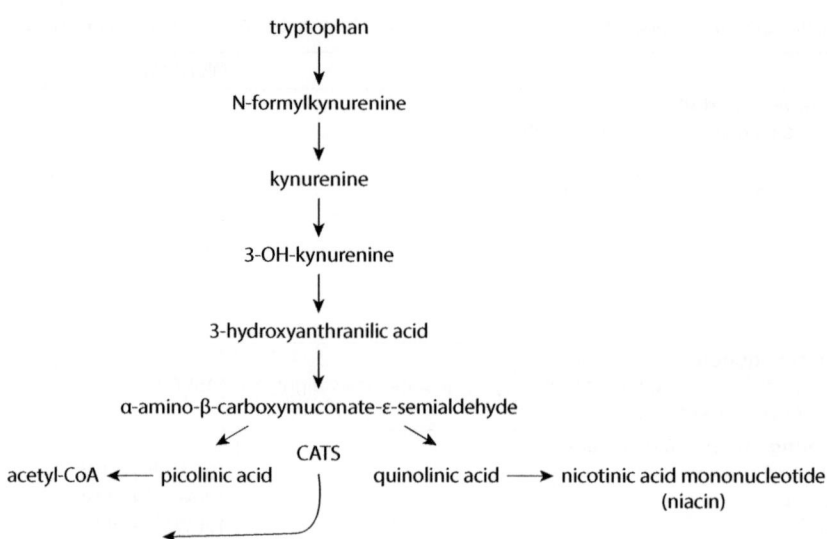

Fig. 11.8. Synthesis of niacin from tryptophan. In cats, the pathway favours picolinic acid and acetyl Co-A formation with little niacin synthesis.

signs with time (months or years), because the rate of synthesis is low and the relative metabolic demand in cats is high due to their high protein turnover. Taurine deficiency produces the clinical signs of feline central retinal degeneration and ultimately blindness, dilated cardiomyopathy and ultimately cardiac failure, and reproductive failure in queens due to fetal failure.

4. Niacin or nicotinamide is an essential vitamin acquired in animals from two sources: dietary niacin and synthesis from the essential amino acid, tryptophan (Fig. 11.8).

   Cats tend to produce picolinic acid instead of quinolinic acid, the intermediate in niacin synthesis. The rate of niacin synthesis in cats is insufficient for normal metabolism and additional preformed niacin is required.

5. Cats require more sulphur amino acid (methionine plus cysteine) than dogs because they produce the sulphur amino acid felinine and N-acetylfelinine, which are pheromones excreted in urine, particularly in tom cats. Sulphur amino acids in cats then are required for hair and skin production, muscle growth, felinine production, taurine production and for the extensive methylation reactions in rapid metabolism, making the sulphur amino acids the first-limiting amino acids in cat diets.

6. Arginine is required for the urea cycle (Fig. 11.9) and in cats this cycle is critical to the disposal of the large amounts of ammonia generated by the high rate of protein catabolism.

   Within hours of consuming an arginine-free diet, cats develop hyperammonaemia, emesis (vomiting), muscle spasms, ataxia, hyperaesthesia,

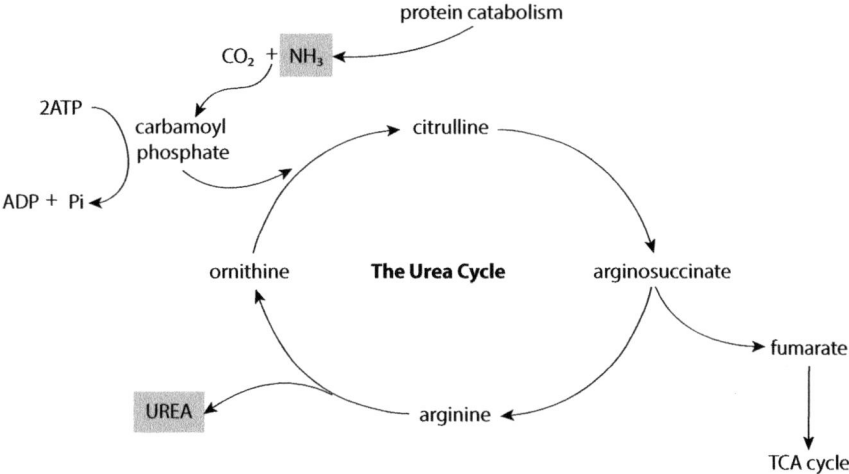

Fig. 11.9. The urea or arginine/ornithine cycle is particularly important in cats to detoxify the large amounts of ammonia generated by extensive protein catabolism.

tetanic muscle contractions, coma and death. Cats are particularly sensitive to arginine deficiency because they cannot synthesise ornithine and citrulline in the intestinal epithelium. This citrulline travels to the kidneys to produce the arginine required there and in other tissues of the body, so failure to produce intestinal ornithine/citrulline leads to a deficit in arginine for the urea cycle and for body tissue metabolism and growth.

Provided cats are fed a protein source largely based on animal proteins they are unlikely to be deficient in the essential amino acids methionine, cysteine, lysine, threonine and tryptophan. Diets high in plant-based proteins require careful amino acid attention to ensure no deficits of these amino acids and taurine.

The protein requirements for cats are indicated in Table 11.11.

Table 11.11. Protein requirements for cats for maintenance, growth, gestation and lactation

Life stage	Crude protein (g/kg DM)[a]		Crude protein (g/1000 kcals)[b]	
	NRC (2006)	AAFCO (2014)	NRC (2006)	Dzanis (1994)
Adult cats maintenance	200	200	50	65
Kittens post weaning	225	300	56.3	75
Late gestation	213	300	53	75
Peak lactation	213	300	53	75

[a]Assumes an energy density of 4000 kcals/kg. For feeds that are not 4000 kcals/kg, multiply this value by the energy density of the food and divide by 4000.
[b]To calculate the amount of protein/1000 kcals by the ME requirement estimated from Table 10.8 and divide by 1000

Vitamin and mineral requirements of cats

The nutrient requirements of cats, like those for dogs, are expressed on a per unit ME basis (Table 11.12).

The AAFCO standard for dog and cat foods is the benchmark for diet quality

Given the complexity of establishing nutrient requirements for dogs, and the enormous number of pet food manufacturers in the marketplace, the Association of American Feed Control Officials (AAFCO) prepared the AAFCO Dog and Cat

Table 11.12. Nutrient requirements for growth, late gestation and peak lactation of cats

Vitamin or mineral	Units	Nutrient requirements (units/kg DM) assuming ME = 4000 kcals/kg	
		Growth, late gestation and peak lactation (minimum)	Adult maintenance (minimum)
Calcium	g	8.0	2.9
Phosphorus	g	5.0	2.6
Ca:P ratio		1.25	1.2
Potassium	g	3.6	5.2
Sodium	g	2.0	0.68
Chloride	g	3.0	0.96
Magnesium	g	0.6	0.4
Iron	mg/kg	70	80
Copper	mg/kg	12.4	5.0
Manganese	mg/kg	7.2	4.8
Zinc	mg/kg	96	74
Iodine	µg/kg	1800	1400
Selenium	µg/kg	300	300
Vitamin A	µg retinol/kg	2000	1000
Vitamin D	µg/kg	7	7
Vitamin E	mg/kg	31	38
Vitamin K	mg/kg	1.0	1.0
Thiamine	mg/kg	6.3	5.6
Riboflavin	mg/kg	4.0	4.0
Pantothenic acid	mg/kg	5.75	5.75
Niacin	mg/kg	40	40
Pyridoxine	mg/kg	2.5	2.5
Folic acid	mg/kg	0.75	0.75
Vitamin B12	µg/kg	22.5	22.5
Choline	mg/kg	2550	2550

Source: NRC (2006)

Food Nutrient Profiles to establish practical minimum and some maximum levels of nutrients for dog and cat foods made form commonly available, non-purified, complex ingredients. AAFCO does not regulate, test or approve pet food standards. AAFCO establishes the nutritional standards for complete and balanced pet foods. It is the responsibility of the pet food manufacturers to formulate their products in line with the AAFCO standards (Dzanis 1994).

Types of dog food

The dog food market is enormous, with many tens of billions of dollars spent each year on various dog foods. From simple options a few decades ago, dog owners are now faced with many choices of foods differing in price, texture, water content, ingredients, processing methods, additives, nutrient composition and increasingly with claims of special characteristics (organic, biodynamic, vegetarian, natural, contains antioxidants, premium, super premium, science-based, and so on). Others are formulated heavily on the basis of price points in the marketplace. Some of the formulations are based around convenience to the owner (e.g. dry foods are easy to carry and store) and many are extruded to appear as human foods. Of course the dog has no interest in the food looking like a miniature bone or hamburger, but it is the owner that is the target market. In general, dog foods can be classified on the basis of texture/moisture (dry, semi-moist, wet canned), market segment (generic, supermarket, premium, super premium) or therapeutic. In addition there is an increasing interest in raw food diets, vegetarian diets and homemade diets (see Table 11.13).

Issues in dog and cat nutrition
Coprophagy (faeces eating) in dogs

Coprophagy is common in dogs, affecting almost 50% of the domestic canine population at some point in their lives (Boze 2010). It is a behaviour of great concern to dog owners, with 6% of dogs returned to a shelter being relinquished due to coprophagy (Wells and Hepper 2000). Some owners consider the practice so abhorrent they consider euthanasia of their pet. Coprophagy in dogs is considered a behavioural problem with perceived, but not evidenced, negative impacts on the health of both pet and owner. Coprophagy in dogs is considered by many to indicate a nutritional problem (Meriwether and Johnson 1980; Read and Harrington 1981; Hart and Hart 1985). However, there is no evidence that coprophagy is indicative of a nutritional issue (Boze 2010). Coprophagic dogs do display depraved appetite in which non-food items are consumed (pica), and both behaviours are associated with anxiety disorders, suggesting that coprophagy is a behavioural rather than nutritional disorder. This is supported by the fact that providing regular vitamin or enzyme supplements has no effect on the incidence of

Table 11.13. Dog food market segmentation

Dog food types	Typical composition (as fed)	Typical composition (dry matter basis)	Advantages	Disadvantages
Dry food	84–90% DM 16–30% CP 2800–4050 kcal/kg	18–32% CP 3000–4500 kcal/kg	• Economic, light to carry • No pathogens • No refrigeration needed • Crunchy food reduces tartar • Usually AAFCO compliant	• May include fillers • May induce more allergies
Semi-moist	70–85% DM 17–20% CP 2550–2800 kcal/kg	20–28% CP 3000–4000 kcal/kg	• May be unbalanced • Usually AAFCO compliant	• Contains simple sugars, salts, humectants, glycerol • Contains preservatives
Canned	25% DM 7–13% protein 875–1250 kcal/kg	28–50% CP 3500–5000 kcal/kg	• Palatable • Economic • Long unopened shelf life • Good for dogs with poor teeth • Good for dogs with renal problems • Usually AAFCO compliant	• Prone to contamination • More dental issues • Loose stools in some dogs
Homemade	Variable		• May be more palatable • Less heat damage (more digestible?) • Allows control over ingredients • No preservatives, colourings, flavourings	• Probably unbalanced unless properly formulated by a nutritionist • Pathogens? • Short half life • Expensive
Raw (BARF or raw meat-based)	Variable		• Better teeth • May be better stools • Allows control over ingredients • No preservatives, colourings, flavourings	• Potential bowel obstructions • Expensive • Pathogens?

coprophagy (Boze 2010). There is no evidence that changing a dog's diet decreases its tendency to eat faeces (Wells 2003), nor does feeding frequency relate to the incidence of coprophagy (Boze 2010). Dogs with exocrine pancreatic insufficiency (see 'Exocrine pancreatic insufficiency in dogs' later) eat faeces but this is associated with undigested fats and other nutrients in the faeces and is unrelated to coprophagy in normal dogs. Other reasons for coprophagy in dogs include searching by castrated males for testosterone (Boze 2010), expression of anxiety-related disorders, boredom, attention-seeking (Wells 2003), evolutionary scavenging behaviour or a continuation of the practice whereby females eat puppy faeces to remove odours attracting predators. The fact that the three most effective treatments for coprophagia in dogs related to human interventions (removal of faeces, positive reward and distracting the dog) suggest that coprophagia can be controlled by behaviour modification (Boze 2010).

The 'humanisation' and 'premiumisation' of dog feeding practices

In affluent countries such as the USA, Australia, the United Kingdom and many European countries, more than 60% of households own a pet and the majority of these are fed commercial, nutritionally balanced foods. This widespread use of nutritionally balanced diets has been a contributing factor to the increased longevity of pets (Kraft 1998). However, there is an increasing interest in the use of 'non-commercial' foods such as homemade foods, vegetarian and raw foods. Some of the motivations for pet owners to use these foods are listed in Table 11.14.

These trends in dog food preferences in Western countries can be attributed to two driving forces: the desire to treat dogs as human family members ('**humanisation**') and the desire to feed dogs the best quality feeds possible ('**premiumisation**'). These motivations are directly matched to human food motivations and the move towards organic, natural, wholesome, vegetarian, unprocessed, homemade, high-'quality' ingredients. As pets moved from working roles (e.g. farm dogs) to 'members of the family', logically owners do not want to feed a member of the family a by-product of the human food industry that the humans in the family won't eat. The second reason for moving away from commercial diets is the desire to prepare food as a social activity for a loved-one, much as we do for human members of the family. Preparing the dog's diet also gives the owner perceived control over the animal's health and wellbeing. Finally, the pet food industry is dominated by large, multinational organisations in contrast to the increasing trend in human food purchases towards local, fresh, organic farmer's market produce. Occasional food-related pathologies such as the ataxia and tetraplegia induced in cats after consumption of an imported, irradiated

Table 11.14. Motivations for pet owners to feed homemade, raw or vegetarian foods

Motivation
'Want to pamper my pet'
'Processed foods are poorer quality'
'Commercial foods are less wholesome'
'Raw foods are more natural, healthy and organic'
'Alternative foods are medically better for my animal'
'Supports my philosophical value system' (vegetarianism, concern for the environment, animal rights)
'Dogs are carnivores and cant digest the grains and starches used as fillers in commercial dog foods'

Source: Adapted from Schelsinger and Joffe (2011)

commercial cat food into Australia, and pet food recalls after melamine contamination, reinforce the perceptions of unhealthy commercial diets.

There are two diametrically opposed views on dog nutrition currently being proposed. One is that we should feed dogs a heavily meat-based diet to reflect the canid's evolution from the carnivorous ancestral wolf. This argument stresses strongly the view that dogs are not designed to digest starchy grains and plant-based foods. The other is that dogs should be fed vegetarian diets, for the same ethical and philosophical reasons some people advocate vegetarianism (concerns about factory farming of animals, animal rights, environmental concerns about animals, fewer famed animals means more food for people, and so on). These two views are now discussed.

Raw food and homemade diets for dogs: the evidence?
There is a strong trend in some quarters away from grains in dog foods reflecting a belief that dogs are carnivores and evolved to eat raw meat and bones. Raw meat-based diets include BARF® diets (originally **B**ones **A**nd **R**aw **F**ood and now **B**iologically **A**ppropriate **R**aw **F**oods), which are based on the concept that dogs evolved to eat raw foods These diets include 'raw whole animal protein, raw vegetables and fruit, uncooked eggs, yoghurt, kelp and healthy fresh herbs'. The most common forms of raw meat-based diets are fresh, frozen and freeze-dried diets available commercially. Some are intended as complete, balanced rations while others are intended as supplements and are not nutritionally balanced. There is an increasing number of internet-based recipes based on BARF® 'principles' and some are formulated by veterinary general practitioners, dog breeders, owners and trainers. Some meet AAFCO nutrient guidelines, but many homemade recipes rely on rotation of ingredients to provide the nutrient balance.

The claims are that 'many dogs cannot tolerate modern processed pet food' and that 'the BARF® diet will result in reductions in obesity, arthritis, allergies, autoimmune disease, diabetes, IBD and other gastrointestinal conditions, kidney disease and many more'. Claims are made of shinier coats and healthier skin, improved smelling breath and reduced stool volume and odour. Evidence for these claims is of low-quality and largely based on testimonials, case studies or poor-cohort studies.

There is no evidence of nutritional **benefit** of raw foods to dogs but there is some (poor-quality) evidence of a nutritional **risk** (Schlesinger and Joffe 2011). Nutritional osteodystrophy has been reported in puppies fed a bones and raw food diet from 3 weeks of age (DeLay and Laing 2002) but this was a case study only. Nutritional secondary hyperparathyroidism has been reported in German shepherd puppies fed a diet of rice and raw meat (Kawaguchi *et al.* 1993), as a result of excessive phosphorus in the diet. Analysis of homemade and commercial BARF® diets in Germany revealed that 76% of 'BARF' rations tested showed at least one nutritional imbalance (Dillitzer *et al.* 2011). Freeman and Michel (2001) found that all of the homemade and raw-meat based diets they tested had multiple nutrient imbalances. Calcium to phosphorus ratios in homemade diets are often not optimal and vitamin levels were highly variable. Dillitzer *et al.* (2011) found that 60% of raw meat-based diets had major nutritional imbalances. Many raw meat-based diets are high in fat (probably producing the shiny coats) but potentially producing obesity, and of risk to animals predisposed to pancreatitis. In another study, 200 homemade diets were evaluated and 95% had at least one essential nutrient below the AAFCO minima (Stockman *et al.* 2013).

In addition to nutritional concerns about homemade and BARF® diets there are well-founded concerns about the microbiological safety of raw foods. There are many reports of the presence of pathogenic bacteria and particularly *Salmonella* spp. in raw meat-based diets, and evidence that these bacteria are shed in faeces posing a potential zoonotic threat to dog owners (Finley *et al.* 2007; Freeman *et al.* 2013). Freezing BARF® diets does not eliminate the pathogens so they remain a threat to the dog and their owners. It is difficult to eliminate *Salmonella* using conventional disinfection of food bowls. Other zoonotics may also reside in raw feed products including the parasite *Toxoplasma gondii*.

> A high proportion (60–95%) of homemade diets for dogs are nutritionally inadequate and many raw meat-based diets contain pathogenic bacteria.

Vegetarian diets for dogs?

Many dog owners are attracted to the idea of feeding their dogs vegetarian diets for the same ethical reasons they choose this diet themselves. The dentition of dogs, combined with the finding that dogs have developed genes associated with starch digestion not present in the ancestral wolf (Axelsson *et al.* 2013), point strongly to the fact that dogs are well equipped for digesting grains and vegetables as well as meat (Brown 2009). Other aspects of digestion and metabolism such as a lower protein requirement than carnivores, the ability to convert β-carotene to vitamin A, linolenic acid to arachidonic acid, and taurine from sulphur amino acids, means dogs can obtain these essential nutrients from plants while cats cannot (Brown 2009).

> The only relevant questions regarding BARF® and vegetarian diets, as is the case for all diets, are:
> 1. Does the diet meet the nutritional needs of the dog?
> 2. Is the diet sufficiently palatable to ensure adequate intake?
> 3. When the diet is fed, is the animal in good health?
>
> Source: Brown (2009).

Provided a nutritionally balanced, digestible diet can be formulated that is palatable to dogs, and maintains health of the dog in its physiological state (pregnancy, lactation, growing, exercising, working, etc.), there is no nutritional reason not to adopt such a ration if so desired (Brown 2009).

It is possible to formulate a diet without meat that meets the nutritional requirements of dogs, but plant-based diets suffer the following potential problems (Brown 2009):

- They may have high levels of **non-starch polysaccharides**, which are indigestible or are digested in the large intestine with potential flatulence problems.
- Plant proteins are often **low in lysine**, the first-limiting amino acid for dogs.
- Many potential plant sources of protein such as soybean, contain **anti-nutritive factors** such as trypsin inhibitors, lectins, tannins and phytates.
- Plant proteins may be contaminated with **mycotoxins**.
- Some plant protein sources are **unpalatable**.

> Formulating a nutritionally adequate vegetarian diet for dogs is not easy and is beyond the ability of most owners.

Obesity in dogs

Obesity, defined as a dog 15–20% overweight, is the most common nutritional disorder in dogs. Dogs classified as overweight or obese account for ~40% of the canine populations in most developed countries, with obesity in ~5–8% of all dogs (Sloth 1992; Colliard *et al.* 2006; McGreevy *et al.* 2005). Obesity is a major risk factor in the development of serious medical conditions including insulin resistance, hip dysplasia, pancreatitis, respiratory distress, heat intolerance, renal pathology and osteoarthritis in dogs (Edney and Smith 1986; Laflamme 2006). The strongest associations of obesity are with diabetes and osteoarthritis (Laflamme 2006). The longevity of dogs is directly linked to body condition score. Dogs with BCS scores above 5 on a 9-point scale, lived to a median age of 11.2 years, while their restricted-fed counterparts lived to a median age of 13 years: a 16% increase in longevity (Kealy *et al.* 2002). The risk factors for obesity (Table 11.15) include breed, female gender, neutering and age (Edney and Smith 1986). Susceptible breeds include labrador retriever, cairn terrier, cavalier King Charles spaniel, cocker spaniel, dachshund, basset hound and beagle. There are no clear associations between obesity and overweight dogs and their diet types (Edney and Smith 1986; Laflamme 2006). Feeding once-per-day is associated with obesity incidence as is the feeding of additional treats, living in a single dog household and lack of exercise (Robertson 2003).

Table 11.15. Factors contributing to overweight and obese dogs

Risk factor	Likely effect	References
Neutering	• Partly confounded with age • Reduced energy metabolism • Reduced energy utilisation in mating/mate searching behaviours • Hormonal changes associated with increased intake	Colliard et al. (2006); Laflamme (2006); McGreevy et al. (2005); Robertson (2003); German (2006); Lund et al. (2006); Suarez et al. 2012)
Sex (females > males)	• Hormonal effects on intake/energy utilisation?	
Fed snacks/treats	• Increased total energy intake	
Fed once-daily	• Reduced energy loss from thermogenesis of digestion	
Single-dog household	• Less exercise/play with other dogs	
Age	• Metabolic rate decrease • Exercise decrease • Pituitary endocrinopathies?	
Lack of exercise	• Reduced energy utilisation	
Breed (retrievers, cavalier King Charles spaniels, beagles, basset hounds, cocker spaniels, dachshunds, at risk breed)	• Metabolic rate? • Hormonal appetite drive? • Gut microbiome?	
Endocrinopathies (hypothyroidism, hyperadrenocorticism)	• Reduced metabolic rate • Increased circulating glucose/adipogenesis	
Rural > urban	• Perceived 'exercise' so no planned exercise? • Access to many 'foods' • Greater age (confounded?)	
Obese owners	• Reduced exercise	
Lower socioeconomic group	• Poorer access to footpaths, parks, facilities? • Feed unbalanced rations (cheaper)?	
Price of pet food (lower > higher)	• Unbalanced diets?	

In preventing and reversing obesity in dogs, the first challenge is to teach owners how to assess the body condition of their dog and to learn to appreciate when their dog is overweight or obese. Interestingly, many dog owners do not realise their dog is overweight or obese and it appears that this tendency to see overweight dogs is 'normal' is increasing and at the same time dogs at ideal condition scores are now seen as malnourished and their owners can be subjected to abuse. Owners often score their dogs 1 score lower than experts in a 5-point system and assign score 4 and 5 in a 5-point system as ideal weight and condition. Having convinced an owner that their dog needs an obesity-management strategy,

the steps in the feeding and exercise plan described earlier, should be
implemented.

Food allergies and adverse food reactions in dogs

Adverse reactions to foods can take several forms: (1) food allergies (FA), in which
immunological mechanisms and hypersensitivity reactions occur; and (2) food
intolerances (FI), which might reflect an enzyme deficiency in the animal (e.g.
lactase), a toxin in the food, or a pharmacological response (e.g. drug-like effect of
a food component such as histidine in poorly preserved fish is converted to
histamine in the gut). Food allergies are defined as an aberrant, adverse immune
response to a particular food substance, usually a protein (Gaschen and Merchant
2011), and characterised by an IgE-antibody dependent response with cutaneous
disease. The most common food allergens in dogs are in beef meat, dairy products,
wheat, eggs and chicken meat. Most allergens are water-soluble glycoproteins
10–70 kDa in size and are relatively stable to heat, acid and proteases. The bovine
IgG heavy chain and the enzyme phosphoglucomutase in beef meat and milk are
the targets of IgE in dogs.

Only a very low proportion of dogs develop FA in relation to the large
number of antigens presented to dogs daily, because there are several
mechanisms operating to stop allergies occurring including digestive enzymes,
bile, the mucus layer, the tight junctions of the gut epithelium, and cells and
factors from the innate and adaptive immune systems that render dietary
antigens less immunogenic (Gaschen and Merchant 2011). Breakdown of these
protective mechanisms results in adverse allergic reactions in the gut and skin
resulting from an interaction between IgE and mast cells, which degranulate,
releasing their vasoactive and proinflammatory cytokines. In dogs the
response is characterised by dermatitis, gastroenteritis and urticaria/
angioedema. In dogs there is an overlap of signs between food allergy and
atopic dermatitis.

In one study (Harvey 1993), food allergies occurred in young animals (up to
50% of allergy cases were less than 1 year old) and at any time of the year. Dogs can
have been on the problem diet for months or even years showing no signs of
adverse food reaction. Several tests are available to identify food allergies including
serological tests and skin prick tests, but the gold standard test is to do an
elimination trial as follows:

1. **Replace the current feed with a low-allergen feed.** This can be a homemade
   feed but this raises all the issues raised under the section on homemade and
   BARF® diets earlier. The advantage of homemade diets for allergen testing is
   they can be made without preservatives, colourings, flavourings, antioxidants
   and processing (e.g. heat extrusion), and it is possible that some allergens are

created during processing, or added to, commercial diets. If a homemade exclusion diet is to be made, it is essential that a nutritionist be involved in the formulation, or an approved recipe is used, because more than 90% of homemade elimination diets are unbalanced (Roudebush and Cowell 1992). Many homemade diets are deficient in Ca and/or have an inappropriate Ca:P ratio. The protein source should be a meat of an animal unrelated to the problem diet (e.g. typically horse, lamb, fish, turkey, kangaroo) or a legume protein (e.g. beans). The carbohydrate source should be potato, sweet potato, rice or pasta. The ingredients should be cooked, mixed in the correct ratios, and frozen in meal-sized sealable plastic bags. Intake rates should be estimated on the basis of the ME content of the diet, which can be calculated from the ingredient list. The new feed should be introduced slowly over 4 days. Alternatively use a commercially available prescription therapeutic diet that has been specifically formulated to be low in antigens. It is vital that the elimination diet is not supplemented by the dog or its owners from other sources! Feed for at least 9–10 weeks.

2. **Feed the original diet to confirm return of the problem.**
3. **Reintroduce the restricted diet as above.**
4. **Re-feed the original diet.**

If the above program produces partial or complete improvement (1), relapse (2), partial or complete improvement (3), and relapse again (4), then an adverse food reaction is confirmed and a strategic feeding program needs to be devised as follows:

1. Develop a complete, balanced, minimal antigen food as a homemade feed (with a nutritionist's advice). This diet should contain a low ratio of omega-6 to omega-3 fatty acids to reduce inflammation and pruritus.
2. Use a commercial minimal antigen therapeutic diet (expensive).
3. Identify the offending ingredient using single antigen testing on the restricted diet above and formulate a specific antigen-free diet.

It is essential that the dog not be provided any other sources of food other than the low-antigen diet.

Exocrine pancreatic insufficiency in dogs

Exocrine pancreatic insufficiency (EPI) is a syndrome of dogs characterised by inadequate production of the exocrine pancreatic enzymes by the acinar cells of the pancreas (see review by Westermarck and Wiberg 2012). The critical role played by these enzymes (proteases, lipases, amylases, peptidases, nucleases) in digestion becomes apparent in dogs with EPI. Clinical signs include: weight loss despite polyphagia (high appetite); a large quantity of faecal output, which can be loose and smelly; and fat is usually present in the faeces (steatorrhea). Fat appears in the faeces because it is not digested in the small intestine under the action of the pancreatic lipase. Coprophagia is common due to the presence of undigested fat in the faeces. EPI has a genetic basis and is more common in certain dog breeds (German shepherds, rough-coated collies, chow chows and cavalier King Charles spaniels). The clinical signs of EPI are the same as those associated with atrophy of the acinar cells of the exocrine pancreas and of chronic pancreatitis (Westermarck and Wiberg 2012). Treatment for EPI involves supplementation with exogenous pancreatic enzymes. EPI is also associated with deleterious changes in the intestinal microbiome, which is not alleviated by provision of exogenous enzymes (Isaiah et al. 2017). The large reserve secretory activity of the exocrine pancreas means maldigestion is not apparent until more than 90% of the secretory function is lost.

Feline hepatic lipidosis

Feline hepatic lipidosis (FHL) is the most common liver disease of cats and is characterised by accumulation of excessive triglycerides in the hepatocytes and a subsequent increase in liver weight of 50%. The aetiology of FHL is complex and involves a combination of negative energy balance (due to anorexia) and subsequent alterations to the pathways of synthesis, uptake, degradation and secretion of fatty acids by the liver. The anorexia is commonly brought on by **stress**, but not always in obese cats (Center 1993), and secondary FHL can be induced by other pathologies such as inflammatory bowel disease, pancreatitis, neoplasia, renal failure or diabetes mellitus. Secondary lipidosis accounts for the majority (95%) of FHL cases (Valtolina and Favier 2017). In primary FHL, changes in diet or stresses associated with changes in the environment (new pets, change of homes, etc.) can induce the anorexia. FHL develops from 2–14 days after the onset of the anorexia.

Hepatic lipidosis appears to be a particular problem for carnivores due to their unique reliance on fat and protein metabolism. Cats are unable to produce the long-chain polyunsaturated fatty acids (LCPUFA) linoleic and linolenic, but also produce insufficient arachidonic acid. The LCPUFA and specifically the omega-3 fatty acids protect against hepatic lipidosis by favouring fatty acid oxidation over triglyceride storage (Mahfouz et al. 1984). They also direct glucose away from fatty acid synthesis to glycogen synthesis. They also up-regulate genes involved in fatty acid oxidation. Cats have poor ability to adapt to protein levels in their diet (see 'Protein requirements for cats for maintenance, growth, gestation and lactation' earlier), so a prolonged period of anorexia induces essential amino acid deficiency and protein malnutrition with subsequent effects on lipoprotein synthesis. The sequence of events thought to induce FHL is summarise in the Box 11.2.

Box 11.2. Feline hepatic lipidosis

1. Anorexia is induced by stress or as a secondary result of another pathology.
2. This causes mobilisation of high levels of triglycerides, which enter the hepatocytes and are stored, rather than oxidised, due to reduced oxidation (associated with low long-chain polyunsaturated fatty acids).
3. The rapid onset of protein malnutrition and essential amino acid deficiencies, particularly of arginine and taurine, compromises the secretion of triglycerides from the liver as very low density lipoproteins. Carnitine deficiency may also play a role in reducing the mitochondrial uptake of fatty acids for oxidation.

Clinical signs and nutritional treatment of FHL

Anorexia for 7 days or more, depression, weight loss, muscle wasting, increased circulating liver enzymes, anaemia, jaundice, increased blood urea nitrogen, dehydration and ptyalism (drooling) are characteristic of FHL and usually preceded by a stressful event or illness. After stabilisation of electrolyte balance and cessation of vomiting, rapid nutritional intervention is essential. Feeding via nasogastric tube or gastrostomy of a high-protein, high-energy diet (e.g. commercial recovery diets are adequate), supplemented with carnitine, taurine and vitamins is recommended in regular and frequent feeds throughout the day (Center 2005). Most cats require 3–6 weeks of intense dietary therapy before appetite returns. Gradual withdrawal of the tube feeding and reintroduction of a highly palatable diet follows.

References

AAFCO (2014) *AAFCO Dog and Cat Food Nutrient Profiles*. Association of American Feed Control Officials, Champaign, IL, USA.

Axelsson E, Ratnakumar A, Arendt M-L, Maqbool K, Webster MT, Perloski M, *et al.* (2013) The genomic signature of dog domestication reveals adaptation to a starch-rich diet. *Nature* **495**, 360–364. doi:10.1038/nature11837

Bermingham EN, Thomas DG, Cave NJ, Morris PJ, Butterwick RF, German AJ (2014) Energy requirements of adult dogs: a meta-analysis. *PLoS One* **9**(10), e109681.

Boze B (2010) A comparison of common treatments for coprophagy in *Canis familiaris*. *Journal of Applied Companion Animal Behavior* **2**, 22–29.

Brosey BP, Hill RC, Scott KC (2000) Gastrointestinal volatile fatty acid concentrations and pH in cats. *American Journal of Veterinary Research* **61**, 359–361. doi:10.2460/ajvr.2000.61.359

Brown WY (2009) Nutritional issues regarding vegetarianism in the domestic dog. *Recent Advances in Animal Nutrition* **17**, 137–143.

Butterwick RF, Hawthorne AJ (1998) Advances in dietary management of obesity in dogs and cats. *The Journal of Nutrition* **128**, 2771S–2775S.

Center SA (1993) Feline hepatic lipidosis. *Veterinary Annals* **33**, 244–254.

Center SA (2005) Feline hepatic lipidosis. *Veterinary Clinics of North America: Small Animal Practice* **35**, 225–269. doi:10.1016/j.cvsm.2004.10.002

Chauncey HH, Henriques BL, Tanzer JM (1963) Comparative enzyme activity of saliva from the sheep, hog, dog, rabbit, rat and human. *Archives of Oral Biology* **8**, 615–627. doi:10.1016/0003-9969(63)90076-1

Colliard L, Ancell J, Benet J-J, Paragon B-M, Blanchard G (2006) Risk factors for obesity in dogs in France. *Journal of Nutrition* **136**, 1951S–1954S. doi:10.1093/jn/136.7.1951S

Corva PM, Medrano JF (2000) Diet effects on weight gain and body composition in high growth (hg/hg) mice. *Physiological Genomics* **3**, 17–23. doi:10.1152/physiolgenomics.2000.3.1.17

Darwin CR (1890) *Journal of Researches into the Natural History and Geology of the Countries Visited During the Voyage Round the World of H.M.S. Beagle under the Command of Captain Fitzroy, R.N.* First Murray illustrated edition. John Murray, London, UK.

DeLay J, Laing J (2002) Nutritional osteodystrophy in puppies fed a BARF diet. *AHL Newsletter* **6**, 23.

Dierenfeld ES, Alcorn HL, Jacobsen KJ (2002) *Nutrient Composition of Whole Vertebrate Prey.* Rodentpro.com, Inglefield, IN, USA, <http://www.rodentpro.com/qpage_articles_03.asp>.

Dillitzer N, Becker N, Kienzle E (2011) Intake of minerals, trace elements and vitamins in bone and raw food rations in adult dogs. *British Journal of Nutrition* **106**, S53–S56. doi:10.1017/S0007114511002765

Driscoll C, Macdonald DW, O'Brien SJ (2009) From wild animals to domestic pets, an evolutionary view of domestication. *Proceedings of the National Academy of Sciences of the United States of America* **106**, 9971–9978. doi:10.1073/pnas.0901586106

Dzanis DA (1994) The Association of American Feed Control Officials dog and cat food nutrient profiles: substantiation of nutritional adequacy of complete and balanced pet foods in the United States. *Journal of Nutrition* 124, 2535S–2539S.

Edney AT, Smith PM (1986) Study of obesity in dogs visiting veterinary practices in the United Kingdom. *The Veterinary Record* **118**, 391–396. doi:10.1136/vr.118.14.391

Finley R, Ribble C, Aramina J, Vandermeer M, Popa M, Litman M, *et al.* (2007) The risk of salmonellae shedding by dogs fed *Salmonella*-contaminated commercial raw food diets. *The Canadian Veterinary Journal. La Revue Veterinaire Canadienne* **48**, 69–75.

Freeman LM, Michel KE (2001) Evaluation of raw food diets. *Journal of the American Veterinary Medical Association* **218**, 705–709. doi:10.2460/javma.2001.218.705 (Erratum published in *Journal of the American Veterinary Medical Association* **218**, 1716).

Freeman LM, Chandler ML, Hamper BA, Weeth LP (2013) Current knowledge about the risks and benefits of raw meat-based diets for dogs and cats. *Journal of the American Veterinary Medical Association* **243**, 1549–1558. doi:10.2460/javma.243.11.1549

Gaschen FP, Merchant SR (2011) Adverse food reactions in dogs and cats. *Veterinary Clinics of North America: Small Animal Practice* **41**, 361–379. doi:10.1016/j.cvsm.2011.02.005

German AJ (2006) The growing problem of obesity in dogs and cats. *Journal of Nutrition* **136** (Supplement), 1940S–1946S.

German AJ, Holden SL, Bissof T, Morris PJ, Biourge V (2009) Use of starting condition score to estimate changes in body weight and composition during weight loss in obese dogs. *Research in Veterinary Science* **87**, 249–254. doi:10.1016/j.rvsc.2009.02.007

German AJ, Holden SJ, Mather NJ, Morris PJ, Biourge V (2011) Low-maintenance requirements of obese dogs after weight loss. *British Journal of Nutrition* **106**, S93–S96.

Goff JP (2015) Gastrointestinal motility. In *Duke's Physiology of Domestic Animals.* 13th edn. (Eds WO Reece, HH Erikson, JP Goff and EE Uemura) p. 467. John Wiley and Sons, Ames, IA, USA.

Hart BL, Hart LA (1985) *Canine and Feline Behavioural Therapy.* Lea and Febiger, Philadelphia, PA, USA.

Harvey RG (1993) Food allergy and dietary intolerance in dogs: a report of 25 cases. *The Journal of Small Animal Practice* **34**, 175–179. doi:10.1111/j.1748-5827.1993.tb02647.x

Hayes ER, Elliot R (1942) Distribution of the taste buds on the tongue of the kitten, with particular reference to those innervated by the chorda tympani branch of the facial nerve. *Journal of Comparative Neurology* **76**, 227–238. doi:10.1002/cne.900760204

Hill RC (1998) The nutritional requirements of exercising dogs. *Journal of Nutrition* **128**, 2686S–2690S.

Hinchcliff KW, Reinhart GA, DiSilvestro R, Reynolds A, Blostein-Fujii A, Swenson RA (1996) Oxidant stress in sled dogs subjected to repetitive endurance exercise. *American Journal of Veterinary Research* **61**, 512–517.

Isaiah A, Parambeth JC, Steiner JM, Lidbury JA, Suchodolski JS (2017) The fecal microbiome of dogs with exocrine pancreatic insufficiency. *Anaerobe* **45**, 50–58. doi:10.1016/j.anaerobe.2017.02.010

Kawaguchi K, Braga IS, III, Takahashi A, Ochiai K, Itakura C (1993) Nutritional secondary hyperparathyroidism occurring in a strain of German shepherd puppies. *The Japanese Journal of Veterinary Research* **41**, 89–96.

Kealy RD, Lawler DF, Ballam JM, Mantz SL, Biery DN, Greeley EH, *et al.* (2002) Effects of diet restriction on life span and age-related changes in dogs. *Journal of the American Veterinary Medical Association* **220**, 1315–1320. doi:10.2460/javma.2002.220.1315

Kraft W (1998) Geriatrics in canine and feline internal medicine. *European Journal of Medical Research* **3**, 31–41.

Laflamme DP (2006) Understanding and managing obesity in dogs and cats. *The Veterinary Clinics of North America. Small Animal Practice* **36**, 1283–1295. doi:10.1016/j.cvsm.2006.08.005

Lavy E, Goldberger D, Friedman M, Steinberg D (2012) pH values and mineral content of saliva in different breeds of dogs. *Israel Journal of Veterinary Medicine* **67**, 244–248.

Li X, Li W, Wang H, Cao J, Maehashi K, Huang L, *et al.* (2005) Pseudogenization of a sweet-receptor gene accounts for cats'indifference toward sugar. *PLOS Genetics* **1**, e3. doi:10.1371/journal.pgen.0010003

Loftus JP, Yazwinski M, Milizio JG, Wakshlag JJ (2014) Energy requirements for racing endurance sled dogs. *Journal of Nutritional Science* **3**, 1–5.

Lund EM, Armstrong PJ, Kirk CA, Klausner JS (2006) Prevalence and risk factors for obesity in adult dogs from private US veterinary practices. *International Journal of Applied Research in Veterinary Medicine* **4**, 177–186.

NRC (2006) *Nutrient Requirements of Dogs and Cats.* National Research Council, The National Academies Press, Washington, DC, USA.

Mahar KM, Portelli S, Coatney R, Chen EP (2012) Gastric pH and gastric residence time in fasted and fed conscious Beagle dogs using the Bravo® pH system. *Journal of Pharmacological Sciences* **101**, 2439–2448. doi:10.1002/jps.23159

Mahfouz MM, Smith TL, Kummerow FA (1984) Effect of dietary fats on desaturase activities and the biosynthesis of fatty acids in rat-liver microsomes. *Lipids* **19**, 214–222. doi:10.1007/BF02534800

McGreevy PD, Thomson PC, Pride C, Fawcett A, Grassi B, Jones B (2005) Prevalence of obesity in dogs examined by Australia veterinary practices and the risk factors involved. *The Veterinary Record* **156**, 695–702. doi:10.1136/vr.156.22.695

Meriwether D, Johnson MK (1980) Mammalian prey digestibility by coyotes. *Journal of Mammalogy* **61**, 774–775. doi:10.2307/1380339

Read DH, Harrington DD (1981) Experimentally induced thiamine deficiency in Beagle dogs: clinical observations. *American Journal of Veterinary Research* **42**, 984–991.

Robertson ID (2003) The association of exercise, diet and other factors with owner-perceived obesity in privately owned dogs in metropolitan Perth WA. *Preventive Veterinary Medicine* **58**, 75–83. doi:10.1016/S0167-5877(03)00009-6

Robinson PP, Winkles PA (1990) Quantitative study of fungiform papillae and taste buds on the cat's tongue. *The Anatomical Record* **226**, 108–111.

Rogers QR, Morris JG, Freedland RA (1977) Lack of hepatic enzymatic adaptation to low and high levels of dietary protein in the adult cat. *Enzyme* **22**, 348–356. doi:10.1159/000458816

Roudebush P, Cowell CS (1992) Results of a hypoallergenic diet survey of veterinarians in North America with a nutritional evaluation of homemade diet prescriptions. *Veterinary Dermatology* **3**, 23–28. doi:10.1111/j.1365-3164.1992.tb00139.x

Russell K, Lobley GE, Millward DK (2003) Whole-body protein turnover of a carnivore, *Felis silvestris catus*. *British Journal of Nutrition* **89**, 29–37. doi:10.1079/BJN2002735

Sagawa K, Li F, Liese R, Sutton SC (2009) Fed and fasted gastric pH and gastric residence time in conscious Beagle dogs. *Journal of Pharmacological Sciences* **98**, 2494–2500. doi:10.1002/jps.21602

Schelsinger DP, Joffe DJ (2011) Raw food diets in companion animals: a review. *The Canadian Veterinary Journal. La Revue Veterinaire Canadienne* **52**, 50–54.

Sloth C (1992) Practical management of obesity in dogs and cats. *The Journal of Small Animal Practice* **33**, 178–182.

Suarez L, Pena C, Carreton E, Juste E, Bautosta-Castano I, Montoya-Alonso JA (2012) Preferences of owners of overweight dogs when buying commercial pet food. *Journal of Animal Physiology and Animal Nutrition* **96**, 655–659. doi:10.1111/j.1439-0396.2011.01193.x

Stockman J, Fascetti AJ, Kass PH, Larsen JA (2013) Evaluation of recipes of home-prepared maintenance diets for dogs. *Journal of the American Veterinary Medical Association* **242**, 1500–1505. doi:10.2460/javma.242.11.1500

Thes M, Koeber N, Fritz J, Wendel F, Dillitzer N, Dobenecker B, *et al.* (2016) Metabolisable energy intake of client-owned adult dogs. *Journal of Animal Physiology and Animal Nutrition* **100**, 813–819.

Valtolina C, Favier RP (2017) Feline hepatic lipidosis. *The Veterinary Clinics of North America. Small Animal Practice* **47**, 683–702. doi:10.1016/j.cvsm.2016.11.014

Vila C, Maldonado JE, Wayne RK (1999) Phylogenetic relationships, evolution and genetic diversity of the domestic dog. *The Journal of Heredity* **90**, 71–77. doi:10.1093/jhered/90.1.71

Vila C, Savolainen P, Maldonado JE, Amorim IR, Rice JE, Honeycutt RL, *et al.* (1997) Multiple and ancient origins of the domestic dog. *Science* **276**, 1687–1689. doi:10.1126/science.276.5319.1687

Wells DL (2003) Comparison of two treatments for preventing dogs eating their own faeces. *The Veterinary Record* **153**, 51–53. doi:10.1136/vr.153.2.51

Wells DL, Hepper DG (2000) Prevalence of behaviour problems reported by owners of dogs purchased from an animal rescue shelter. *Applied Animal Behaviour Science* **69**, 55–65. doi:10.1016/S0168-1591(00)00118-0

Westermarck E, Wiberg M (2012) Exocrine pancreatic insufficiency in the dog: historical background, diagnosis, and treatment. *Topics in Companion Animal Medicine* **27**, 96–103. doi:10.1053/j.tcam.2012.05.002

Witzel AL, Kirk CA, Henry CA, Toll PW, Brejda JJ, Paetau-Robinson I (2014) Use of a novel morphometric method and body fat index system for estimation of body composition in overweight and obese dogs. *Journal of the American Veterinary Medical Association* **244**, 1279–1284. doi:10.2460/javma.244.11.1279

12

Horse nutrition

Key points

- Horses are hindgut fermenters that evolved to consume high-fibre diets in small but frequent meals ('a little, but often').
- Sixty-five per cent of the total digesta contents of a horse is in the large intestines.
- Medium-quality hay or pasture is the foundation of good equine nutrition.
- Nuchal or cresty neck fat score is a good indicator of a horse's propensity to equine metabolic syndrome (EMS), characterised by insulin resistance, laminitis, osteochondrosis, hyperlipaemia, pituitary pars intermedia dysfunction (equine Cushing's Disease) and systemic inflammation.
- The Digestible Energy, crude protein, vitamin and mineral requirements of horses for maintenance, growth, gestation, lactation and exercise can be estimated from a series of simple equations related to horse bodyweight, average daily gain, stage of pregnancy and stage of lactation.
- Development orthopaedic disease affects as many as 50% of foals in early life. DOD is multifactorial involving genetics, growth rate (energy intake), dietary calcium/phosphorus ratio, micromineral intakes and endocrine factors.
- Laminitis (founder) is a multifactorial disease with a strong nutritional component (high soluble carbohydrate intakes from pasture or grains) interacting with phenotypes characterised by insulin resistance, hyperinsulinaemia, chronic inflammation, and general and regional obesity.
- Exertional rhabdomyolysis syndromes relate to defects in glycogen storage (PSSM) or muscle calcium regulation (RER) that respond to reductions in dietary soluble carbohydrates and increases in dietary oils in conjunction with well-managed exercise regimes.

Introduction

Horses belong to the Order Perissodactyla, which also contains donkeys, zebras, rhinoceroses and tapirs. Members of the Perissodactyla have an odd number of toes and are hindgut fermenters. The diet of the ancestral horse comprised grasses and other forages obtained by grazing and browsing. The quality and availability of these forages varied greatly throughout the year, with periods of nutritional inadequacy interspersed with periods of nutritional abundance. Physiologically, horses adapted to have higher appetites in periods of plenty and to store excess energy as adipose tissue, which is drawn upon in times of scarcity. The insulin sensitivity of horses varies throughout the year to allow this efficient storage of energy. However, this efficient storage of energy leads to obesity when high-quality feed is available constantly. This is exacerbated in domesticated horses by their lower level of activity relative to their ancestors. Many of the diseases of horses are associated with this nutritional/metabolic phenotype. Termed 'equine metabolic syndrome' or EMS, it is analogous to the so-called 'metabolic syndrome' in humans and is characterised by obesity, insulin resistance, hyperinsulinaemia, laminitis and increased regional adiposity with or without obesity (Frank *et al.* 2010). It is likely that a much wider spectrum of pathologies, such as chronic inflammation, oxidative stress, prothrombosis and perturbed lipid metabolism, are associated with this syndrome, as they are in human metabolic syndrome.

The native forages on which horses evolved contained soluble sugars, cell wall contents, soluble proteins and lipids, but relatively little starch. These forages were available throughout the day, allowing the horse to graze at regular, frequent and short intervals ('a little, but often'). This rule should guide us as to how we should feed the modern, domesticated horse. Many domesticated horses are 'slug-fed' intermittently on diets containing high levels of starch and low moisture content. Many of the disorders associated with horses are nutritional in origin or have a strong nutritional component (colic, laminitis, gastric ulceration, obesity, hyperlipaemia, developmental orthopaedic disease and certain exercise-induced myopathies). The origins of these problems lies in the horses' digestive functioning and the solutions lie in designing better nutritional and exercise programs for horses, as discussed in this chapter.

Digestive anatomy and physiology of horses
Dentition, chewing and saliva in horses

Most adult (>5 years) female horses have 36 permanent teeth as follows (with no canines and no first premolar (wolf teeth). Male adults may have as many as 44 teeth (with four small canines and two wolf teeth (first premolar) as follows:

	Incisors	Canines	Premolars	Molars
Upper	3	0–1	3–4	3
Lower	3	0–1	3	3

The wolf teeth (first premolar) are only present in the upper jaw and mainly in males. The canines are small and only present in males. Unlike ruminants, horses have incisors on both upper and lower jaws. The strong mobile upper lip grasps forages and presents them to the incisors, which shear off the forage. The tongue then moves the material back to the 'cheek teeth' (or premolars) and molars for extensive mastication. The jaw movements of horses are vertical and horizontal, producing a grinding motion across the large premolar/molar surfaces. Feed particles are comminuted to ~1.6 mm (Meyer *et al.* 1985), and then swallowed with large quantities of saliva. Horse teeth continue to erupt throughout life, and on concentrate feeds, which require significantly less chewing than forages, the teeth can develop sharp points that must be 'floated' or filed back regularly. Equine saliva is secreted from three main pairs of salivary glands: the parotid, mandibular and sublingual glands. Horses produce up to 40 L of saliva/day with a pH of 9.0. Equine saliva contains virtually no amylase, so its major function is lubrication and gastric buffering. The electrolyte concentration in equine saliva is similar to that found in canines and porcines and is a mixture of sodium, potassium, chloride, calcium, magnesium, bicarbonate and phosphate (Alexander 1966).

Gross anatomy of the equine digestive tract

The horse is a hindgut (post-gastric) fermenter and this is reflected in the relative capacities of the stomach, small intestines and large intestines (Table 12.1). The total digesta capacity in the adult horse is ~200 L with only 35% (70 L) in the stomach and small intestines and 65% (130 L) in the caecum and colon.

The simple stomach of horses differs substantially from that of other monogastric animals in that almost half of the epithelium is non-glandular, stratified squamous (protective) epithelium, similar to that of the oesophagus (Fig. 12.1). The glandular region contains parietal cells, which secrete acid (HCl), zymogen cells, which secrete pepsin and lipase, and enterochromaffin cells, which secrete

Table 12.1. Capacity of the digestive compartments of the adult horse (500 kg)

Compartment	Proportion of total (%)	Volume (L)
Stomach	8	16
Small intestines	27	54
Caecum	16	32
Colon	49	98
TOTAL		200

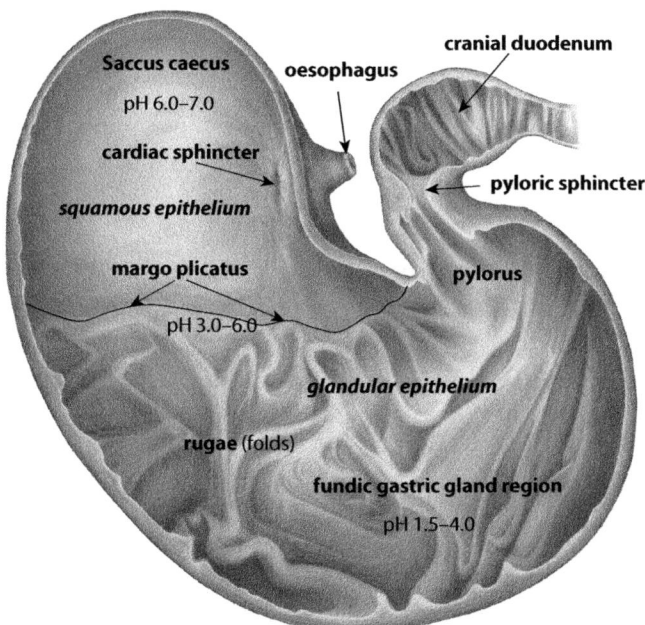

Fig. 12.1. The equine stomach is relatively small and is divided into a glandular region below, and a squamous region above, by the line of margo plicatus. The pH in the squamous region is usually above 6 because there are no parietal (acid) cells and high-pH saliva enters via the oesophagus. The pH in the mid-section around the margo plicatus is around 3.0–6.0. In the region of the proper (fundic) gastric glands, the pH drops to as low as 1.5. Both lower regions have lower pH the longer the gap between feeds.

histamine. The pyloric mucosa also produces the hormone gastrin. The squamous and glandular regions of the horse's stomach are separated by the line of margo plicatus, a clear demarcation of the two epithelia (Fig. 12.1).

The opening of the oesophagus into the stomach is guarded by a strong muscular sphincter, the cardiac sphincter, which rarely relaxes, so horses rarely vomit, and any gases generated by fermentation in the saccus caecus region must pass down the gastrointestinal tract, potentially causing pain (colic). The pyloric sphincter is also tight and regulates the emptying of the stomach digesta into the duodenum. The residence time of feed in the equine stomach varies with the diet type, with liquids emptying more rapidly than solids. Small, solid meals have a half-life of ~90 min (Lohmann *et al.* 2000). The secretion of gastric acid into the equine stomach is continuous, although the rate declines after feeding.

The small intestine in horses is relatively short, being ~25 m total in an adult 500 kg animal. The duodenum is ~1 m in length and the ileum ~0.7 m, leaving the jejunum at more than 23 m. The bile duct and pancreatic duct enter the duodenum ~15 cm posteriorly to the pyloric sphincter and a second pancreatic duct enters the duodenum close to the diverticulum. Horses have no gall bladder, so bile secretion is continuous.

The large intestine commences at the end of the ileum at the ileo-caeco-colic junction (Fig. 12.2). The caecum in horses lies in the right half of the abdomen. It is ~1 m long and comprises a base, a curved tapering body and a blind-ending apex deflected cranioventrally. A longitudinal layer of muscle runs in four bands along the caecum wall, visible externally as 'taenia', between which the caecal wall is puckered into four rows of sacculations or 'haustrae' (Fig. 12.2). The colon of the horse comprises three major parts: a large ascending colon, which is arranged in two U-shaped loops lying on top of each other; a short transverse colon; and a long descending colon. The first two parts are known as the large colon because they have a large luminal diameter, and the last part is known as the small colon because the diameter is considerably smaller. The two loops of the ascending colon are formed by three sharp flexures. The right dorsal colon runs cranioventrally until it reaches the xiphoid region where it is deflected caudally at the sternal flexure and becomes the left ventral colon. At the pelvis it is then deflected dorsally

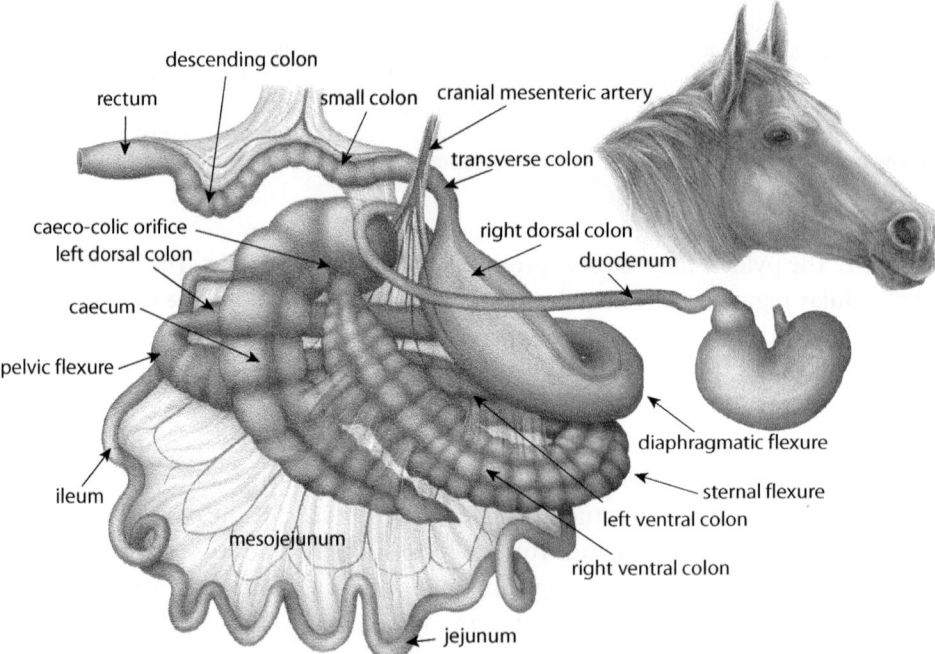

Fig. 12.2. The equine gastrointestinal tract. The gastric stomach of horses holds only 8% of the total GI tract contents, the small intestines only 27%, the caecum 16% and the colon 49%, so the large intestines hold 65% of the total tract contents and the stomach and intestines only 35%. The colon is characterised by three distinct flexures: the sternal flexure, the pelvic flexure and the diaphragmatic flexure. The large intestinal wall has rows of longitudinal muscle called taenia, which form folds called haustrae that slow up the rate of passage and ultimately shape the faecal output as water is resorbed.

and cranially at the pelvic flexure and becomes the left dorsal colon. The left dorsal colon then runs cranially until it reaches the diaphragm and becomes the right dorsal colon after the 360° turn at the diaphragmatic flexure. The right dorsal colon is wide but short, and ends at the much smaller diameter transverse colon. The descending colon is similar in diameter to the jejunum and is ~2–4 m long ending at the rectum at the pelvis. The descending colon is suspended by the mesentery at the duodenocolic fold (Fig. 12.2).

Digestion in the stomach, and small and large intestines in horses

Although pepsinogen is activated in the equine stomach by HCl, the extent of protein digestion in the stomach appears to be negligible and certainly many-fold less than the proteolysis occurring in the small intestine (Kern *et al.* 1974). Similarly for gastric lipase activity. Lipase will release free fatty acids from dietary triacylglycerols but the lipase activity in gastric mucosae is low compared with pancreatic lipase activity in the small intestine (Merritt and Julliand 2013). There is microbial activity in the equine gastric stomach, particularly in the upper sections where the pH (6.0–7.0) is conducive to fermentation. The fermentation acids, lactic acid and volatile fatty acids (VFA) are present, depending on the prevailing pH, but the extent to which they are absorbed and contribute to total energy supply is probably small.

As in other animals, the small intestine is lined by villi and microvilli for absorption of the end products of intestinal enzymic activity. Equine pancreatic juice contains bicarbonate, sodium, chloride, potassium, α-amylase and lipase. The disaccharidases maltase, sucrase and lactase are secreted in the brush border of the intestinal villi. Lactase activity diminishes with age in horses, so large quantities of dietary lactose in adult horses cause digestive upsets. Generally, the proteolytic enzyme activity of the equine pancreatic secretions is low and the contribution of intestinal protein digestion and amino acid absorption is unclear. Pancreatic lipase activity in horses is high and it appears that horses can digest fat quite efficiently, which can be useful in treating certain myopathies and for fuelling endurance performance (see 'Exertional rhabdomyolysis syndromes' later).

The extent of starch digestion before large intestinal fermentation is critical given the adverse effects of high starch overload into the large intestines. The concentration of α-amylase in horse pancreatic juice is low compared with other species, but the activity of disaccharidases is similar. Presumably the activity of pancreatic amylase on starches and the activity of brush border disaccharidases on sugars provides the horse with large quantities of intestinal hexoses, but there appears to be significant spill over of undigested carbohydrates to the large intestines. There is microbial activity in the small intestine of horses and this will also influence starch availability in the large intestines.

Overload of soluble carbohydrate into the large intestines is a major problem for horses. 'Slug feeding' of grains should be avoided and medium-quality roughage should be available at all times.

The caecum and colon of horses contain bacteria, archaea, ciliated protozoans and anaerobic fungi, many of which are similar to those prevailing in the forestomachs of the pre-gastric fermenters. Bacteria make up more than half the dry weight of the faeces. Bacterial numbers in the large intestines number $0.5-5 \times 10^9$/g. Ciliated protozoans number 10^3-10^4/mL contents and fungi 10^3-10^4/mL contents. The bacteria include the cellulolytic bacteria *Ruminococcus flavefaciens*, *Fibrobacter succinogenes* and the glycolytic and amylolytic bacteria streptococci, lactobacilli and enterococci. Lactate-utilising bacteria are also present, including *Megasphaera* and *Veillonella* species. The numbers of bacteria are greatest in the caecum and ventral colon. Horses are less efficient than ruminants at digesting high-fibre feeds, but this probably reflects the more rapid transit time of feed in the horse.

The main products of microbial fermentation in the large intestines of horses are acetate and butyrate. Propionate and lactate are produced if starch is entering the large intestines. The VFA are absorbed across the caecal and colonic epithelia, presumably by similar mechanisms to VFA absorption for the rumen (see Fig. 3.21). The removal of undissociated acids, and the reciprocal movement of bicarbonate into the lumen, buffer the pH of the intestinal lumen, thereby preventing catastrophic drops in luminal pH.

Protein digestion in horses commences in the stomach under the action of pepsin activated from pepsinogen by acid. The pepsin begins breaking peptide bonds and this continues in the small intestine under the action of pancreatic proteases, releasing amino acids and dipeptides. The amino acids are absorbed across the small intestinal epithelium by amino acid transporters (see Table 2.2) specific to groups of amino acids (e.g. transporter A transports gly, pro, ala, gln, his, met; ASC transports ala, ser, cys, thr, gln; imino transports pro; L transports gly, ala, val, leu, ile, met, phe, trp, ser, thr, cys, tyr, asn, gln; T transports phe, trp, tyr; X_{AG} transports asp, glu;, and y^+ transports arg, lys, his). Protein and non-protein nitrogen that is not digested in the small intestines enters the large intestine where it contributes to microbial protein synthesis, but there is no evidence of significant amino acid absorption across the large intestinal epithelia. The 'quality' of protein for horses depends on the combination of the amino acid profile of the protein and its digestibility. The pre-caecal digestibility is particularly important because only the protein digested in the small intestines contributes to the available amino acid pool for the animal. On average, the apparent whole-tract digestibility of proteins in horses is ~79% and pre-caecal digestibility averages 51% (NRC 2007). A feeding system that used ileal digestibility of proteins, such as that used in pig nutrition, would be ideal for horses given the large effect of the hindgut on overall tract digestibility. However, there are insufficient data on feeds related to their 'ileal digestibility' to implement such as scheme at present.

Lysine is the first-limiting amino acid for horses and a level of 4.3% is the minimum required by growing animals.

The energy, protein, vitamin and mineral requirements of horses

Assessing the bodyweight and body condition score of horses

To formulate diets for horses we require direct measure of the bodyweight of the animal or an indirect assessment from morphometric measurements. The fatness of the horse can also be useful in guiding nutritional decisions and commonly this is gauged by body condition scoring (see next section).

Estimating body condition score in horses

A commonly used method for estimating the fatness or 'condition' of a horse is the Henneke system (Henneke *et al.* 1983). This system uses scores from 1 to 9 as indicated in Table 12.2 and assessed by palpation of subcutaneous fat at specified sites.

Estimating cresty neck scores in horses

Neck adiposity (crest fatness) can also be scored (Table 12.3).

The cresty neck score appears to be a useful indicator of the propensity of a horse to develop equine metabolic syndrome (EMS – see 'Laminitis' later), and has been likened to waist measurements in humans as an indicator of adiposity and propensity to metabolic syndrome in humans. Fatty crests can develop even in horses that are not generally obese and this may be a good indicator of insulin resistance (Johnson *et al.* 2009).

> Horses with equine metabolic syndrome have a prominent and characteristic fatty nuchal crest, show signs of insulin resistance, laminitis, osteochondrosis, hyperlipaemia, PPID (equine Cushing's disease) and systemic inflammation (Johnson *et al.* 2009).

The implications of obesity and insulin resistance in horses are considered in 'Common nutritional and metabolic diseases of horses' later.

Table 12.2. Body condition scoring in horses (after Henneke *et al.* 1983)

Score	Visual	Description
1 Poor		• Extremely emaciated • Spinous processes, ribs, tailhead, hooks (tuber coxae, hip joints), pins (tuber ischia, lower pelvic bones) prominent • Bone structure of withers, shoulders and neck easily noticeable • No palpable fatty tissue
2 Very thin		• Emaciated • Slight fat over lumbar spinous and transverse processes • Spinous processes, ribs, tailhead, hooks and pins prominent • Withers, shoulders and neck structures faintly discernible
3 Thin		• Fat built up halfway on spinous processes • Transverse processes cannot be felt • Slight fat cover over ribs • Spinous processes and ribs easily discernible • Tailhead is prominent • Individual vertebrae cannot be easily identified • Hook bones rounded but easily discernible • Pin bones not distinguishable • Withers, shoulders and neck are accentuated
4 Moderately thin		• Negative crease along back • Faint outline of ribs discernible • Fat can be felt around tailhead • Hook bones not discernible • Withers, shoulder and neck not obviously thin
5 Moderate		• Back is level • Ribs can't be visually distinguished but can be easily felt • Fat around tailhead is spongy • Withers appear rounded over spinous processes • Shoulders and neck blend smoothly into body
6 Moderately fleshy		• Slight crease down back • Fat over ribs feels spongy • Fat around tailhead feels soft • Fat beginning to be deposited along sides of withers, behind shoulders and along neck
7 Fleshy		• May have crease down back • Individual ribs can be felt but with noticeable fat between • Fat around tailhead soft • Fat is deposited along withers, behind shoulders and along neck
8 Fat		• Crease down back • Difficult to feel ribs • Fat around tailhead is very soft • Areas along withers and behind shoulders are filled with fat • Noticeable thickening of neck • Fat deposited along inner thighs
9 Extremely fat		• Obvious crease down back • Patchy fat over ribs • Bulging fat over tailhead, along withers, behind shoulders, along neck • Fat along thighs may rub together • Flank is filled with fat an flush with rest of body

Table 12.3. Nuchal or cresty neck scoring system

Score	Visual	Description
0		No visual appearance of a crest (tissue apparent above the ligamentum nuchae). No palpable crest.
1		No visual appearance of a crest, but slight filling felt with palpation.
2		Noticeable appearance of a crest, but fat deposited fairly evenly from poll to withers. Crest easily cupped in one hand and bent from side to side.
3		Crest enlarged and thickened, so fat is deposited more heavily in middle of neck than towards poll and withers, giving a mounded appearance.
4		Crest grossly enlarged and thickened, and can no longer be cupped in one hand or easily bent from side to side. Crest may have wrinkles/creases perpendicular to top line.
5		Crest is so large it permanently droops to one side.

Source: After Carter *et al.* (2009)

Table 12.4. Digestible Energy (MJ/day) and Crude Protein (g/day) requirements for maintenance of horses at three levels of activity

Activity level	DE (MJ/day)	CP (g/day)
Low (stalled, sedentary lifestyle, old arthritic)	0.13W (kg)	1.08W (kg)
Moderate (normal paddock activity, walking)	0.14W (kg)	1.26W (kg)
High (highly strung, nervous horses, young horses, stallions, 'bad doers')	0.16W (kg)	1.44W (kg)

Source: NRC (2007)

Digestible Energy and crude protein requirements of horses

The NRC (2007) *Nutrient Requirements of Horses* (6th Edition) provides the theoretical basis for formulating diets for horses of varying ages, bodyweights, growth rates, pregnancy, lactation and exercise. This system uses a Digestible Energy (DE) system for evaluating both the animal's requirements for energy and the provision of that energy in feeds. Although it is acknowledged that Net Energy (NE) systems provide the closest value to the animals' requirements, in practical terms it is difficult to measure and account for the components that reduce available energy from DE to NE. In DE systems, these components are estimated and DE is derived from NE taking into account these deductions. The NRC (2007) uses crude protein requirements calculated using the factorial method of adding up the various storages and losses of protein depending on the physiological state of the animals (age, growth, gestation, lactation, exercise). Tables 12.4–12.8 summarise the requirements derived in NRC (2007).

Digestible Energy and crude protein requirements for maintenance in horses

Table 12.4 shows the maintenance requirements for DE and CP in horses under three activity scenarios. For example, a 500 kg horse stabled continuously would require 65 MJ DE/day (0.13 × 500) and 540 g CP/day (1.08 × 500). A typical forage (say grass hay, cool season mature) contains 8.5 MJ DE/kg dry matter, and 10.8% CP. To meet the DE requirement, the horse would have to eat 7.6 kg DM/day (= 9.1 kg as fed). The CP intake would be 820 g at this intake. Maximum intake of this horse would be ~10 kg dry matter (2% of bodyweight), so this is feasible.

Digestible Energy requirements (MJ/day) for growing and exercising horses

The composition of gain and the efficiency of growth in growing horses, changes with the age of the animal, so the DE requirements for each kilogram of gain change with age (Table 12.5).

Table 12.5. Digestible Energy requirement for weight gains in horses of different ages

Age of horse (months)	DE for gain (MJ/kg gain)
6	36
12	56
18	71
24	79

Source: NRC (2007)

> To estimate the DE requirement for a 2-year old, 400 kg horse growing at 0.5 kg/day, choose the activity level scenario from Table 12.4 (e.g. moderate), estimate the DE requirement for maintenance (0.14W = 0.14 × 400 = 56 MJ) then add the growth requirement = 79 × 0.5 kg/day = 39.5. Total DE requirement = 95.5 MJ/day.

Feeding a horse to change its body condition score requires estimates of the relationship between bodyweight change and condition score and between weight change and DE requirement. The desired rate of change in condition score also determines how much DE above maintenance is required to change the score. Table 12.6 shows the extra DE required to achieve a change in condition score of a 500 kg horse from 4–5 out of 9 on the Henneke scoring system (Table 12.2).

For growing horses, the CP requirement is adjusted for changes in the efficiency of utilisation of dietary protein. Young animals use dietary protein with 50% efficiency, but this drops off to 30% by 12 months of age. This is accounted for in the following equation:

$$CP \text{ requirement (g/day)} = 1.44W + (ADG \times 0.2) / (E \times 0.79)$$

where: W = bodyweight (kg); ADG = average daily gain (g/day); E = efficiency of protein use (4–6 months = 0.50; 7–8 months = 0.45; 9–10 months = 0.40; 11 months = 0.35; >12 months = 0.30. The term 1.44W is the maintenance CP requirement assuming the high value in Table 12.4.

Table 12.6. Increase in DE required to change the condition score of a 500 kg horse from 4 to 5 (Henneke system)

Time to achieve an increase in condition score of 1 (days)	DE above maintenance (MJ/day)[a]
60	25
90	17
120	13
150	10
180	8

[a]Estimate maintenance from Table 12.4 and add the selected value from this table

Table 12.7. Digestible Energy requirements for different levels of exercise in horses

Exercise level	DE (MJ/day)	CP (g/day)
Light (recreational riding, show horses) (1–3 h/week)	0.16W (kg)	1.35W (kg)
Medium (school horses, beginning of training, frequent show horses, polo, farm work) (3–5 h/week)	0.20W (kg)	1.44W (kg)
Heavy (farm work, polo, show horses frequent and strenuous, medium-level eventing, middle stages of race training)	0.22W (kg)	1.53W (kg)
Very heavy (racing – thoroughbred, quarter horse, standardbred, endurance, elite 3-day event) (6–12 h/week slow or 1 h/week speed work)	0.29W (kg)	1.61W (kg)

Source: NRC (2007)

DE and CP requirements of exercising horses

The DE and CP requirements of horses exercising at different levels is shown in Table 12.7. A 500 kg horse undergoing moderate exercise at the beginning of a training period will require 100 MJ DE/day (500 × 0.20) and 720 g CP/day (500 × 1.44). The maximum intake of a 500 kg horse is ~10 kg DM/day, so it would have to contain 10 MJ/kg DM DE and 7.2% CP (e.g. an immature forage legume hay).

Digestible Energy and crude protein requirements for gestation and lactation in horses

For gestation, multiply the maintenance requirement estimated from Table 12.7 by the factors in Table 12.8, depending on stage of gestation.

For pregnant mares, the CP requirement is calculated the same as for growth, the assumption being that fetal protein deposition is equivalent to early body gains (E = 0.50):

$$\text{CP requirement (g/day)} = 1.26W \text{ (kg)} + 0.51 \times \text{ADG (g/day)}$$

The CP requirements for lactation can be estimated if one makes some assumptions about milk yield/bodyweight relationships, protein digestibility, milk protein composition and efficiency of use of protein for lactation.

$$\text{CP requirement for lactation (g/day)} = 2.86W$$

Assumptions:

- milk yield = 3.2% of bodyweight
- milk protein = 2%

Table 12.8. Factors used to increase the maintenance DE requirement during pregnancy

Gestation month	Factor (×M)
9th	1.11
10th	1.13
11th	1.20

Source: NRC (2007)

- efficiency of utilisation of protein for milk production = 50%
- CP digestibility = 79%.

For example, a 730 kg mare in early lactation = 730 × 0.14 (Table 12.4) + 0.14 × W (lactation) = 204 MJ/day.

Vitamin and mineral requirements of horses

Table 12.9 shows the requirements of horses with a mature bodyweight of 500 kg. Values are the required concentrations of the nutrients in feedstuffs to meet the

Table 12.9. Requirements of horses at different physiological stages for macrominerals, trace elements and vitamins

	Percentage of feed DM						
	Ca	P	Mg	K	Na	Cl	S
General	**0.2–0.4**	**0.14–0.38**	**0.075–0.15**	**0.25–0.53**	**0.10–0.4**	**0.40–0.81**	**0.15–0.19**
Maintenance	0.2	0.14	0.075	0.25	0.1	0.4	0.15
Exercise	0.4	0.29	0.15	0.53	0.41	0.93	0.19
Growth	0.39	0.22	0.06	0.2	0.08	0.37	0.15
Gestation	0.36	0.26	0.075	0.26	0.11	0.41	0.15
Lactation	0.59	0.38	0.13	0.48	0.13	0.46	0.19
	mg/kg feed DM						
	Co	Cu	I	Fe	Mn	Se	Zn
General	**0.05–0.06**	**10 to 13**	**0.02–0.05**	**40–65**	**40–50**	**0.01–0.013**	**40–50**
Maintenance	0.05	10	0.035	40	40	0.01	40
Exercise	0.06	13	0.044	50	50	0.013	50
Growth	0.05	10	0.034	48	39	0.01	39
Gestation	0.05	13	0.04	50	40	0.01	40
Lactation	0.06	13	0.044	63	50	0.013	50
	×1000 IU						
	A	D	E				
General	**15–30**	**3.3–6.2**	**0.5–1.0**				
Maintenance	15	3.3	0.5				
Exercise	23	3.3	1				
Growth	17	6.2	0.78				
Gestation	30	3.3	0.8				
Lactation	30	3.3	1				

Source: NRC (2007)

animal's daily requirements, assuming the dry matter intake of the animal is at 2% of its bodyweight. The values are typical ranges of concentrations required in feeds, plus variations in required concentrations with growth, exercise, gestation and lactation.

Exercise has very significant effects on the requirements for the macrominerals as a result of the loss of large quantities of electrolytes in sweat. For most minerals and vitamins, the requirements are increased by gestation, lactation and growth.

Common nutritional and metabolic diseases in horses

Like most animals, many of the diseases of horses are associated with poor nutrition. Overfeeding and obesity, unbalanced rations and high levels of starchy cereals in 'slug' feeds, have negative effects on the development of insulin resistance, laminitis, osteochondrosis and colic in horses. Although there is a genetic component to some of these disorders, attention to simple nutritional principles is often the most effective preventative measure. In this section some of the common diseases in which nutrition plays a significant role in the aetiology or treatment are considered in relation to the latest scientific literature.

Developmental orthopaedic disease

Developmental orthopaedic disease (DOD) entails a spectrum of conditions related to disturbances of the musculoskeletal system in general and of the articular and metaphyseal cartilage in particular. It encompasses physitis (inflammation of the metaphyseal growth plate) and osteochondrosis (OC: a disturbance of the process of endochondral proliferation, maturation and ossification of bones). OC is also called dyschondroplasia or osteochondrosis dissecans (OD), the latter reflecting the breaking away of damaged cartilage fragments. As many as 10–50% of foals can be affected by the disease, with most abnormalities apparent early in life (1 month of age). Most lesions regress after this period, but there appear to be 'windows' of susceptibility after which regression is unlikely. The point of no return appears to be at ~5 months of age for the hock and 8–12 months of age for the stifle (van Weeren and Barneveld 1999). The incidence in warmbloods and racing breeds is as high as 25% (van Weeren 2006), so an understanding of the aetiology of the disease is a high priority for the equine industries (Jeffcott 1996). OC is a multifactorial disease involving genotype, growth rate, body size, energy intake, macromineral intake, trace element intake, endocrinopathy and biomechanical trauma (Jeffcott 1996). The very early onset may infer involvement of the maternal diet but little is known of the potential maternal 'programming' effects, particularly on the endocrine system. Rapid growth rate (which may relate to genotype effects), dietary imbalances of calcium/phosphorus and possibly effects of high-starch concentrates on insulin-glucose responses, have all been thought to influence the development of OC.

There is currently considerable debate about the relative roles of genetics, nutrition, exercise, endocrine factors, physical trauma and failure of vascularisation in the aetiology of DOD (van Weeren 2006; Geor *et al.* 2013). Some studies show a strong effect of high-energy intakes and rapid growth rate and DOD incidence (Savage *et al.* 1993), but others show no effect of energy intake but an effect of growth rate (Donabedian *et al.* (2006). It is thought the relationship between energy intake and DOD is via increases in glucose and insulin after feeding, particularly when the diet is high in easily digested carbohydrates. Insulin and the insulin-like growth factors (IGF-1 and IGF-2) directly influence endochondral ossification by stimulating cartilage matrix synthesis (Henson *et al.* 1997). Low plasma IGF-I is associated with OC lesions. Insulin also stimulates the removal of the thyroid hormones form the blood. Thyroid hormones are involved in cartilage cell (chondrocyte) differentiation and vascularisation of the cartilage. At present there is some evidence that high-energy, high-glycaemic diets can induce OC lesions, probably through insulin effects, but other evidence is contradictory. Given the role of calcium and phosphorus in bone growth, these macrominerals are obvious targets for a role in OC. Low levels of calcium in foal diets are associated with a high incidence of OC (Knight *et al.* 1985) so adequate calcium in the ration is advised. Excess calcium is not recommended because it may interfere with absorption of other minerals. Diets with excess phosphorus are associated with a high incidence of OC lesions (Savage *et al.* 1993). Ca:P ratios of 1.4:1 to 3:1 are recommended in addition to attention to absolute levels.

Given the role of copper as a cofactor in lysyl oxidase, the enzyme involved in connective tissue synthesis, a role for copper in OC has been examined. Results are conflicting but in general there appears to be little evidence for a role of copper in the incidence of OC (Vervuert and Ellis 2013), but some evidence that copper status may influence regression of OC lesions. Supplementation of the mare may be required to elevate the liver copper levels in foals.

Developmental orthopaedic disease in foals

- DOD has a genetic component (DOD affects 25% of warmbloods and racing horses).
- High-energy intakes may induce DOD through effects on insulin and thyroid hormones.
- High P intakes and low Ca:P ratios promote DOD.
- Supplementing the mare with copper may improve the regression of OC lesions.

Laminitis

Laminitis is a painful and debilitating disease characterised by failure of the bond between the inner hoof wall and the distal phalanx (the last digit). The subsequent structural collapse of the foot and rotation of the distal phalanx are extremely painful and laminitis represents a major animal welfare issue. The pathophysiological changes associated with laminitis include activation of the matrix metalloproteinases in the hoof lamellae, the presence of markers of inflammation, altered blood flow (increased or decreased), and a metabolic predisposition to insulin resistance and obesity known as equine metabolic syndrome (EMS). It has long been known that overload of starchy grains (cereals) can produce laminitis and pasture-associated laminitis is triggered by overload of rapidly fermentable carbohydrates and the absorption of substances that cause failure of the lamellar structure of the hoof. In support of this theory, oral administration of oligofructose induces laminitis in healthy horses. However, many horses do not succumb to laminitis when the risk factors in pasture are highest, suggesting other predisposing factors, including genetics, are operating. It has now been established that horses with metabolism characterised by regional and general obesity, and insulin-resistance are at increased risk of laminitis when placed on pastures with high soluble (non-structural) carbohydrates. This syndrome is referred to as EMS and is characterised by regional adiposity (particularly around the nuchal crest = cresty neck (see Table 12.3) or general adiposity. Although general obesity is observed in the majority of cases, some affected horses have a leaner overall body condition but regional adiposity, and others are normal in appearance. Affected animals are insulin-resistant and display hyperinsulinaemia, or abnormal glycaemic and insulinaemic responses to oral or IV glucose challenges. They also have a predisposition to laminitis, which is unrelated to common laminitis-inducing factors such as grain overload, colic, colitis or retained placenta (Frank *et al.* 2010). The EMS phenotype is also characterised by dyslipidaemia, hyperleptinaemia, arterial hypertension, altered reproductive cycling in mares and increased systemic markers of inflammation associated with obesity (Frank *et al.* 2010). Although many of these signs are consistent with equine Cushing's disease (ECD), EMS horses are not hirsute, and test negative for ECS in endocrine tests. Horses showing signs of severe EMS may be the tip of the iceberg in that subclinically affected animals may be those that progress to overt pasture-associated laminitis when the pasture conditions become conducive. If this is the case, the term pre-laminitic metabolic syndrome would be appropriate (Huntington *et al.* 2009).

So how are obesity, insulin resistance and laminitis related physiologically? Fat is now regarded as the largest endocrine organ in the body and the adipokines and inflammatory cytokines released from the adipocytes (leptins, resistin, interleukins, TNF-α, IL-1, IL-6, adiponectin, visfatin, apelin, among others) are thought to contribute to the low-grade inflammation apparent in obese individuals (Frank *et al.* 2010). These individuals are predisposed to insulin resistance and hyperinsulinaemia has been shown to directly produce laminitis on otherwise normal ponies (Asplin *et al.* 2007). Further studies of the relationships between regional adiposity (particularly nuchal fat), insulin resistance, inflammatory cascades and susceptibility to pasture-induced laminitis are urgently required to establish clear links between the nutritional/digestive and metabolic pathophysiological causes of laminitis.

Exertional rhabdomyolysis syndromes

Muscle pain and cramping during or after exercise in horses is termed 'tying up', 'Monday morning disease', azoturia or 'set fast' and belongs to a syndrome called exertional rhabdomyolysis. When the syndrome occurs repeatedly it is termed recurrent exertional rhabdomyolysis (RER) in thoroughbreds, standardbreds and Arabians, and polysaccharide storage myopathy (PSSM) in quarter horses, paints, European warmbloods, appaloosas, Morgan horses and draught breeds. RER appears to be a heritable defect in calcium regulation, leading to muscle necrosis. PSSM is a glycogen storage disorder associated with a mutation in the *GYS-1* gene, which codes for the glycogen synthase-1 enzyme. This enzyme has a regulatory inhibitory site that allows the enzyme to switch 'off' in response to high levels of accumulated glycogen in the muscle. A defect in this site means the enzyme is continually switched 'on' and an abnormal form of glycogen (amylase-resistant, periodic acid Schiff (PAS)-positive) polysaccharide accumulates in the skeletal muscle Type-II muscle fibres). High glycogen storage can also result from increased sensitivity to insulin shunting more glucose into the myocytes. In both RER and PSSM, there is evidence that reducing dietary starches and sugars, and increasing dietary fat are beneficial in managing the disorders, but management of exercise regimes in addition to dietary intervention is essential. The clinical signs of PSSM are more frequent in horses receiving diets with moderate to high levels of soluble carbohydrates from cereals. Feeding diets with low soluble carbohydrates and fats comprising 6–25% of the DE, reduces muscle glycogen and reduces pain. There is little evidence of selenium and vitamin E-responsive myopathies in horses, but a small increase in antioxidants including Se/vitamin E may be advisable and unlikely to be harmful (Table 12.10).

Equine gastric ulceration syndrome (EGUS)

A very high proportion of horses are diagnosed with clinically significant gastric ulceration. As many as 93% of racehorses and 37% of pleasure horses (Murray *et al.* 1996), 40% of performing quarter horses (Bertone 2000) and 58% of show horses (McClure *et al.* 1999) have been diagnosed as clinically affected by EGUS. Commonly reported risk factors in racehorses include feeding high-concentrate

Table 12.10. Dietary management of PSSM and RER in horses

PSSM	RER
• Minimise starch and sugar intake (no cereals) to <10% of DE. • Use forages or forage replacers containing <10–12% non-structural (soluble) carbohydrates. • Provide >15% of DE as oils (fats) but manage bodyweight and condition score carefully. • Ensure the ration is balanced for protein, vitamin and mineral content. • Very gradually return to exercise. • Avoid high sugar, high legume pastures. • Treat as if laminitis-prone. • Add vitamin E to oil (1–1.5 IU/mL oil).	• Reduce starches/sugars to <20% of DE. • Use grass hays, not haylage, and avoid legume forages. • Include oils at 15–20% of DE. • Avoid high-sugar, high-legume pastures. • Treat as if laminitis-prone. • Add vitamin E to oil (1–1.5 IU/mL oil).

diets, low quantities of hay in diets, fasting, exercise regimes, meal or 'slug' feeding and the use of non-steroidal anti-inflammatory drugs (Merritt 2003; Lester 2004; Jonsson and Egenvall 2006; Videla and Andrews 2009). Nutritional mismanagement consistently appears as a major contributing factor to the development of EGUS. Horses produce ~1.5 L of gastric juice/h and, unlike other mammals, secrete the juice continuously (Campbell-Thompson and Merritt 1990).

The mucosa of the stomach of horses is divided into a squamous epithelium above, and a glandular mucosa below: the line of margo plicatus (Fig. 12.1). The glandular mucosa is well protected from acid because it secretes bicarbonate and mucus, unlike the squamous epithelium, which does not secrete these protective agents (Ross *et al.* 1981). Most ulcers (80%) therefore occur in the squamous region, often close to the margo plicatus. It appears likely that ulceration occurs when there is an imbalance between protective agents (mucus, bicarbonate, mucosal blood flow, prostaglandins and epithelial integrity) and ulcer-causing agents (HCl acid, pepsin, bile acids and VFA (Luthersson *et al.* 2009)). Risk factors for equine ulceration include active racing and training, increasing age, intermittent feeding, stall confinement, high-concentrate diets, the use of non-steroidal anti-inflammatory agents and possibly the presence of *Helicobacter* bacteria. Table 12.11 shows the major risk factors of various management practices on the frequency and severity of EGUS in horses, along with the potential mechanism operating to generate ulcers.

Prevention of EGUS, then, largely centres on nutritional management, including reducing high-starch intakes in individual meals and per day, provision of medium- to high-quality hay (high in protein) and not straw only, and *ad libitum* access to water. Therapeutic treatment for equine gastric ulceration is largely directed to increasing the gastric pH by suppressing HCl production. Inhibitors of the K^+/H^+ proton pump (e.g. omeprazole) and antagonists of the

Table 12.11. Nutritional risk factors for development of equine gastric ulceration and their potential mechanisms of inducing the syndrome

Risk factor	Potential mechanism
Fasting	• Continued acid secretion with no buffering from saliva and feed components • Increasingly watery contents splash gastric juice onto the squamous mucosa
High-starch rations and grain feeding before roughage feeding	• Rapid production of VFA and lactic acid • More fluid gastric contents • Inadequate saliva production and gastric buffering • Low calcium and other buffering agents
Stabling	• Associated with 'slug feeding' and grain feeding • No continued access to saliva-inducing roughages • No exercise
'Slug' feeding (large infrequent meals)	• Combination of periods of fasting and sudden rapid acid production results in increased acid and reduced dilution and buffering
Straw as the only roughage component	• Damage to the epithelium by lignified materials • Low in calcium and protein, which act as buffers • Alterations to the fibrous mat
Restricted access to water	• Water dilutes the gastric secretions, reducing the acidity
Transport	• Perturbed feed and water intake means less dilution of acid and less feed buffers

Source: Adapted from Luthersson *et al.* (2009)

histamine H2 receptor (e.g. ranitidine) are major treatments based on our knowledge of gastric acid secretion (Fig. 2.5).

References

Alexander F (1966) A study of parotid salivation in the horse. *The Journal of Physiology* **184**, 646–656. doi:10.1113/jphysiol.1966.sp007937

Asplin KE, Sillence MN, Pollitt CC, McGowan CM (2007) Induction of laminitis by prolonged hyperinsulinaemia in clinically normal ponies. *Veterinary Journal (London, England)* **174**, 530–535. doi:10.1016/j.tvjl.2007.07.003

Bertone JJ (2000) Prevalence of gastric ulcers in elite, heavy use Western performance horses. *Journal of Veterinary Internal Medicine* **14**, 366.

Campbell-Thompson ML, Merritt AM (1990) Basal and pentagastrin-stimulated gastric secretion in young horses. *The American Journal of Physiology* **259**, R1259–R1266.

Carter RA, Geor RJ, Staniar WB, *et al.* (2009) Apparent adiposity assessed by standardised scoring systems and morphometric measurements in horses and ponies. *Veterinary Journal (London, England)* **179**, 204–210. doi:10.1016/j.tvjl.2008.02.029

Donabedian M, Fleurance G, Perona G, *et al.* (2006) Effects of fast versus moderate growth rate related to nutrient intake on developmental orthopaedic disease in the horse. *Animal Research* **55**, 471–486. doi:10.1051/animres:2006026

Frank N, Geor RJ, Bailey SR, Durham AE, Johnson PJ (2010) Equine metabolic syndrome. *Journal of Veterinary Internal Medicine* **24**, 467–475. doi:10.1111/j.1939-1676.2010.0503.x

Geor RJ, Coenen M, Harris P (2013) *Equine Applied and Clinical Nutrition: Importance of Nutrition for Health, Welfare and Performance.* Saunders, Elsevier, UK.

Henneke DR, Potter GD, Kreider JL, *et al.* (1983) Relationship between condition score, physical measurements and body fat percentage in mares. *Equine Veterinary Journal* **15**, 371–372. doi:10.1111/j.2042-3306.1983.tb01826.x

Henson FM, Davenport C, Butler L *et al.* (1997) Effects of insulin and insulin-like growth factors I and II on the growth of equine fetal and neonatal chondrocytes. *Equine Veterinary Journal* **29**, 441–447.

Huntington P, Pollitt C, McGowan C (2009) Recent research into laminitis. In *Advances in Equine Nutrition IV.* (Ed. JD Pagan) pp. 293–311. Nottingham University Press, Nottingham, UK.

Jeffcott LB (1996) Osteochondrosis – an international problem for the horse industry. *Journal of Equine Veterinary Science* **16**, 32–37.

Johnson PJ, Wiedmeyer CE, Ganjam VK (2009) Medical implications of obesity in horses – lessons for human obesity. *Journal of Diabetes Science and Technology* **3**, 163–174. doi:10.1177/193229680900300119

Jonsson H, Egenvall A (2006) Prevalence of gastric ulceration in Swedish standardbreds in race-training. *Equine Veterinary Journal* **38**, 209–213. doi:10.2746/042516406776866390

Kern DL, Slyter LL, Leffel EC, *et al.* (1974) Ponies versus steers: microbial and chemical characteristics of intestinal ingesta. *Journal of Animal Science* **38**, 559–564. doi:10.2527/jas1974.383559x

Knight DA, Gabel AA, Reed SM, *et al.* (1985). Correlations of dietary minerals to incidence and severity of metabolic bone disease in Ohio and Kentucky. In *Proceedings of the 31st Annual Convention American Association Equine Practitioners.* pp. 445–461. American Association Equine Practitioners Lexington, KY, USA.

Lester GD (2004) Gastrointestinal diseases of performance horses. In *Equine Sports Medicine and Surgery.* (Eds KW Hinchcliff, AJ Kaneps and RJ Geor) pp. 1037–1043. Saunders Elsevier, Philadelphia, PA, USA.

Lohmann K, Roussel AJ, Cohen ND, *et al.* (2000) Comparison of nuclear scintigraphy and acemenophen absorption as a means of studying gastric emptying in horses. *American Journal of Veterinary Research* **61**, 310–315. doi:10.2460/ajvr.2000.61.310

Luthersson N, Hou Nielsen K, Harris P, Parkin TDH (2009) Risk factors associated with equine gastric ulceration syndrome (EGUS) in 201 horses in Denmark. *Equine Veterinary Journal* **41**, 625–630. doi:10.2746/042516409X441929

McClure SR, Glickman LT, Glickman NW (1999) Prevalence of gastric ulcers in show horses. *Journal of the American Veterinary Medical Association* **215**, 1130–1133.

Merritt AM (2003) The equine gastric stomach: a personal perspective. *Proceedings of the American Association of Equine Practitioners* **49**, 75–102.

Merritt AM, Julliand V (2013) Gastrointestinal physiology. In *Equine Applied and Clinical Nutrition; Health, Welfare and Performance.* (Eds RJ Geor, PA Harris and M Coenen) pp. 3–32. Saunders Elsevier, London, UK.

Meyer H, Coenen M, Gurer C (1985) Investigations of saliva production and chewing in horses fed various feeds. In *Proceedings of the Equine Nutrition and Physiology Society.* pp. 38–41. East Lansing, MI, USA.

Murray MJ, Schusser GF, Pipers FS, Gross SJ (1996) Factors associated with gastric lesions in thoroughbred horses. *Equine Veterinary Journal* **28**, 368–374. doi:10.1111/j.2042-3306.1996.tb03107.x

NRC (2007) *Nutrient Requirements of Horses.* 6th revised edn. National Research Council, The National Academies Press, Washington, DC, USA.

Ross IN, Bahari HM, Turneberg LA (1981) The pH gradient across mucus adherent to mouse fundic mucosa in vivo and the effect of potential damaging agents. *Gastroenterology* **81**, 713–718.

Savage CJ, McCarthy RM, Jeffcott LB (1993) Effects of dietary energy and protein on induction of dyschondroplasia in foals. *Equine Veterinary Journal. Supplement* **16**, 74–79.

van Weeren PR (2006) Etiology, diagnosis and treatment of OC(D). *Clinical Techniques in Equine Practice* **5**, 248–258. doi:10.1053/j.ctep.2006.08.002

van Weeren PR, Barneveld A (1999) The effect of exercise on the distribution and manifestation of osteochondritic lesions in the warmblood foal. *Equine Veterinary Journal* **31**(Supplement), 16–25. doi:10.1111/j.2042-3306.1999.tb05309.x

Vervuert AM, Ellis AD (2013) Developmental orthopedic disease. In *Equine Applied and Clinical Nutrition; Health, Welfare and Performance.* (Eds RJ Geor, PA Harris and M Coenen) pp. 536–548. Saunders Elsevier, London, UK.

Videla R, Andrews FM (2009) New perspectives in equine gastric ulcer syndrome. *The Veterinary Clinics of North America. Equine Practice* **25**, 283–301. doi:10.1016/j.cveq.2009.04.013

13

Pig nutrition

Key points

- The wild pig evolved as a simple-stomached omnivore adapted to a wide range of feeds.
- The secretions of the gastric stomach (pepsin, lipase, rennin), the exocrine pancreas (proteases, lipases, carbohydrases, nucleases), liver (bile) and the brush border (carbohydrases, peptidases) produce simple products for absorption (hexoses, amino acids, fatty acids) by simple diffusion, facilitated diffusion or carrier-mediated transport.
- Fermentation in the large intestines provides 20–30% of the energy for the pig (depending on the diet fibre content).
- Regulating feed intake in pigs is critical to managing the phases of the production cycle (grower, finisher, gestation and lactation) to ensure high milk yields, healthy piglets and rapid growth of weaners.
- Pig diets are formulated on the basis of Digestible Energy and standardised ileal digestible amino acids. Lysine is the first-limiting amino acid for pigs and other amino acids levels are expressed relative to available lysine.
- Pastures are nutritionally inadequate for young growing pigs, lactating sows, pregnant sows and even dry sows. Even high-quality pastures are inadequate for pigs. Free-range pigs eat less, grow slower and have reduced dressing percentages, even when given access to balanced grain-based rations.
- The amount of feed additives and exogenous enzymes in pig feeds will increase as alternative feedstuffs are increasingly used.

Introduction

Pigs belong to the Order Artiodactyla and the Family Suidae. They are even-toed ungulates (hooved animals). Both Asian and European domestic pig breeds diverged from the same Eurasian wild boar ancestor ~500 000 years ago and were domesticated ~9000 years ago (Giuffra et al. 2000). Asian pig genetics were introgressed into European pig breeds in the 18th and 19th centuries to create the modern European pig breeds from hybrid origins (Giuffra et al. 2000). The total number of pigs worldwide has increased from 566 million (1967/69) to 767 million (2019) and is projected to reach 1062 million by 2030. Per capita consumption of pork is dominated by China (41.0 kg pa), the European Union (40.8 kg pa), Montenegro (40.3 kg pa), Taiwan (39.8 kg pa), Serbia and South Korea (37 kg pa), Belarus (33.5 kg pa), the USA (29.2 kg pa), Vietnam (25.6 kg pa) and Norway (25.2 kg pa) (National Pork Board USA 2016). Production of pigs 100 years ago was combined with other enterprises, and involved grazing in free-range systems. In developed countries, production is now highly industrialised with highly productive genotypes, careful management of reproduction and close attention to the provision of diets to match requirements and production levels. Nutrition is a key driver in maximising the reproductive performance of sows and growth rate of weaned piglets. Optimising nitrogen and phosphorus nutrition is a main means of reducing odour and effluent pollution from piggeries. Nutrition also influences the quality of pork through pre-slaughter effects on post-slaughter biochemistry.

Much of our knowledge of pig nutrition has come from the use of pigs as a model for human nutrition. Studies using neonatal pigs have been instrumental in elucidating the role of colostrum in intestinal development, the role of growth factors in intestinal growth and the impact of protein/energy malnutrition on development of neonates. Impacts of maternal malnutrition during pregnancy on growth retardation and its sequelae have been studied in pigs and further work on the developmental origins of health and disease benefits from the use of pigs as models.

Digestive anatomy and physiology of pigs
The mouth and salivary glands

The mouth of the pig is long and varies with breed. At birth piglets have eight deciduous teeth, four incisors and four canines. Adult pigs have 44 teeth, which are complete by 18 months of age, with the following dental formula:

	Incisors	Canines	Premolars	Molars
Upper	3	1	4	3
Lower	3	1	4	3

The canines in boars continue to grow throughout life, becoming the fighting tusks.

The pig has three major salivary glands – the parotid glands, sublingual glands and the mandibular or submaxillary glands – plus several minor glands (labial, lingual, buccal and palatine). The parotid and mandibular glands have major ducts draining into the oral cavity. The minor glands have numerous small openings into the oral cavity. Saliva is a mixture of water containing electrolytes, mucus, enzymes (α-amylase, lysozyme) and immunoglobulin-A. Like other mammals, the role of saliva in pigs is to lubricate the food to allow smooth passage through the oesophagus, begin starch digestion to glucose, protect against disease (immunoglobulins) and to recycle electrolytes to the intestines. The volume and composition of saliva varies with the type of feed consumed. Dry feeds or a high feed intake induce a high volume of saliva secretion. As the rate of saliva production increases, the concentration of electrolytes, dry matter and amylase decreases (Kidder and Manners 1978). The control of saliva output and composition is through the autonomic nervous system, so parasympathetic stimulation ('rest and digest' arm of the autonomic nervous system) increases the rate of secretion. It is unlikely that salivary amylase plays a significant role in starch breakdown in the pig, relative to the extensive amylolytic activity of pancreatic amylase.

The gastrointestinal tract of the pig

In contrast to the horse, the digestive tract of pigs is directed towards gastric and small intestinal digestion rather than large intestinal fermentation. This is reflected in the relative capacities of the various segments (Table 13.1).

The gastric stomach of the pig

The pig has a simple, gastric stomach (Fig. 13.1), which is relatively much larger than that of other mono-gastrics. The oesophagus enters the stomach at the cardia. An unusual blind diverticulum occupies the fundic region on the dorsal aspect. It

Table 13.1. Absolute lengths, absolute capacities and relative capacities of the segments of the gastrointestinal tract of a 200 kg adult pig

Segment	Length (absolute) (metres)	Capacity (absolute) (litres)	Capacity (relative) (%)
Stomach		8	29
Small intestine	18.3	9.2	33
Caecum	0.2	1.55	6
Colon and rectum	5.0	8.70	32

Source: Hill (1970)

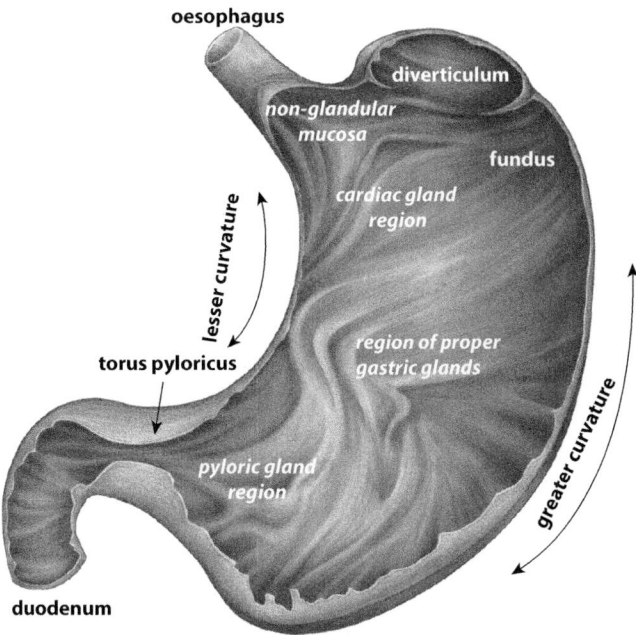

oesophagus

diverticulum

non-glandular mucosa

fundus

cardiac gland region

lesser curvature

torus pyloricus

region of proper gastric glands

greater curvature

pyloric gland region

duodenum

Fig. 13.1. The gastric stomach of the pig. Note the gastric diverticulum: an unusual 'outpocket' of the fundus. It is this sort of 'blind sac' development that led to the multi-chambered ruminant and camelid stomach. The major regions of the mucosa are the cardiac gland region, the region of proper gastric glands and the pyloric gland region.

appears as though the pig started down the evolutionary road towards a multi-chambered stomach and pre-gastric fermentation, but settled for a mono-gastric system. The body of the stomach contains regions with varying glandular epithelia. The cardiac gland region in pigs is much larger than that of other mono-gastrics. This region produces a large amounts of mucus (Kararli 1995). The gastric gland region and pyloric gland region contain parietal and chief cells, producing gastric acid and pepsin (Kararli 1995). The mucosal wall of the pig stomach is highly folded into rugae or folds. The pyloric antrum leads to the pylorus, which ends in a tight sphincter (pyloric sphincter or torus pyloricus) that controls gastric emptying into the duodenum (see Dyce *et al.* 2010).

The small and large intestines of pigs

The entire gastrointestinal tract of the pig is shown in Fig. 13.2.

The stomach empties into the duodenum through the pyloric sphincter. The small intestines comprise the duodenum, jejunum and ileum (16–21 m long in the

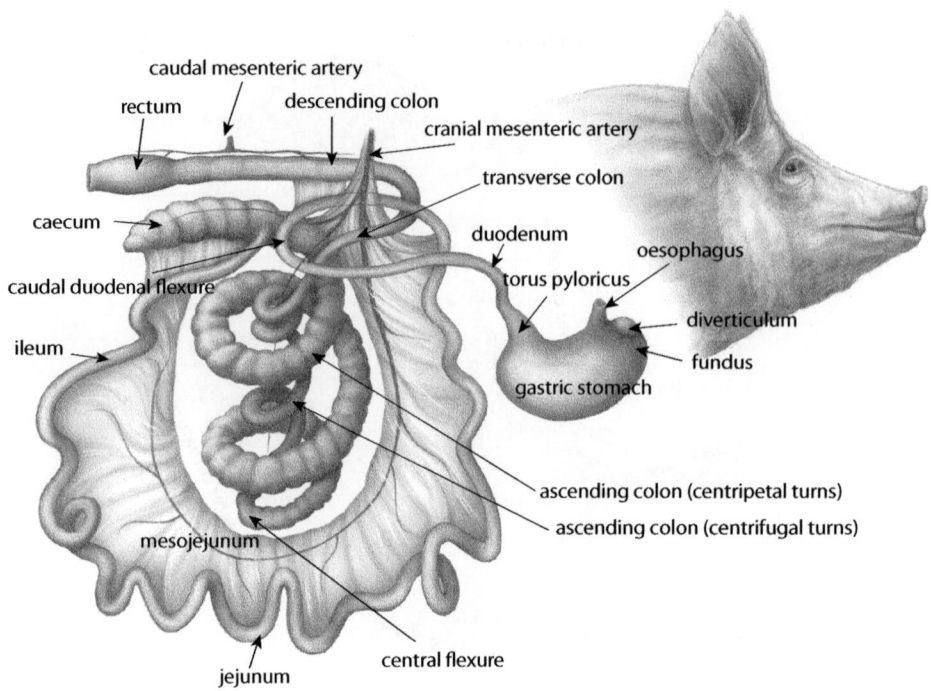

Fig. 13.2. The digestive tract of a pig. Note the relatively large gastric stomach and the cone-shaped spiral of the large intestine. The ascending colon spirals in a clockwise direction, reverses direction at the central flexure, then spirals counter-clockwise until the colic flexure. The ascending colon and the caecum are haustrated (gathered into folds by a longitudinal muscle).

adult pig) (Table 13.1). The large intestines comprise a caecum and a colon, which spirals in two directions in a cone shape. The caecum in pigs is a cylindrical blind sac with the blind apex facing caudo-ventrally. It has three taenia (rows of longitudinal muscle) and three corresponding rows of sacculations (haustrae).

The colon of pigs has the usual three parts (ascending, transverse and descending colon) but the ascending part is greatly elongated and coiled to form a cone-shape, with the base of the cone attached to the abdominal roof (Fig. 13.2). The ascending colon leaving the caecum forms centripetal clockwise turns (from above) to the apex of the cone ventrally. It then reverses turns at the central flexure and spirals counter clockwise tightly and inside the clockwise centripetal coils. The transverse colon then turns sharply and enters the descending colon, which is suspended by mesentery to the abdominal wall.

The small intestine is almost 80% of the length of the entire gut and has a capacity of almost 10 L. The caecum is relatively small in the pig compared with ruminants and equids. The relative lengths of the small intestine segments are 5% (duodenum), 90% (jejunum) and 5% (ileum). The wall of the intestines comprises four major layers: the mucosa, the submucosa, the muscularis and the serosa (Fig. 2.2). The mucosa comprises three sublayers: the muscularis mucosa, the lamina propria and the epithelium. The muscularis mucosa has two muscle layers (circular and longitudinal). The lamina propria contains blood vessels, lymphocytes, lymph nodules called Peyer's patches, and neurons bound together by connective tissues. The epithelium is a simple columnar epithelium containing absorptive cells, goblet cells (mucus) and entero-endocrine cells, which originate from stem cells proliferating in the base of the crypts between the villi, the finger-like projections into the lumen. The crypts of Lieberkühn contain the proliferating replacement cells for the epithelium and are also the main source of the secretions. As the cells progress up the villi, the microvilli on the luminal surface elongate and the secretion of digestive enzymes begins. The microvilli increase the surface area of the epithelium by ~40-fold. The folds of Kerckring, the villi and microvilli increase the total surface area for absorption and secretion by ~1000 times relative to a simple tube.

Feed intake regulation in pigs

Feed intake in pigs is important because growing pigs and lactating sows are usually given free access (*ad libitum*) to feed under the assumption that these animals will consume sufficient feed to maximise production. For newly weaned pigs, a decrease in feed intake at weaning reduces energy intake below the maintenance requirement producing a growth 'check' that is economically damaging. Non-lactating gilts or sows, on the other hand, are usually restricted-fed because high levels of feeding during gestation results in reduced intake during lactation, which will reduce milk yield and piglet growth. Understanding the factors that control feed intake is important because the total supply of nutrients to the pig depends on nutrient density multiplied by intake. Any factor reducing intake will reduce nutrient supply and therefore production. Feed intake is influenced by a large number of factors including animal genotype, live weight, age, parity, nutritional history, stage of lactation, temperature, diet energy density, diet protein content, nutrient deficiencies and the feeding system being used.

Genotype and voluntary intake

Genetic selection for increased feed conversion efficiency in pigs, and possibly selection for leanness, appears to indirectly select for reduced feed intake (Fowler *et al.* 1976). Close and Cole (2000) note a reduction of ~15–30% in voluntary Digestible Energy intake of growing pigs between 1963 and 1988. This may also be the case in the sows of the more-efficient genotypes. Breed differences in appetite, although confounded by other variables such as litter size, suggest that there is a strong genetic component to voluntary intake, which will translate to feed composition requirements.

Live weight, age and parity and voluntary intake

Live weight, age and parity are related, but heavier pigs have higher maintenance requirements and will consume more feed. Voluntary feed intake increases with parity up to Parity 6. Litter size has a large effect on intake because more piglets means greater sucking stimulus and greater milk yield. Milk yield is a major driver of voluntary intake. Sows suckling 16 piglets consume 15% more feed than those suckling four piglets (Close and Cole 2000). In general the intake increases by 0.2–0.4 kg/piglet/day.

Nutritional history and voluntary feed intake

Sows that are overfed during gestation will consume less feed during lactation. Overfed sows during pregnancy deposit fat, which is then used during lactation to support milk production in place of additional feed. Such sows lose more weight during lactation. This is important because genetic selection has already reduced their lactational intake (as discussed) and, although energy demands may be met, other essential nutrients may be too low in conventional diets to support lactation. Feed utilisation efficiency during lactation is higher than during pregnancy, so it makes economic sense to restrict feed during gestation and increase feed intake during lactation. The depressing effect of gestation feed intake on intake during lactation becomes more pronounced at Parity 2 than Parity 1. Body fatness is the signal that reduces intake during lactation. However, restricted feeding in late gestation may not be wise given the rapid growth of the fetuses over the last month. It is common practice to increase intake over the last 2–5 weeks of gestation and this does not reduce intake during lactation.

Stage of lactation and voluntary feed intake

Feed intake follows the pattern and level of milk production. Typically intakes range from 3–4 kg/day on day 1 and 5–6 kg/day on day 7. Voluntary intake reaches a maximum on about day 15 of lactation and thereafter declines. Typical DE intakes are 63 MJ/day on day 3 to 78 MJ/day on day 17.

Temperature and feed intake

The lower critical temperature (LCT) and upper critical temperature (UCT) of pigs are the lower and upper limits of the thermoneutral zone. Below the LCT

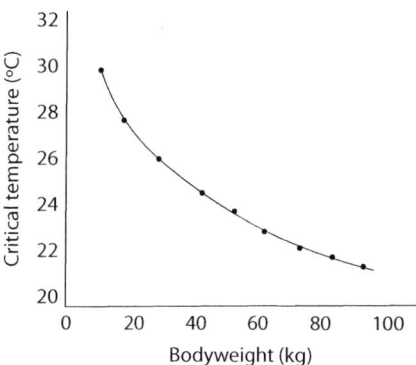

Fig. 13.3. Effect of bodyweight on upper critical temperature above which feed intake will be reduced. Large pigs require lower temperatures to maintain feed intake.

intake increases and nutrient requirements are met, but temperatures above the UCT cause a reduction in feed intake as the animal attempts to maintain homeostasis of core body temperature. The reduced nutrient intake is a problem because insufficient energy and other nutrients may be obtained. The UCT depends on the bodyweight of the pig, the relative humidity, airflow and environmental housing. The effect of bodyweight of the pig on the critical temperature at which daily feed intake is reduced is shown in Fig. 13.3. As pigs increase in weight, the critical temperature above which intake and growth rate will decline decreases.

Wetting or misting systems to reduce the negative impacts of heat are now used to reduce this problem. Increasing the fat content of diets can reduce the negative effects of reduced feed intake in hot environments because fat has a lower heat increment than carbohydrate or protein.

Energy density of the diet and voluntary feed intake

Being mono-gastrics, pigs tend to eat to meet their energy requirement up to the point at which gut fill prevents further consumption. In other words, as the fibre content of a feed is incrementally increased and its DE content decreased, the pig will consume incrementally more of the feed to maintain a relatively constant DE intake. While this holds true generally, in practice the compensating mechanisms to changes in energy density are incomplete. For example, increasing the DE content of feeds can result in an increase in total DE intake (the reduction in feed intake does not occur). Similarly, as DE density decreases, the increase in intake expected can be limited by gut fill, so DE intake declines. This relationship is shown in Fig. 13.4.

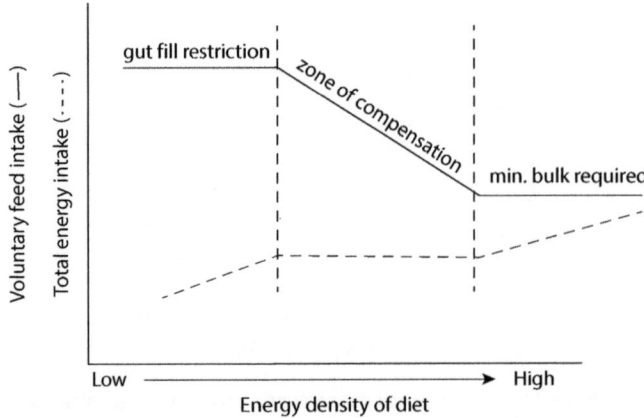

Fig. 13.4. Pigs 'eat for energy' and compensate when fed low-energy diets by eating more or, conversely, eating less when fed high-energy diets. However, there is a limit to compensation on very low-energy diets imposed by gut fill, and on very high-energy diets because a minimum amount of feed is required. Source: Adapted from data from Close and Cole (2000).

Weaning

Reductions in intake post weaning are common and the main cause of post weaning lag in growth. Adjustments to the nutrient density of the diet can compensate to some extent for this reduction. Typically a 21-day-old pig will grow at 250–300 g/day, but immediately after weaning the voluntary feed intake drops and is insufficient for maintenance.

Maximal voluntary feed intake in pigs can be achieved by following the simple steps outlined in Box 13.1.

Box 13.1. Practical tips to maximise feed intake in pigs

- Ensure the feed is fresh, palatable and contains no mycotoxins, odours, moulds or bitter flavours.
- Ensure the feed is nutritionally balanced to meet the needs of the particular class of pig.
- Feed several times/day or to appetite.
- Do not overfeed in pregnancy.
- Ensure fresh water is always available.
- Increase gut capacity to maximise lactation intake, by feeding high soluble fibre during pregnancy.
- Separate gestation and lactation diets are essential.
- Avoid temperatures above 20°C.
- Ensure adequate feeding space.

The energy and protein requirements of pigs

Maximising the efficiency of conversion of feed to pig meat is critical to the profitability of pig-producing enterprises, because feed costs represent 55–70% of the total variable costs in the business. The objective is to achieve the desired reproductive outputs, growth rates and carcase composition for the target market, at minimal feed cost. This is best achieved using feed formulation programs that account for the complex interactions between pig genotype, environment, management system, sow parity, stage of gestation, target live weight, body composition and the available feed ingredients. The nutritional requirements have been well established since 1944 by the National Research Council editions of *Nutrient Requirements of Swine*. The latest edition of this publication was 2012 (NRC 2012), which is the basis of the following

Table 13.2. Digestible Energy, crude protein and lysine requirements of selected groups of growing, gestating and lactating pig

Class bodyweight (kg)	Feed DE (MJ/kg)	Intake (g/day)[a]	Bodyweight gain (g/day)	CP (%)	Lysine (%)
Grower/finisher					
5–7	14.9	280	210	23.7	1.7
7–11	14.9	493	335	23.7	1.53
11–25	14.6	953	585	20.9	1.4
25–50	14.2	1582	758	18	1.12
50–75	14.2	2229	900	15.5	0.97
75–100	14.2	2636	917	13.2	0.84
100–135	14.2	2933	867	13.2	0.71
Gestation					
140 (0–90 days)	14.2	2130	578	11.9	0.61
140 (90–115 days)	14.2	2530	543	14.7	0.8
185 (0–90 days)	14.2	2210	472	9.6	0.45
185 (90–115 days)	14.2	2610	408	12	0.62
Lactation					
175 (Parity 1)	14.2	5950	0	16.6	0.86
210 (Parity >1)	14.2	6610	0	16.1	0.83
Dry sow/boar	**13**	**2500–3000**	**0**	**8.8**	**0.47**

[a]Includes wastage at 5%

Source: Adapted from NRC (2012)

requirement tables. Computer feed formulation software is used to estimate the precise requirements of specific classes of pigs, so these tables are only a general guide to the required levels of DE, CP and lysine.

Mineral and vitamin requirements of pigs

The mineral requirement of pigs at various physiological stages are similar to those of other animals and include the macrominerals (Ca, P, Mg, Cl, S, K, Na) and trace elements (Fe, Zn, Cu, Se, I, Mn, Cr). Sulphur requirements are considered to be met by the normal provision of the essential sulphur-containing amino acids and are therefore not included here. The roles of these essential minerals in metabolism are the same as those described previously (Table 1.11). Most changes to the requirements since the last edition (NRC 1998) have been to increase the daily intakes of minerals to account for the following:

- selection for rapid growth
- selection for rapid lean tissue gain and feed conversion efficiency, which has reduced voluntary feed intake thereby requiring higher dietary concentrations to achieve the same daily intake
- increasing use of unconventional feedstuffs, which has increased the probability of adverse mineral interactions and bioavailability
- increasing understanding of the factors affecting the bioavailability of minerals.

Table 13.3 shows the mineral requirements of growing pigs fed *ad libitum* a 90%DM diet (NRC 2012).

The mineral and vitamin requirements of breeding sows are given in Table 13.4.

Table 13.3. Mineral and fat-soluble vitamin requirements of growing pigs at different stages of growth. Typical ME densities of diet and body gain at each stage are indicated.

	5–7 kg	7–11 kg	11–25 kg	25–50 kg	50–75 kg	75–100 kg	100–135 kg
ME density (MJ/kg)	14.2	14.2	14.0	13.8			
Bodyweight gain (g/day)	210	335	585	758	900	917	867
Ca (%)	0.81	0.76	0.67	0.62	0.55	0.49	0.43
P (total)	0.66	0.61	0.57	0.53	0.49	0.45	0.41
P (available)[a]	0.39	0.34	0.27	0.25	0.21	0.2	0.17
Na (%)	0.4	0.35	0.28	0.1	0.1	0.1	0.1
Cl (%)	0.5	0.45	0.32	0.08	0.08	0.08	0.08
Mg (%)	0.04	0.04	0.04	0.04	0.04	0.04	0.04
K (%)	0.3	0.28	0.26	0.23	0.19	0.17	0.17

Table 13.3. (Continued)

Cu (mg/kg)	6	6	5	4	3.5	3	3
I (mg/kg)	0.14	0.14	0.14	0.14	0.14	0.14	0.14
Fe (mg/kg)	100	100	100	60	50	40	40
Mn (mg/kg)	4	4	3	2	2	2	2
Se (mg/kg)	0.3	0.3	0.25	0.2	0.15	0.15	0.15
Zn (mg/kg)	100	100	80	60	50	50	50
Vit A (IU/kg)	2200	2200	1750	1300	1300	1300	1300
Vit D (IU/kg)	220	220	200	150	150	150	150
Vit E (IU/kg)	16	16	11	11	11	11	11
Vit K (IU/kg)	0.5	0.5	0.5	0.5	0.5	0.5	0.5

[a]Apparent total tract digestibility
Source: NRC (2012)

Table 13.4. Mineral and vitamin requirements of gestating and lactating sows with typical ME density of diets and typical feed intakes (assumes diet is 90%DM and includes feed wastage)

Nutrient requirement	Gestation	Lactation
ME density (MJ/kg)	13.8	13.8
Feed intake (g/day)	2210	6280
Mineral and vitamin		
Ca (%)	1	1
P (total)	0.8	0.8
P (available)[a]	0.35	0.35
Na (%)	0.15	0.2
Cl (%)	0.12	0.16
Mg (%)	0.06	0.06
K (%)	0.2	0.2
Cu (mg/kg)	10	20
I (mg/kg)	0.14	0.14
Fe (mg/kg)	80	80
Mn (mg/kg)	25	25
Se (mg/kg)	0.15	0.15
Zn (mg/kg)	100	100
Vit A (IU/kg)	4000	2000
Vit D (IU/kg)	800	800
Vit E (IU/kg)	44	44
Vit K (IU/kg)	0.5	0.5

[a]Apparent total tract digestibility
Source: NRC (2012)

Feeding programs for different classes of pigs
Feeding the young pig

It is essential that piglets grow rapidly during the suckling and weaning phases because at slow growth rates the relative amount of energy going to maintenance is substantially higher than at high growth rates. For example, the DE required for maintenance of a 5 kg piglet is 1.6 MJ/day. If the piglet is growing at 250 g/day (requiring an additional 4.4 MJ/day), then maintenance makes up 27% of the daily energy intake. If that piglet is only growing at 125 g/day the maintenance cost is now 42% of the total energy intake. The slower growing piglet will also take twice as long to meet the market weight, so the total DE required to reach market weight is greatly increased. Given feed costs make up ~60% of total variable costs, slow growth is economically damaging.

The two main growth phases of young pigs is the suckling phase and the weaning phase. Figure 13.5 shows the typical growth paths of young piglets during the suckling and the weaning phases.

The increase in growth rate during the early suckling period reflects the establishment of lactation milk yield in the sow and formation of nursing order. Teats at different position along the body vary in milk yield and solids contents, so piglets adopting superior teats (cranial and middle) are heavier at weaning than those on the caudal teats (at the back of the sow). The adoption of superior teats

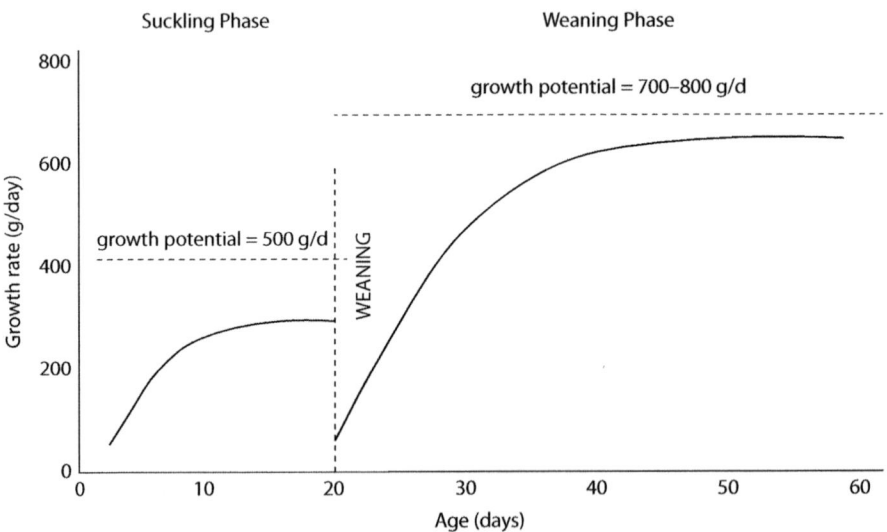

Fig. 13.5. Growth paths of young pigs during the suckling and post weaning phases. Note the growth interruption at weaning and the time to return to genetic potential post-weaning. Data derived from 265 piglets from 23 litters of a high-growth genotype. Source: Based on data from Dividich and Le Seve (2001).

may itself reflect the initial birthweight of the piglet, because heavier piglets will outcompete for the better teats.

Transition to solid feed in young pigs

In feral pigs the transition of young pigs to solid feeds takes place over many weeks, which allows the piglet to gradual reduce milk intake and to replace this with solids, which it gradually learns are nutritionally acceptable. Modern piggeries abruptly wean piglets as early as 10 days, but more commonly at 21 days of age. In addition to not having a fully developed gut (small intestine particularly) to handle solid feed, there are behavioural and immunological challenges all of which result in reduced nutrient intake and depressed growth (Fig. 13.6). Pigs that fail to adapt to the new regime early in the post weaning period do not thrive and are more susceptible to subsequent disease. Several strategies are used to reduce this post-weaning growth check including liquid feeding, gruel feeding and introducing creep feed pre-weaning (Mavromichalis 2006).

Creep feeding and maximising feed intake post weaning

Creep feeding is the practice of offering the piglet access to a high-quality, nutrient-dense, highly palatable solid feed while it is still sucking milk from the sow. The objective is to speed up the development of the digestive capability (enzyme activity, gut morphology) of the piglet so at weaning the animal is able to transition to the solid fed with a smaller growth check. The evidence for creep feeding making a difference to weaning weights is not strong. Certainly in young pigs the intake of creep feed is insignificant but then suddenly rises at about day 25. In some piglets, however, their intake of the creep feed remains very low even after day 28. Creep feeding only works if the weaning age is sufficiently late that the intake of solid feed at weaning is ~500 g/day (Pluske et al. 1995). Early-weaned piglets are unlikely to achieve this rate of solids intake and should not be fed creep feeds because they can develop hypersensitivity reactions to the feed. If creep feed is used, it should be made from high-quality ingredients containing cooked cereals, milk proteins, lactose, fishmeal and animal plasma proteins (if allowed). The creep diet should be palatable, readily digestible and introduced at 7–10 days of age. The creep feed should be offered in small amounts initially on flat trays close to the piglets but inaccessible to the sow. Pellets and blends of pellets with mash can be very effective creep feeds. Access to fresh water is critical because the young piglet has yet to learn how to regulate water intake after a period of milk feeding.

It is critical that newly weaned pigs are kept at temperatures close to their lower critical temperature to ensure they do not use energy to keep warm but also do not suffer intake depression due to heat. Provision of a hygienic environment is critical to avoid disease. Provision of continuous lighting increases intake because piglets stay in groups in the dark (Bruininx *et al.* 2001). It is important that the weaning diet contains easily digestible components such as cooked cereals, caseins, fishmeal, lactose and sucrose. Provision of appetisers, additives (such as immunoglobulins, antimicrobial agents, organic acids, zinc oxide, copper sulphate) or provision of a gruel (made of feed and water or milk replacer mixed in bowl feeders) can greatly improve intake and growth performance of newly weaned piglets (van Dijk *et al.* 2001). Although these nutrient-rich diets containing complex components (plasma protein and lactose) are expensive, they are only fed for a short period of transition and therefore contribute relatively little to the total feed costs.

Weaner feeding

Weaners should be fed diets *ad libitum* until 20–25 kg live weight. The rations should be high in DE (14.5–14.9 MJ/kg), because the young piglets have a limited intake capacity at this stage. The ration should also be high in crude protein and lysine (Table 13.2). Feed must be kept fresh and troughs checked regularly for soiling. Water quality is critical to the early growth of piglets. Water for piglets should be free of microbial contamination and low in total dissolved solids (<1000 ppm).

Grower/finisher feeding

Growing pigs should be fed *ad libitum* on a high-energy (14.5–15 MJ DE/kg), 20–24% CP diet until ~45–60 kg, to maximise the growth of lean tissue. Dividing the growth phase into several periods on a bodyweight basis allows provision of diets more aligned to the animals' nutrient requirements (e.g. three or four grower/finisher phases). This is particularly useful in reducing the costly protein component of rations because the protein requirement declines with bodyweight (Table 13.2). It also reduces the total excretion of nitrogen and phosphorus in effluent. Split-sex feeding also allows more targeted feeding because gilts require more protein than barrows.

Dry sows

Dry sows require ~2.5–3.0 kg of grain-based feed depending on their bodyweight (Table 13.2). Sows overfed during pregnancy have a reduced feed intake during lactation, which is detrimental to piglet growth. They should be fed to maintain a good level of body condition but not to over fatness.

Feeding systems for gilts and sows

Feeding strategies for the breeding herd have changed markedly over the past few decades. Twenty years ago it was assumed that gilts would accumulate sufficient fat reserves before farrowing to maintain her through pregnancy and lactation. Gilts

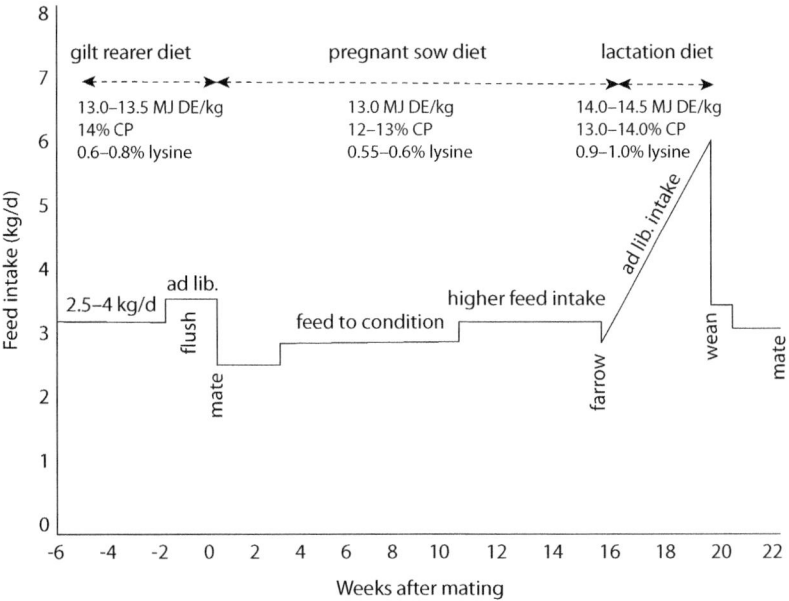

Fig. 13.6. A typical feeding program for first-parity gilts. Source: Adapted from Close and Cole (2000).

were often mated at 36–38 weeks weighing 125–130 kg. Modern piggeries mate gilts at ~28–30 weeks when gilts weigh ~110–115 kg. Selection for lean growth has also reduced the fat reserves in gilts and possibly their voluntary feed intake, all of which makes carrying the pregnancy and then lactation difficult. Piglets are now weaned at 3–5 weeks of age and the sows are expected to rebreed during peak lactation when the energy deficit is highest. She cannot consume sufficient food during peak lactation to maintain weight. The longer the lactation, the greater the chance she can regain lost weight, so shorter lactations reduce this opportunity. This is a particular problem for young sows (King and Williams 1984). It is customary to feed gilts *ad libitum* for 14 days before service to maximise ovulation rate (Fig. 13.6).

Most feeding strategies for pregnant and lactating sows are based on controlled and limited weight gain during pregnancy and then maintaining body condition during lactation by allowing a high level of feeding during suckling (Fig. 13.6).

Typically during pregnancy a diet containing 13.0 MJ DE/kg and 12–13% CP with a moderate level of lysine is fed. The requirement for other amino acids is assumed to be the same as for 'ideal' protein. For lactating sows, the protein requirement depends on milk yield. Typically a lactating sow diet contains 14.0–14.5 MJ/kg DE and 13–14% CP with a high level of lysine. Intakes are of the order of 5–6 kg of a high-grain diet. First parity sows may require a higher density of DE and protein to account for their lower appetite than older sows.

A typical feeding strategy for a first parity gilt in an ideal environment would involve the phases indicated in Fig. 13.6.

Feeding systems for free-range systems

In some countries there is a trend towards production of free-range swine products in response to consumer demands for less-intensive production systems on the grounds of perceived animal welfare benefits and possible improvements in product quality. From a nutritional viewpoint, access to pastures and forage plants is problematic because pigs are poor at digesting fibre, and pastures change markedly in nutritive value throughout the seasons. Pastures are inadequate for young growing pigs, lactating sows, pregnant sows and even dry sows. Even high-quality pastures are inadequate for pigs. Even when given access to balanced grain-based rations, free-range pigs eat less, grow slower and have reduced dressing percentages due to increased gut volume (Hoffman *et al.* 2003).

Feed additives to stimulate gut health in young pigs
Exogenous feed enzymes to increase the feeding value of pig diets

Feed enzymes are worth more than US$650 million per year and save the global feed market US$3–5 billion per year (Kiarie *et al.* 2013). Phytases make up 60% of the market, carbohydrases 30% and proteases and lipases the remaining 10%. They have been widely adopted because the diets fed to pigs in commercial settings are relatively narrow and based around a high-cereal grain component. The evolution of wild pigs produced a digestive system that allowed the pig to grow, reproduce, produce milk and protect their young. Two major changes have occurred since this evolution: (1) the growth potential of the pig has been greatly increased; and (2) the diet of the pig has changed greatly from a highly variable diet with no ingredient making up more than ~20–30% of the total, to diets based largely on one or two grains and a protein source or two. Most pig diets are either maize, barley or wheat based, with a protein source such as soybean meal. Each of these ingredients contains antinutritive factors (ANFs), which can reduce performance. Indeed ANFs can be the factor limiting the inclusion of otherwise nutritionally valuable components, so any means of reducing ANF levels as they pass through the digestive tract can be economically valuable.

The major non-starch polysaccharides (NSPs) in maize (cellulose, arabinoxylans), wheat (arabinoxylans, xyloglucan), barley (β-glucans), oats (β-glucans), soybean meal (galacturonans, arabinans, galactomannan), rapeseed meal (arabino-β-1,4-galactan) and sorghum (arabinoxylans, cellulose, hemicelluloses) are susceptible to degradation by carbohydrases such as xylanase, β-glucananase, cellulase, α-amylase, β-mannanase, α-galactasidase and pectinase. By degrading these NSPs, which can comprise up to 20% of the grain or meal, the digestibility of the feed is increased and the viscosity of the digesta is reduced, thereby increasing the rate of passage. Other effects include potential prebiotic effects of the oligosaccharides released by NSP digestion. This may be beneficial to the gut microbiota and gut health (Kiarie *et al.* 2013). Typical modes of action of feed enzymes are summarised in Box 13.2.

Box 13.2. Modes of action of feed enzymes on mono-gastric digestion and health

- hydrolysis of chemical bonds in feeds that are not degraded by the animal's enzymes
- increased availability of starches, amino acids and minerals by breaking nutrient-encapsulating cell wall polysaccharides
- breakdown of anti-nutritional factors such as phytic acid, releasing phosphorus
- solubilisation of insoluble non-starch polysaccharides, which are then available through hindgut fermentation
- provision of added levels of endogenous enzymes such as amylase, protease and lipase to supplement the young animal's own enzymes
- improved gut microbes by stimulating oligosaccharide release from non-starch polysaccharides.

Phytic acid (myo-inositol 1,2,3,4,5,6-hexakis phosphate) is the storage form of phosphorus in cereal grains and oilseeds. By degrading the phytate, the P is released for utilisation by the animal. Phosphorus is the third most expensive ingredient in pig diets so increasing the availability of P using phytase is cost effective. Phytic acid also exerts antinutritive effects on the ileal digestibility of proteins and it increases the production of mucins. The reduced mucin production may beneficially influence the gut microbiota (Kiarie *et al.* 2013).

Many feed additives are aimed at promoting better gut health as follows:

- enhancing the pig's immune response with immunoglobulins, omega-3 fatty acids, yeast-derived β-glucans
- reducing the pathogen load in the gut with organic and inorganic acids, high levels of zinc oxide, essential oils, herbs and spices, prebiotics, bacteriophages and antimicrobial peptides
- stimulating the establishment of beneficial gut microbes with pro- and prebiotics
- stimulating digestive function with various feed additives like organic acids (butyric, gluconic, lactic), amino acids (glutamine, threonine, cysteine) and nucleotides.

Table 13.5. Common feed additives used in pig nutrition

Additive	Effect	Stage
Antibiotic growth promotants	Daily gain Feed/gain	Starter Grower/finisher
Probiotics	Daily gain Feed/gain	Starter Grower/finisher
Copper sulphate	Daily gain Feed/gain	Starter
Zinc oxide	Daily gain Feed/gain	Starter
Yucca plant extract	Daily gain Feed/gain	Grower/finisher
Mycotoxin binders	Daily gain	Grower/finisher
Acidifiers	Daily gain Feed/gain	Starter
Phytase	Daily gain Feed/gain	Grower/finisher
Carnitine	Fat utilisation Daily gain Feed/gain Litter size Birthweight	Starter Grower/finisher Gestation
Chromium (organic)	Lean gain Litter size	Grower/finisher Gestation
Conjugated linoleic acid	Lean gain Feed/gain	Grower/finisher
Ractopamine	Daily gain Feed/gain Lean	Finisher

Table 13.5 lists some of the most common feed additives used in pig production.

Some of these additives such as ractopamine, which is a β-adrenergic agonist, increases protein deposition but also reduces feed intake. This means a higher protein content of the ration is required to meet the increased demand and decreased supply of amino acids.

Post-weaning diarrhoea in early-weaned pigs

Post-weaning diarrhoea (PWD) is a major source of economic loss and welfare compromise in piggeries. In well-run piggeries, mortality associated with PWD is typically 0.5% but in poorly managed piggeries mortality can be as high as 7%. Obviously microbial pathogens are responsible for severe diarrhoea, but it is increasingly recognised that nutritional factors play a major role in the pathogenesis (Sissons 1993). As antibiotics are being removed from pig feeds, it is becoming increasingly important to identify the role of non-microbial factors in the development of, and treatment of, PWD. These include the sudden withdrawal of whole milk, poor maturation of digestive function, the presence of antinutritive factors in post-weaning diets, and the role of dietary antigens in creating a hypersensitivity reaction in the piglet.

The early-weaned piglet undergoes profound changes in intestinal morphology (Hampson 1986), epithelial function (Hampson and Kidder 1986) microbial colonisation of the intestines (Katouli et al. 1999) and immunity (Blecha et al. 1983). Some relate to the abrupt change in feed from sow's milk to solids, which results in a sudden reduction in nutrient intake. Young pigs are well suited to milk as a feed source and its removal, particularly in very young pigs (less than 4 weeks of age), creates problems. Milk also contains several protective components such as antibodies and binding proteins that prevent attachment of pathogenic organisms to enterocyte receptors. Lactobacilli in milk competitively exclude pathogens (Deprez et al. 1986), so removal of milk creates a sudden and dramatic change in the intestinal microbiota. These are exacerbated further by the removal of in-feed antibiotics in pig enterprises globally and the scarcity of skim milk as a feed additive (Sissons 1993). Replacement of skim milk with alternative proteins such as soybeans, canola (rapeseed), peas and beans is problematic because these contain antinutritive factors such as enzyme inhibitors, lectins and dietary antigens, which are implicated in the development of diarrhoea. It appears that local intestinal hypersensitivity to dietary antigens may play a very significant role in the pathogenesis of PWD (Miller et al. 1984; Newby et al. 1985). The presence of dietary antigens in creep feed, for example, appears to play a role in sensitising the newly weaned piglet to those antigens (Miller et al. 1984).

There are several dietary approaches to control PWD in piglets. Restricting the intake of feed during the early weaning period to match the delayed immature digestive function will ensure the gut is not overwhelmed. Adding fibre to the diet also appears to reduce PWD due to reduced nutrient density and reduced intake to match digestive ability. Adding fats to the weaner diet may reduce PWD by reducing the rate of gastric emptying, thereby slowing the supply of soluble carbohydrate to the intestines, which would create a hyperosmotic condition (Sissons 1993). Processing feeds (heat, alcohol or pre-digestion) to reduce the antigenicity of proteins may also reduce the incidence of PWD. The potential for using probiotics and prebiotics to reduce pathogen colonisation has also been successful (Giang et al. 2010). Alternatives to antibiotics to prevent PWD include probiotics (various microbes such as *Lactobacillus, Bifidobacterium, Streptococcus* and *Bacillus*), oligosaccharides, short-chain fatty acids, zinc compounds, plants extracts and natural clays. Many of these show promise as means of reducing PWD in early-weaned piglets (Vondruskova *et al.* 2010).

References

Blecha F, Pollmann DS, Nichols DA (1983) Weaning pigs at an early age decreases cellular immunity. *Journal of Animal Science* **56**, 396–400. doi:10.2527/jas1983.562396x

Bruininx EMAM, van der Peet-Schwering CMC, Schrama JW, Vereijken PFG, Vesseur PC, Everts H, *et al.* (2001) Individual feed intake characteristics and growth performance of group-housed weanling pigs: effects of sex, initial body weight, and body weight distribution within groups. *Journal of Animal Science* **79**, 301–308.

Close WH, Cole DJA (2000) *Nutrition of Boars and Sows*. Nottingham University Press, Nottingham, UK.

Deprez P, Hende C, Van D, Muyulle E, Oyaert W (1986) The influence of the administration of sow's milk on the postweaning excretion of hemolytic *E. coli* in the pig. *Veterinary Research Communications* **10**, 469–478. doi:10.1007/BF02214010

Dividich J, Le Seve B (2001) Energy requirements of the young pig. In *The Weaner Pig* (Eds MA Varley and J Wiseman) pp. 17–44. CABI Publishing, Wallingford, UK.

Dyce KM, Sack WO, Wensing CJG (2010) *Textbook of Veterinary Anatomy*. 4th edn. Saunders Elsevier, Grand Rapids, MI, USA.

Fowler VR, Bichard M, Pease A (1976) Objectives in pig breeding. *Animal Production* **23**, 365–387. doi:10.1017/S0003356100031482

Giang HH, Viet TQ, Ogle B, Lindberg JE (2010) Growth performance, digestibility, gut environment and health status in weaned piglets fed a diet supplemented with potentially probiotic complexes of lactic acid bacteria. *Livestock Science* **129**, 95–103. doi:10.1016/j.livsci.2010.01.010

Giuffra E, Kijas JMH, Amarger V, Carlborg O, Jeon J-T, Andersson L (2000) The origin of the domestic pig: independent domestication and subsequent introgression. *Genetics* **154**, 1785–1791.

Hampson DJ (1986) Alterations in piglet small intestinal structure at weaning. *Research in Veterinary Science* **40**, 32–40. doi:10.1016/S0034-5288(18)30482-X

Hampson DJ, Kidder DE (1986) Influence of creep feeding and weaning on brush border enzyme activities in the piglet small intestine. *Research in Veterinary Science* **40**, 24–31. doi:10.1016/S0034-5288(18)30481-8

Hill KJ (1970) Developmental and comparative aspects of digestion. In *Duke's Physiology of Domestic Animals*. 8th edn. (Ed. MJ Swenson). pp. 351–362. Cornell Univiversity Press, Ithaca, NY, USA.

Hoffman LC, Styger E, Muller M, Brand TS (2003) The growth and carcass and meat characteristics of pigs raised in a free-range or conventional housing system. *South African Journal of Animal Science* **33**, 166–175.

Kararli TT (1995) Comparison of the gastrointestinal anatomy, physiology and biochemistry of humans and commonly used laboratory animals. *Biopharmaceutics & Drug Disposition* **16**, 351–380. doi:10.1002/bdd.2510160502

Katouli M, Melin L, Jensen-Waern M, Wallgren P, Mollby R (1999) The effect of zinc oxide supplementation on the stability of the intestinal flora withy special reference to composition of coliforms in weaned pigs. *Journal of Applied Microbiology* **87**, 564–573. doi:10.1046/j.1365-2672.1999.00853.x

Kiarie E, Romero LF, Nyachoti M (2013) The role of added feed enzymes in promoting gut health in swine and poultry. *Nutrition Research Reviews* **26**, 71–88. doi:10.1017/S0954422413000048

Kidder DE, Manners MJ (1978) *Digestion in the Pig.* Scientechnica, Bristol, UK.

King RH, Williams IH (1984) The effect of nutrition on the reproductive performance of first-litter sows 1. Feeding level during lactation, and between weaning and mating. *Animal Production* **38**, 241–247. doi:10.1017/S0003356100002233

Mavromichalis I (2006) *Applied Nutrition for Young Pigs.* CABI, Wallingford, UK.

Miller BG, Newby TJ, Stokes CR, Hampson DJ, Brown PJ, Bourne FJ (1984) The importance of dietary antigen in the cause of post weaning diarrhoea in pigs. *American Journal of Veterinary Research* **45**, 1730–1733.

National Pork Board USA (2016) *PorkCheckOff.* National Pork Board (USA), Des Moines, IA, USA, <http://www.pork.org/pork-quick-facts/home/stats/u-s-pork-exports/top-10-pork-producing-countries/>.

Newby TJ, Miller B, Stokes CR, Hampson D, Bourne FJ (1985) Local hypersensitivity response to dietary antigens in early weaned pigs. In *Recent Developments in Pig Nutrition.* (Eds DJA Cole and W Haresign) pp. 211–221. Butterworths, London, UK.

NRC (1998) *Nutrient Requirements of Swine.* 10th revised edn. National Research Council, The National Academies Press, Washington, DC, USA.

NRC (2012) *Nutrient Requirements of Swine.* National Research Council, The National Academies Press, Washington, DC, USA.

Pluske JR, Williams IH, Aheren FX (1995) Nutrition of the neonatal pig. In *The Neonatal Pig: Development and Survival.* (Ed MA Varley) pp. 187–235. CABI, Wallingford, UK.

Sissons JW (1993) Aetiology of diarrhoea. In *Recent Developments in Pig Nutrition 2* (Eds DJA Cole, W Haresign and PC Garnsworthy) pp. 267–284. Nottingham University Press, Nottingham, UK.

van Dijk AJ, Everts H, Nabuurs MJA, Margry RJCF, Beynen AC (2001) Growth performance of weanling pigs fed spray-dried animal plasma: a review. *Livestock Production Science* **68**, 263–274. doi:10.1016/S0301-6226(00)00229-3

Vondruskova H, Slamova R, Trckova M, Zraly Z, Pavlik I (2010) Alternatives to antibiotic growth promoters in prevention of diarrhoea in weaned piglets: a review. *Veterinarni Medicina* **55**, 199–224. doi:10.17221/2998-VETMED

14

Poultry nutrition

Key points

- The avian digestive tract comprises the oesophagus, crop, proventriculus, gizzard, duodenal loop, jejunum, ileum, paired caeca and vent.
- The modern meat-producing bird grows from ~55 g as a day-old chick to 3.0 kg in 5 weeks at a feed conversion ratio (feed/gain) of <1.5.
- Modern layers produce up to 300 eggs/year weighing 55 g each, equating to >16 kg eggs/year or 10 times her bodyweight.
- The nutrient requirements of meat chickens are for high Metabolisable Energy (>13 MJ/kg DM), high crude protein (18–23% CP), highly digestible protein with a balance of the essential amino acids relative to the first-limiting amino acid, lysine.
- Layer birds require high-energy (11.5–12.5 MJ/kg DM), high-protein (16–17%) and a high level of available calcium to available (non-phytate) phosphorus.
- The use of free-range production systems for layers and broiler birds is raising new nutritional issues, including the impact of intake and nutritive value of forages/invertebrates/soil on nutrient balances, and health issues such as gizzard impaction. Seasonality of pasture and forage production and quality in free-range production systems will present an ongoing challenge to nutritionists and producers.

Introduction

The global chicken meat industry is expanding faster than any other meat sector globally and will continue to do so until 2025 (FAO 2018). Most of this growth is in the developing countries where urbanisation and an increase in per capita income is driving the expansion. Global production of chicken meat in 2018 was 123 million tonnes, an increase from 70 million tonnes in 2001 (FAO 2018). By 2025 it is estimated that global chicken meat production will reach 131 million tonnes. The main reason for this global growth in production and consumption is the widespread cultural acceptance of chicken meat, its adaptation to a wide variety of cooking methods and its relative low cost in comparison with other meats. The last has been driven by rapid gains in growth and growth efficiency due to the application of modern genetics and matching the nutrient requirements of the birds very closely to their metabolic requirements. This is reflected in feed conversion ratios (feed consumed/weight gained) now approaching 1.5:1 and continuing on a downward trajectory, which shows no signs of abating. Egg production likewise is growing rapidly and by 2015 about 70 million tonnes of eggs were produced worldwide, equating to some 1338 billion eggs. Nutrition is the key driver of profitability of chicken meat and egg production enterprises but there are some significant challenges facing the future poultry industries including:

1. The trend towards **free-range production systems** in developed countries poses challenges in developing pasture-based systems that do not compromise efficiency or bird health.
2. Finding alternatives to current expensive sources of energy and protein without compromising efficiency.
3. Health issues including ascites, tibial dyschondroplasia and sudden death syndrome.
4. Finding **alternatives to antibiotics** to maintain bird health.

These issues are dealt with in 'Common nutritional and metabolic diseases of poultry' later.

Digestive anatomy and physiology of chickens

The chicken is a mono-gastric animal with a simple but efficient digestive system heavily reliant on intestinal digestion of high-quality feeds. The digestive tract comprises the oesophagus (pre-crop and post-crop), the crop, the proventriculus, the gizzard, the small intestine (duodenum, jejunum and ileum) and the large intestine (paired caeca, colon and cloaca) (Fig. 14.1).

Mouth and salivary glands of the chicken

The beak is a keratinised structure overlying the mandibles and incisive bones. There are no teeth but the hard palate has five transverse rows of large horny

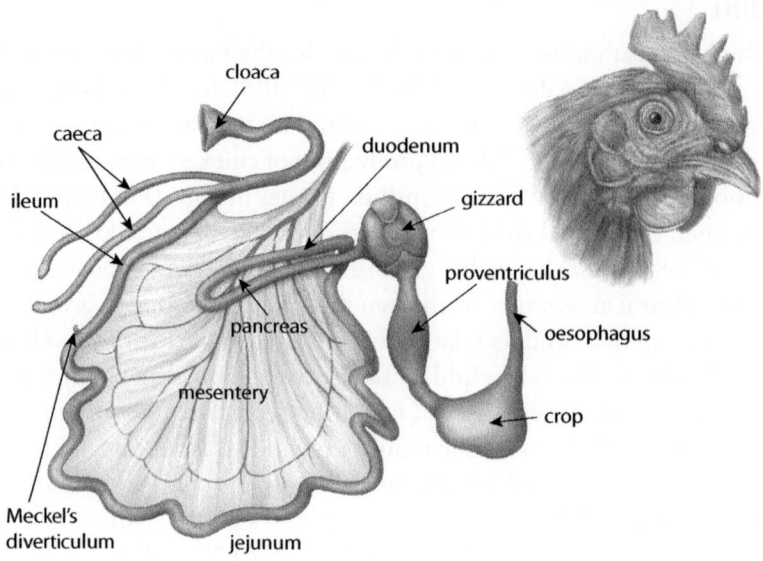

Fig. 14.1. The digestive tract of a chicken. The crop is a dilation of the oesophagus used for storing feed. The proventriculus begins acid/pepsin digestion. The gizzard is a hard, muscular organ for grinding feed. The pancreas lies in the loop of the duodenum. The jejunum and ileum are delineated by Meckel's diverticulum. The caecum in birds is paired either side of the ileum.

papillae that face backwards. The salivary glands (maxillary, palatine, phenopteryoid, submandibular and lingual glands) have ducts that pierce the hard palate. The tongue of birds is pointed to fit the beak shape and the epithelium is thick and horny.

Crop

The pre-crop oesophagus leads to a dilation of the oesophagus known as the crop, which is used in grain-eating birds to store food. This storage function allows birds to eat 'meals' but continually digest food. The epithelium of the oesophagus has a mucus membrane, which lubricates the passage of food. There may be some fermentation occurring in the crop but it is likely to be a minor contributor to total energy production. Impaction of the crop can occur if chickens are fasted and then allowed access to fibrous feeds, or in free-range chickens on fibrous pastures (see 'Nutritional issues in free-range poultry systems' later). The post-crop oesophagus leads to the proventriculus.

Proventriculus

The proventriculus, or glandular stomach, is a dilation of the gut and is theoretically the site of the beginning of protein digestion because it contains pepsin and hydrochloric acid. However, the proventriculus pH is higher than the optimum for pepsin activity so it is unlikely that much proteolysis occurs here.

Gizzard (ventriculus)

The thick, muscular gizzard is a rounded organ covered on the serosal side by connective tissue that is thicker at the centre than at the edges. The circular and longitudinal smooth muscles are thick and produce strong grinding motions during the frequent contractions. The inner surface is covered by a thick, horny layer of tissue that is raised in ridges. Grit is often found in the gizzard, where it acts as a grinding agent akin to teeth in mammals. The acidic chyme from the gizzard enters the duodenum, where bicarbonate from the pancreas neutralises the acid.

Small intestines

The entrance to the gizzard from the proventriculus and the exit to the duodenum are in close proximity. The digesta enters the duodenal loop, the centre of which contains the pancreas. Exocrine pancreatic secretions include proteases, lipases and polysaccharidases, as well as bicarbonate, which neutralises the pH of the chyme. The main body of the small intestine is the jejunum, which progresses from the end of the duodenal loop to the diverticulum of Meckel, the vestigial remnant of the yolk stalk. The ileum then progresses to the beginning of the large intestine, which bifurcates into paired caeca running retrograde to the ileum. The large intestine then ends at the cloaca. Small intestinal activity breaks down proteins and peptides into amino acids, polysaccharides into simple sugars and lipids into fatty acids. Minerals and vitamins are also absorbed here and, with the other products of digestion, enter the portal vein to the liver. Fatty acids and fat-soluble vitamins enter the lacteals in the villi and progress to the thoracic duct at the heart.

Large intestines

The two blind caeca are ~16–18 cm in length and they line either side of the ileum to which they are attached by mesentery. Some fermentation occurs in the caeca and large intestines, but the contribution to total energy supply is limited (Moran 1982). The colon in poultry is short and ends at the cloaca. The urodeum is an expanded area in the caudal large intestine, which contains the distal openings of the ureters from the kidneys. Uric acid is deposited here with the faeces. Retrograde movement of urine from the urodeum and other colonic contents into the caecum presumably supplies the microbes of the large intestine with a source of nitrogen and effectively increases the exposure time of the digesta to the intestinal epithelium without requiring a long large intestine.

Energy, amino acid, vitamin and mineral requirements of meat birds

Metabolisable Energy requirements and target bodyweights for meat chickens

Broiler, or meat chickens have been selected for rapid and efficient growth and are usually fed *ad libitum* to maximise growth rates. Varying the Metabolisable Energy content of broiler rations does not impact growth rate as much as one might expect because chickens are capable of changing their feed intake to maintain total energy available. Broiler nutrition is divided into three growth phases (starter, grower and finisher) because the growth characteristics, and therefore the nutrient requirements, change sufficiently over these short periods of time to make changing the rations worthwhile economically.

The currently accepted nutrient requirements of broiler chickens are given in Table 14.1 and target bodyweights and weekly intake rates for Cobb500 broilers over the first 9 weeks of life are shown in Table 14.2.

Table 14.1. Nutrient requirements of meat chickens (/kg feed at 90% dry matter)

Nutrient	Unit	0–3 weeks	3–6 weeks	6–8 weeks
Metabolisable Energy	MJ/kg DM	12.6	13.3	13.5
PROTEIN	%	22–25	21–23	19–21
Arg	%	1.25	1.1	1
Gly + Ser	%	1.25	1.14	0.97
His	%	0.35	0.32	0.27
Isoleu	%	0.8	0.73	0.62
Leu	%	1.2	1.09	0.93
Lys	%	1.1	1	0.85
Met	%	0.5	0.38	0.32
Met + Cys	%	0.9	0.72	0.6
Phe	%	0.72	0.65	0.56
Phe + Tyr	%	1.34	1.22	1.04
Pro	%	0.6	0.55	0.46
Thre	%	0.8	0.74	0.68
Try	%	0.2	0.18	0.16
Val	%	0.9	0.82	0.7
FATS				
Linoleic acid	%	1	1	1
MACROMINERALS				
Ca	%	1	0.9	0.8

Table 14.1. (Continued)

Nutrient	Unit	0–3 weeks	3–6 weeks	6–8 weeks
Cl	%	0.2	0.15	0.12
Mg	mg	600	600	600
Non-phytate P	%	0.45	0.35	0.3
K	%	0.3	0.3	0.3
Na	%	0.2	0.15	0.12
TRACE MINERALS				
Cu	mg	8	8	8
I	mg	0.35	0.35	0.35
Fe	mg	80	80	80
Mn	mg	60	60	60
Se	mg	0.15	0.15	0.15
Zn	mg	40	40	40
FAT-SOLUBLE VITAMINS				
A	IU	1500	1500	1500
D	ICU	200	200	200
E	IU	10	10	10
K	mg	0.5	0.5	0.5
WATER-SOLUBLE VITAMINS				
Cobalamin	mg	0.01	0.01	0.007
Biotin	mg	0.15	0.15	0.12
Choline	mg	1300	1000	750
Folacin	mg	0.55	0.55	0.502
Niacin	mg	35	30	25
Pantothenic acid	mg	10	10	10
Pyridoxine	mg	3.5	3.5	3
Riboflavin	mg	3.6	3.6	3
Thiamin	mg	1.8	1.8	1.8

Source: NRC (1994)

Amino acid requirements of meat chickens

Poultry have the same requirements as mammals for the essential amino acids phenylalanine, valine, threonine, tryptophan, isoleucine, methionine, histidine, arginine, lysine and leucine (PVT TIM HALL see Table 1.7).

However, birds cannot synthesis arginine sufficiently rapidly to meet the high requirement for the urea cycle, which converts ammonia to urea for excretion, so

Table 14.2. Target bodyweights (g) and weekly feed intake rates (g) for Cobb500 broilers (mixed sex values)

Week	Bodyweight (g)	Weekly feed intake (g)
1	185	167
2	465	542
3	943	1192
4	1524	2137
5	2191	3352
6	2857	4786
7	3506	6379
8	4111	8070
9	4649	9785

Source: *Broiler Performance and Nutrition Supplement* (https://cobb-vantress.com)

chickens have an absolute requirement for arginine, particularly when they are growing rapidly. The essential requirement of a rapidly growing broiler chicken is relatively high because the majority of the body growth is body protein. However, maximising body gain and maximising the yield of economically important tissues, such as breast tissue, are not always the same thing. The lysine requirement for maximal breast development, for example, is greater than the lysine requirement for maximal weight gain (Acar *et al.* 1991). The amounts of amino acids required for maximal efficiency are also higher than that required for maximal gain (NRC 1994). It is generally regarded that the sulphur amino acids, methionine and cysteine, are first-limiting for growth of broiler chickens and that lysine is second-limiting.

Ileal digestible amino acids
The total concentration of an amino acid in a feed may not be a good indicator of the availability of that amino acid for protein synthesis if the digestibility of the amino acid is poor. In some feeds, such as meat meal, cottonseed meal, corn gluten meal and alfalfa meal, the digestibility of lysine and methionine can be as low as 50–75% compared with 85–90% for most other protein sources. Diets are increasingly formulated for poultry based on digestible amino acids relative to lysine and corrected for the impact of endogenous losses of amino acids and for the impact of the microbial activity in the large intestines (Bryden and Li 2010). Tables of standardised ileal digestibility (SID) values are now available for most feedstuffs and can be used to formulate diets relative to the SID requirements (Lemme *et al.* 2004).

Vitamin and mineral requirements of meat chickens

Table 14.3 lists the major clinical signs of vitamin deficiencies in poultry. The clinical signs of mineral deficiency in poultry are summarised in Table 14.4.

Table 14.3. Major clinical signs of deficiencies of the vitamins in poultry

Vitamin	Clinical signs
Vitamin A	• Exudate accumulation in eyes • Pustules in mucous membranes • Ataxia • Increased pressure in cerebrospinal fluid • Slowed growth and delayed closure of epiphyseal plate
Vitamin D	• Demineralised bones • Thin eggshells • Awkward gait
Vitamin E	• Encephalomalacia • Exudative diathesis (capillary leakage) • Anaemia (microcytic) • Gizzard necrosis
Vitamin K	• Petechial subcutaneous haemorrhages • Increased prothrombin time
Thiamine (B1)	• Polyneuritis (mad chick disease) • Anorexia • Paralysis • Arrhythmia
Riboflavin (B2)	• Curled-toe paralysis • Neuropathy of sciatic and brachial nerves • Anorexia • Walking on hocks
Niacin (B3)	• Oral inflammation • Enlarged hocks, bowed legs
Pantothenic acid (B5)	• Dermatitis on feet, eyelids and mouth
Pyridoxine (B6)	• Anorexia • Hyper-excitability • Weakness • Perosis (slipped tendon)
Folic acid	• Poor feathering • Anaemia
Biotin	• Dermatitis of the foot pad (crusty) • Poor feathering
Cobalamin (B12)	• Decreased egg size
Choline	• Perosis

Source: NRC (1994)

Table 14.4. Major clinical signs of deficiencies of the essential mineral elements in poultry

Mineral	Deficiency signs
MACROMINERALS	
Calcium	• Rickets • Thin eggshells • Demineralised bones • Tetany, death
Phosphorus	• Rickets • Demineralised bones • Cage layer fatigue
Magnesium	• Anorexia • Nervous tremors • Tetanic convulsions • Thin eggshells
Chloride	• Neuropathy (spasms when startled) • Dehydration (increased haematocrit)
Potassium	• Weakness • Seizures • Demineralised bones • Thin egg shells
Sodium	• Anorexia • Dehydration • Increased haematocrit • Demineralised bones
TRACE ELEMENTS	
Copper	• Anaemia (microcytic) • Aneurisms (poor connective tissue crosslinking) • Leg weaknesses • Bone deformity
Iodine	• Goitre • Delayed hatching time
Iron	• Anaemia (microcytic)
Manganese	• Perosis (slipped tendon) • Deformed skeletal development • Thin eggshells • Tetany, death
Selenium	• Exudative diathesis • Muscular dystrophy
Zinc	• Poor feathering • Enlarged hocks • Scaly dermatitis

Source: NRC (1994)

Common feedstuffs used in meat chicken diets

The main ingredients used in broiler (meat chicken) diets are:

- wheat
- sunflower meal
- maize (corn)
- soya meal
- full fat soya meal
- canola (rapeseed) meal
- oils and fats
- limestone
- phosphate
- salt
- sodium bicarbonate
- minerals and vitamins
- other additives (exogenous enzymes, mycotoxin binders).

Energy, amino acid, vitamin and mineral requirements of growing pullets and laying hens

Modern egg-producing enterprises aim to produce a large number of large eggs that have strong egg shells and that meet the market requirements for shell colour, yolk colour and albumen consistency, and for the lowest possible feed costs. Individual hens can produce 300 eggs/year, but typical flock averages are more like 275/year. Given an average egg weight of 55 g, this equates to 15 kg of eggs/year/hen, which is more than 10 times her bodyweight. The nutritional requirement to achieve this has to meet the high energy, protein, carbohydrate, lipid, mineral and vitamin demands of egg synthesis.

Egg formation in hens

Figure 14.2 shows the left oviduct of a laying hen. The ovary contains all the developing follicles (containing yolk) in different stages of development. Some have not yet commenced growth (primary follicles), some are actively growing, some are at maturity and ready to be released from the ovary and some are atretic (having discharged their yolk previously). The yolk released from the follicle enters the infundibulum, which channels it to the magnum: a process taking ~15 minutes. The yolk remains in the magnum for ~3 h during which time ~40% of the albumen is added to the outside of the yolk. The developing egg enters the isthmus where it spends ~75 minutes during which more albumen (~20%) and the inner and outer shell membranes are synthesised. The first calcium carbonate crystals are laid down on the outer shell membrane, forming the basis of egg shell synthesis in the shell gland or uterus. The developing egg spends ~18–20 h in the shell gland during which time the egg shell is synthesised.

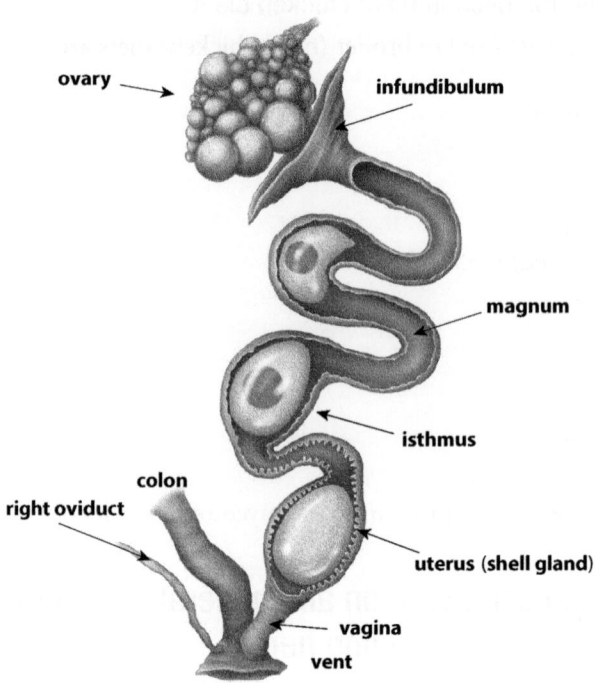

Fig. 14.2. Left oviduct of a laying hen. The right oviduct of hens is non-functional.

Composition of eggs

A typical egg weighing 58 g contains egg yolk (~2.8 g protein, 191 mg cholesterol, 0.63 g carbohydrates, 4.66 g total fat, all fat-soluble vitamins, vitamin D, many B-group vitamins and most essential minerals), albumen (the highest biological value protein in nature) and shell (mainly calcium carbonate and a protein matrix). These high levels of essential nutrients must be present in the bird's diet or be mobilised from body stores during egg formation.

Calcium and phosphorus requirements for egg production

The calcium required for synthesis of the egg shell, calcium carbonate, comes from the diet, medullary bone on the inner cavity of the long bones, and the skeleton. Approximately 2.5 g of calcium are required to form the egg shell. The hen can only obtain ~2.0 g of calcium per day from the diet, so the remainder must come from the bone stores. The supply of calcium from the different sources is finely tuned to the requirement, to the extent that eggs laid in the morning have thinner and weaker shells than those laid in the afternoon because the dietary supply of calcium overnight is less than during the day (Washburn and Potts 1975). Overnight, the dietary calcium supply is low at a time when the shell gland is actively synthesising shell. The calcium deficit is derived from hydroxyapatite (Ca_{10} $(PO_4)_6 (OH)_2$), a calcium storage mineral in medullary bones. This reliance on calcium phosphate stores, when a large quantity of calcium is required, produces a large quantity of phosphorus that must be excreted (Etches 1987). Balancing the total amount of dietary calcium and phosphorus, the timing of the supply of calcium and phosphorus relative to the timing of oviposition, the availability of the dietary calcium and phosphorus, and the ratio of available calcium to available phosphorus are major challenges for the egg industry (Li *et al*. 2016). The carbonate required to form hydroxyapatite for egg shells comes from the blood, being a component of the acid-base homeostasis reaction:

$$CO_2 + H_2O \leftrightarrow H_2CO_3 \leftrightarrow H^+ + HCO_3^-$$

Weak shells can result from hens being overheated, which causes panting and excessive loss of CO_2. This reaction is catalysed by carbonic anhydrase, which is a zinc-mediated enzyme. Zinc deficiency therefore results in weak egg shells.

Phosphorus in hen diets is often locked up in phytate: a storage form of phosphorus in cereal grains. Phytate combines with divalent and trivalent cations (calcium, zinc, cobalt, copper, iron, magnesium and nickel), reducing their availability to the animal (Pallauf and Rimbach 1997). Insoluble calcium phytates greatly reduce the availability of calcium for egg shell synthesis. Phytate also has antinutritional effects on digestive enzymes such as trypsin and amylase, and reduces fat digestion (Leeson *et al*. 1993). For these reasons, poultry feeds are often supplemented with exogenous phytase enzymes to increase nutrient availability.

Achieving the correct ratio of calcium to phosphorus is critical to strong eggshell formation. Ratios of 10:1 are quoted for laying birds, but, given the difficulty of knowing the available calcium and available phosphorus in the diet, future requirements should be expressed as available Ca and available P (Li *et al*. 2016).

Table 14.5 shows the composition of albumen, yolk, egg contents and whole eggs (including shells).

Nutrient requirements and target bodyweights for growing pullets

Pullets hatching at ~40 g bodyweight reach sexual maturity at about week 20 when bodyweight is ~1400 g. Good nutrition is required during this phase of growth

Table 14.5. Composition (%) of egg albumen, egg yolk, egg minus shell and whole eggs (including shell)

Component	Egg white	Yolk	Egg	Whole egg
Water	88.0	48.0	73.7	65.6
Protein	11.0	17.5	12.9	11.8
Fat	0.2	32.5	11.5	11.0
Ash	0.8	2.0	1.0	11.7

because it sets the hen up for higher total egg production during the laying period. Table 14.6 shows the nutrient requirements of growing pullets up to the point of lay.

Given a properly balanced ration for the birds at each stage of development and growth, the nutrient intake can be governed by achieving the rates of feed intake shown in Table 14.7.

Table 14.6. Minimum nutrient requirements of growing layer birds

Nutrient	Units	Starter 0–6 weeks	Grower 6–12 weeks	Developer 12–15 weeks	Pre-layer 15 weeks–production
Protein	%	20.0	17.50	15.50	16.50
Metabolisable Energy	MJ/Kg	11.5–12.4	11.5–12.6	11.3–12.4	11.4–12.4
Metabolisable Energy	Kcal/Kg	2750–2970	2750–3025	2700–2970	2725–2980
	Kcal/Lb	1250–1350	1250–1370	1225–1350	1235–1350
Lysine	%	1.10	0.90	0.66	0.80
Methionine	%	0.48	0.41	0.32	0.38
Methionine + Cystine	%	0.82	0.71	0.58	0.65
Tryptophan	%	0.20	0.19	0.18	0.19
Threonine	%	0.73	0.55	0.52	0.55
Calcium	%	1.00	1.00	1.00	2.75[a]
Average phosphorus	%	0.45	0.43	0.42	0.40
Sodium	%	0.18	0.18	0.18	0.18
Chloride	%	0.18	0.18	0.18	0.18

[a]At least 30–65% of the added limestone should have a minimum particle size of 2250 microns

Table 14.7. Bodyweights and feed intake level targets for pullets during the pre-lay growth stages

Age (weeks)	Bodyweight (g)	Feed consumption (g/bird/day)	Age (weeks)	Bodyweight (g)	Feed consumption (g/bird/day)
1	70	13	10	870–970	56
2	115	20	11	960–1080	61
3	190	25	12	1050–1117	66
4	280	29	13	1130–1250	70
5	380–390	33	14	1210–1310	73
6	480–500	37	15	1290–1370	75
7	580–620	41	16	1360–1430	77
8	680–750	46	17	1500–1540	80
9	770–860	51			

Source: CRC Poultry Hub and HyLine Management Guide online (http://www.poultryhub.org/nutrition/nutrient-requirements/nutrient-requirements-of-meat-chickens-broilers/)

Metabolisable Energy, protein, amino acid, vitamin and mineral requirements of laying hens

Most breeders recommend feeding layers on the basis of digestible amino acids and the 'ideal amino acid' ratios as outlined in Table 14.8. Note all the essential amino acids are referred relative to the first-limiting amino acid, lysine at 100.

Table 14.9 lists the current nutrient requirements for laying hens.

Table 14.8. Digestible amino acid requirements of layer birds

Amino acid	Relative to digestible lysine
Lysine	100
Methionine	43
Methionine + cysteine	84
Threonine	68
Tryptophan	23
Arginine	101
Isoleucine	94
Valine	101

Source: NRC (1994)

Table 14.9. Nutrient requirements of laying birds

Nutrient	Unit	Laying hens
Metabolisable Energy	MJ/kg DM	12.5
PROTEIN	%	17
Arg	%	0.8
His	%	0.19
Isoleu	%	0.58
Leu	%	0.85
Lys	%	0.75
Met	%	0.37
Met + Cys	%	0.65
Phe	%	0.48
Phe + Tyr	%	0.95
Pro	%	
Thre	%	0.53
Try	%	0.16
Val	%	0.65
MINERALS		
Ca	%	3.5
Cl	%	0.15

(Continued)

Table 14.9. (Continued)

Nutrient	Unit	Laying hens
Mg	mg	500
Non-phytate P	%	0.4
K	%	0.4
Na	%	0.15
Cu	mg	2
I	mg	0.3
Fe	mg	45
Mn	mg	35
Se	mg	0.1
Zn	mg	35
VITAMINS		
A	IU	8800
D	ICU	1100
E	IU	7
K	mg	2
B12	mg	0.007
Biotin	mg	0.11
Choline	mg	1100
Folacin	mg	0.4
Niacin	mg	25
Pantothenic acid	mg	6
Pyridoxine	mg	3.3
Riboflavin	mg	4.5
Thiamin	mg	2

Source: NRC (1994)

Common nutritional and metabolic diseases in poultry
Nutritional issues in free-range poultry systems

Free-range poultry systems are defined as those in which the birds have access to areas 'mainly covered with vegetation' and such systems are becoming increasingly adopted in many countries. In Australia, free-range egg production (44%) now exceeds caged egg production (42%) and in the UK, free-range layers account for ~10% of the total. Free-range chicken meat production (15% of total) is also increasing in Australia and the USA, but less so in Europe and the UK. These developments are important because the remarkable success of the modern poultry industries (chicken meat and layer) is the result of application of intense genetic selection for growth rate and efficiency, and finely tuned nutrition. The latter involves the use of feeds that meet the Net Energy requirements of the birds at each stage of growth or egg production, high-protein rations with appropriate balances of the essential amino acids that are digested in the intestines (including addition of synthetic amino acids), and the use of exogenous enzymes that achieve small, but cost-effective, increases in efficiency of digestion. To then place the birds in an environment in which the quantity and nutritive value of a component of the feed are unknown, creates uncertainty in a previously predictable system (Walker and Gordon 2003). Forages may also contain antinutritive factors that depress performance despite low intakes. To further complicate the situation, the nutritive value and quantity of forage available varies with season, rainfall, temperature, radiation input, fertiliser history and sward characteristics. Foraging birds also consume significant quantities of soil and invertebrate organisms of unknown composition. The requirements of the foraging birds also change with the energy demands for foraging. Also, dealing with temperatures outside of the thermoneutral zone must now be taken into account. Laying birds will also be influenced by changes in daylength. Although infectious diseases are outside the scope of this review of nutrition, many farms are visited by wild birds attracted by feed provided outside the barn and increasing the risk of infectious mortalities.

One of the key drivers towards free-range production systems is the consumer perception that free-range birds experience better welfare, and that the quality and taste of free-range products is better than housed birds. There is little evidence to support these claims and further research in this space is required.

Some of the major problems associated with free-range poultry production systems are now considered. The key questions are: (1) how much forage do the birds consume?; (2) what is the nutritive value of this forage?; and (3) what problems are encountered in free-range systems and how might they be avoided?

Intake and nutritional value of free-range forages
Intake of forages by free-range birds is relatively small, ranging from ~1–5% of total dry matter intake in meat birds and 5–10% in layer birds (Singh and Cowieson 2013). If the composition of the forage is equivalent to that of the concentrate ration fed indoors, this would represent a significant reduction in costs of production, because establishing and maintaining pastures is cheaper than concentrates, and feed costs comprise 60–70% of total costs of production. Layer birds can consume as much as 30–40 g of dry matter per day in the form of worms, herbage and insects, and as much as 7 g/day of soil.

The degree of selectivity of free-range birds is not known, so it is difficult to determine the nutritive value of their forage intake. The content of potassium in grasses is high and K influences the dietary electrolyte balance, which has been carefully formulated for the bird's concentrate diet. Dietary electrolyte balance is critical to maintenance of blood pH, so potential problems with metabolic acidosis and alkalosis in free-range birds is possible. Blood acid-base balance involves regulation of CO_2 losses, which impact on egg shell ($CaCO_3$) formation and egg shell quality. K also influences the absorption of magnesium, so high forage K intake may induce hypomagnesaemia in the birds, exacerbating bone and leg problems.

Forages provide more than a nutrient source and the use of appropriate pasture and shrub species may be beneficial for shade, protection from predators, soil protection, and protection from wind. Appropriate choice of species (e.g. herbs such as oregano, rosemary and thyme) may also be used to provide antioxidants, or tannin-containing plants may be useful adjuncts to antibiotics and anti-protozoal therapeutics.

The use of pastures in poultry nutrition may also provide benefits to products. For example, there are reports of increased levels of omega-3 fatty acids in the meat of pasture-fed birds (Ponte *et al.* 2008a) and better sensory properties of meat (Ponte *et al.* 2008b), and possibly eggs.

The energy and protein contents of grasses and legume forages varies greatly with forage species, growth phase and season. Typical values for temperate forage legumes and grasses at different stages of growth are shown in Table 14.10.

Table 14.10. Nutritive value of typical forage legumes and grasses at different phases of growth

Forage type	Phase of growth	ME (MJ/kg DM)	CP (% DM)	Other
Forage legumes (clovers)	Phase 1, active, young, green	13–14	26–30	High minerals; high soluble carbohydrates; potential oestrogens
Forage legumes (clovers)	Phase 3, senescence	9–10	12–16	High minerals; moderate soluble carbohydrates
Phalaris (*P. aquatica*)	Phase 1, active, young, green	11–12	25–30	High soluble carbohydrates
Phalaris (*P. aquatica*)	Phase 3, senescence	7–8	8–10	High cellulose, hemicellulose; low minerals
Mixed meadow sward	Phase 1, active, young, green	11–11.5	15–20	Moderate soluble carbohydrates
Mixed meadow sward	Phase 3 senescence	6–7	6–8	High cellulose, hemicellulose; low minerals

These variations will have large impacts on the nutrition and performance of free-range birds if they consume the forages at significant levels. For example, if a meat bird consumes, say, 10% of its total intake as the mixed meadow sward, the average M/D of its ration will drop from 13.4 MJ/kg DM to 12.7 MJ/kg DM. The crude protein content of its ration will drop from 23% to 21.4%. A daily intake of 12.7 MJ/day and 21.4% CP is insufficient for maximum growth and minimum FCR. Consumption of the legume during early growth, on the other hand, will have little impact on the bird's overall nutrition unless antinutritive factors are present or the protein is badly unbalanced. Invertebrates have a high ME value (12.5 MJ/kg DM) and very high crude protein values (60%) and particularly lysine (4%).

Impaction of the crop, proventriculus and gizzard

Grass impaction of the gastrointestinal tract can be a significant problem for free-range birds, leading to high mortalities (as high as 1% per week in free-range layers) (Singh and Cowieson 2013). The grass fibres became entangled into long, twisted plugs, which prevent the onward passage of the material to the duodenum.

Bone and joint disorders in meat chickens

Studies of meat chickens have shown that up to 30% of birds at 40 days of age can show signs of poor locomotion and ~3% were unable to walk at all (Knowles *et al.* 2008). It is commonly stated that rapid growth rates are responsible for the leg problems and there is evidence of a genetic component to disorders such as tibial dyschondroplasia, but nutritional interventions can prevent the pathology (Leach and Nesheim 1972). Other leg problems are a direct result of a nutritional deficiency or imbalance.

Tibial dyschondroplasia (TD or osteochondrosis) is a cartilage abnormality of the bones known in birds as the tibiotarsus and the tarsometatarsus. It appears between days 14–25 and the first signs are a reluctance of birds to move and a stilted gait. At the ends of these bones there is extensive, abnormal accumulation of cartilage in the growth plates of the metaphyseal region. Instead of calcifying, the chondrocytes (cartilage cells) in the growth plates remain uncalcified while the bird is still growing rapidly. The immature cartilage becomes highly cross-linked with collagen, which makes it resistant to vascularisation and resorption. When the growth rate of the bird slows, this cartilage is resorbed and the trabecular bone finally forms. Although the cartilage remains soft, the leg is prone to bending backwards under the bodyweight of the bird, and leg fractures, lameness and necrosis of the region can occur. The condition is extremely painful for the bird, and can result in starvation or trampling by other birds. TD is a common disorder affecting ~1% of all rapidly growing broiler chickens, turkeys and ducks.

TD appears to be multifactorial, with a strong genetic component associated with selection for rapid growth rates. The inheritance is via a sex-linked recessive allele (Sheridan *et al.* 1978). Other risk factors are copper deficiency, excessive levels of the sulphur amino acids, cysteine and homocysteine, metabolic acidosis (Orth and Cook 1994) and low calcium/phosphate ratios (Edwards and Veltmann 1983). There have been reports of an effect of high-chloride diets on the incidence of TD (Veltmann and Jensen 1981).

Provision of 1,25-dihydroxycholecalciferol (1,25-(OH)2D3) in rations prevents many cases of tibial dyschondroplasia, suggesting that selection for high growth rate has been accompanied by a defect in vitamin D metabolism (Edwards 2000).

Osteoporosis, rickets and weak egg shells
Meat chickens are typically fed diets containing 10 g calcium and 4.5 g phosphorus/kg diet (i.e. a Ca:P ratio of ~2.2:1). Deficiencies of calcium or phosphorus or of vitamin D3, or imbalances of these essential nutrients (particularly low Ca:P), result in bone disorders. In growing birds, inadequate skeletal calcification results in rickets, and in laying birds weak egg shells and osteoporosis result. Rickets is due to inadequate calcium at the cellular level to allow normal bone mineralisation. Rickets is often due to inadequate vitamin D3 availability due to dietary deficiency, poor absorption or poor potency of the D3 supplement in the ration. Rickets is characterised by poor growth rate, enlargement and widening of the ends of the epiphyseal plates in the long bones (proliferating and hypertrophic zones), thickening of the cortical bone, softening of the bone and enlargement of the parathyroid (Itakura *et al.* 1978).

Slipped tendon (perosis)
Perosis or slipped tendon in chickens is caused by a deficiency of manganese either due to a frankly deficient diet, or to a manganese deficiency induced by high levels of calcium and phosphorus in the diet. Perosis is characterised by enlargement of the tibiometatarsal joint, bending of the tibiotarsal bone, thickening and shortening of the leg bones and slippage of the calcaneal tendon off the condyle, resulting in splaying of the leg.

References
Acar N, Moran ET, Jr, Bilgili SF (1991) Live performance and carcass yield of male broilers from two commercial strain crosses receiving rations containing lysine below and above the established requirement between six and eight weeks of age. *Poultry Science* **70**, 2315. doi:10.3382/ps.0702315

Bryden WL, Li X (2010) Amino acid digestibility and poultry feed formulation: expression, limitations and application. *Revista Brasileira de Zootecnia* **39**, 279–287. doi:10.1590/S1516-35982010001300031

Edwards HM (2000) Nutrition and skeletal problems in poultry. *Poultry Science* **79**, 1018–1023. doi:10.1093/ps/79.7.1018

Edwards HM, Veltmann JR (1983) The role of calcium and phosphorus in the aetiology of tibial dyschondroplasia in young chicks. *The Journal of Nutrition* **113**, 1568–1575. doi:10.1093/jn/113.8.1568

Etches RJ (1987) Calcium logistics in the laying hen. *The Journal of Nutrition* **117**, 619–628. doi:10.1093/jn/117.3.619

FAO (2018) *World Food and Agriculture – Statistical Pocketbook 2018.* Food and Agriculture Organisation of the United Nations, Rome.

Itakura C, Yamasaki K, Goto M, Takahashi M (1978) Pathology of experimental vitamin D deficiency rickets in growing chickens. *Avian Pathology* **7**, 491–513. doi:10.1080/03079457808418305

Knowles TG, Kestin SC, Haslam SM, Green LE, Butterwirth A, Pope SJ, *et al.* (2008) Leg disorders in broiler chickens: prevalence, risk factors and prevention. *PLoS ONE* **3**, e1545.

Leach RM, Nesheim MC (1972) Further studies on tibial dyschrondoplasia (cartilage abnormality) in young chicks. *The Journal of Nutrition* **102**, 1673–1680. doi:10.1093/jn/102.12.1673

Lemme A, Ravindran V, Bryden WL (2004) Ileal digestibility of amino acids in feed ingredients for broilers. *World Poultry Science* **60**, 421–435.

Leeson S, Summers JD, Caston L (1993) Response of brown egg strain layers to dietary calcium or phosphorus. *Poultry Science* **72**, 1510–1514. doi:10.3382/ps.0721510

Li X, Bryden WL, Zhang D (2016) 'Available phosphorus requirement of laying hens'. Final report for the Australian Egg Corporation Ltd. Publication No 1UQ101A, AECL, Sydney.

Moran ET, Jr (1982) *Comparative Nutrition of Fowl and Swine: The Gastrointestinal System.* Office for Educational Practice, University of Guelph, Guelph, Ontario, Canada.

NRC (1994) *Nutrient Requirements of Poultry.* 9th edn. National Research Council. National Academy Press. Washington, DC, USA.

Orth MW, Cook ME (1994) Avian tibial dyschondroplasia: a morphological and biochemical review of the growth plate lesion and its causes. *Veterinary Pathology* **31**, 403–414. doi:10.1177/030098589403100401

Pallauf J, Rimbach G (1997) Nutritional significance of phytic acid and phytase. *Archiv für Tierernaehrung* **50**, 301–319.

Ponte PIP, Prates JAM, Crespo JP, Crespo DG, Moura JL, Alves SP, *et al.* (2008a) Restricting the intake of a cereal-based feed in free range-pastured poultry: effects on performance and meat quality. *Poultry Science* **87**, 2032–2042. doi:10.3382/ps.2007-00522

Ponte PIP, Rosado CMC, Crespo JP, Crespo DG, Moura JL, Alves SP, *et al.* (2008b) Pasture intake improves the performance and meat sensory attributes of free range broilers. *Poultry Science* **87**, 71–79. doi:10.3382/ps.2007-00147

Sheridan AK, Howlett CR, Burton RW (1978) The inheritance of tibial dyschondroplasia in broilers. *British Poultry Science* **19**, 491–499. doi:10.1080/00071667808416505

Singh M, Cowieson AJ (2013) Range use and pasture consumption in free-range poultry production. *Animal Production Science* **53**, 1202–1208. doi:10.1071/AN13199

Veltmann JR, Jensen LS (1981) Tibial dyschondroplasia in broilers: comparison of dietary additives and strains. *Poultry Science* **60**, 1473–1478. doi:10.3382/ps.0601473

Walker A, Gordon S (2003) Intake of nutrients from pasture by poultry. *Proceedings of the Nutrition Society* **62**, 253–256. doi:10.1079/PNS2002198

Washburn KW, Potts PL (1975) Effect of strain and age on the relationship of oviposition time to shell strength. *British Poultry Science* **16**, 599–606. doi:10.1080/00071667508416235

Index

Printed and bound by CPI Group (UK) Ltd, Croydon, CR0 4YY

25/06/2025

14694447-0001